New Trends in Probability and Statistics – Volume 2

Analytic and Probabilistic Methods in Number Theory

NEW TRENDS IN PROBABILITY AND STATISTICS
Volume 2

Analytic and Probabilistic Methods in Number Theory

Proceedings of the International Conference in Honour of J. Kubilius, Palanga, Lithuania, 24-28 September 1991

Editors
F. Schweiger and E. Manstavičius

TEV
VILNIUS, LITHUANIA

///VSP///
UTRECHT, THE NETHERLANDS
TOKYO, JAPAN

VSP BV
P.O. Box 346
3700 AH Zeist
The Netherlands

TEV Ltd.
Akademijos 4
Vilnius
Lithuania

First published in 1992
ISBN 90-6764-094-8

CIP-DATA KONINKLIJKE BIBLIOTHEEK, DEN HAAG

New trends in probability and statistics
Utrecht : VSP; Vilnius : TEV.
Vol. 2: Analytic and probabilistic methods in number theory: Proceedings of the International Conference in Palanga in honour of J. Kubilius / ed. E. Manstavičius, F. Schweiger.
ISBN 90-6764-094-8 bound
NUGI 811
Subject headings: probability theory / number theory

Typeset by TEV Ltd., Vilnius
Printed in Lithuania by Spindulys, Kaunas

CONTENTS

IV. MULTIPLICATIVE NUMBER THEORY

V. DISTRIBUTION OF ADDITIVE FUNCTIONS

FOREWORD

Having now entered the winter of my life I can look back upon an interesting and varied life. My early years were ones of rich experience developing into an industrious adulthood containing much change.

Much of my life has been devoted to my nation suffering the hardest times. Fate has willed that I, like most intellectuals of the older generation, could not devote myself to my own speciality. It was a case of sharing most of my mature years between scientific, creative work and other activities. I was making my bread both as a white-collar worker, as a politician and as a social activist. I felt myself a mathematician only after having escaped from prison like a Sunday gentleman. It often happened that mathematics became the subject to deal with in the evenings (or at night, when younger), on Sundays or holidays. In fact I do not know whether I should feel regret about not having written numerous articles and books. However, today when I see how much has been done, and not without my support, by my pupils and colleagues, I do not have any regrets.

I would like to express my cordial thanks to the Rectorate of Vilnius University and all my pupils and colleagues for arranging this fiesta of my jubilee. I also wish to thank those who came and took part in this grand rally, including the scientific conference in Palanga. I regard it as an honour not only to myself but to Vilnius University and to Lithuania, now reborn as an independent state.

Jonas Kubilius

PREFACE

The Lithuanian sea resort Palanga was the site of the conference 'Analytic and Probabilistic Methods in Number Theory' held to honour Professor Jonas Kubilius on the occasion of his 70th birthday. Professor Kubilius not only served as a rector of Vilnius University for more than three decades he has also been an eminent scholar who contributed so much to the flourishing of mathematics in Lithuania. To him and his students we owe some fine work in number theory, probability theory and the links between these areas. Therefore, it seems appropriate to publish the proceedings of the conference in a series which is devoted to promoting research and new perspectives in probability theory and its wide field of applications and relations to other fields.

In 1949 J. Kubilius presented in his first work, a partial solution of a famous conjecture of Kurt Mahler in metrical number theory. It was Kubilius' student W. Sprindzhuk who later gave complete proof. Kubilius and his students have also contributed to analytic number theory, especially to the problem of the distribution of prime ideals in algebraic number fields. However, the most impressive work has been done on the statistical theory of arithmetical functions which almost created a new research area called Probabilistic Number Theory. A monograph devoted to this topic was translated into English in 1964 and became very influential. His former teacher, academician Yu. V. Linnik, once evaluated Kubilius' work as follows:

"Jonas Kubilius' research is a great contribution to science where two domains of mathematics, probability theory and number theory, intersect. Only some considerations and isolated facts had existed in this field before. Now he has developed a comprehensive and far-reaching theory. The essential parallelism of the number theory and that of probability has been established. This can even be considered as having philosophical significance."

The mathematical depth of his work has also received recognition by several honorary degrees awarded to Professor Kubilius.

The 65 participants of the Palanga conference came from 15 countries. The lectures, including plenary ones delivered by E. Wirsing, A. Schinzel and A. N. Andrianov plus short presentations, covered a broad scale of recent research in number theory, especially in its analytic and probabilistic branches. This volume contains most of the articles presented in Palanga as well as some additional papers from researches who were unable to attend the conference. However, to meet the standards of mathematical journals all papers were given to referees. Therefore, we wish to express our sincere thanks not only to the participants of the conference ands the contributors to the proceedings, but also to the referees whose careful work was a great help in editing this book. Last but not least we appreciate the accurate work done by the copy editors Z. Kryžius and V. Stakėnas, which allowed the rather different papers to be presented as one consistent book.

E. Manstavičius and F. Schweiger, editors.

Part One

DIOPHANTINE APPROXIMATIONS AND TRANSCENDENTAL NUMBER THEORY

New Trends in Probab. and Statist., pp. 3–7
F. Schweiger and E. Manstavičius (Eds.)
1992 VSP/TEV

POLYNOMIALS OF MINIMAL DIOPHANTIC RELATIVE DEVIATION FROM ZERO ON THE INTERVAL [0,1]

E. APARICIO BERNARDO*
Department of Mathematics, University of the Basque Country,
E48080 Bilbao, Spain

ABSTRACT

Here the concept of relative diophantic deviation from zero is introduced and a sequence of polynomials with integer coefficients and small relative diophantic deviation from zero on the interval $[0,1]$ is built.

The problem of the structure of the polynomials with integer coefficients, not simultaneously equal to zero, with the minimal uniform diophantic deviation from zero on the interval $[0,1]$, has been proposed by A. O. Gelfond and discused by the author of the present paper in (Aparicio Bernardo, 1954, 1988). As it is pointed out in (Aparicio Bernardo, 1988) this problem is closely connected with the problem of the calculation of the own minimal deviation from zero. In this case, it is important to obtain polynomials with small uniform diophantic deviation from zero.
From now on we shall employ the following notations:

$$\rho_n^{-n} = \min_{P_n \in H_n^*} \max_{0 \leqslant x \leqslant 1} |P_n(x)| \tag{1}$$

where H_n^* is the class of the polynomials of degree n with integer coefficients, not simultaneously equal to zero, and

$$\rho = \lim_{n \to \infty} \rho_n. \tag{2}$$

For the existence of this limit see (Aparicio Bernardo, 1987).
The best upper bound for ρ,

$$\rho \leqslant 2,37686,$$

has been obtained by A. D. Gorshkov's method, see (Aparicio Bernardo, 1981, 1984). To obtain a lower bound, other methods are used. We define the relative deviation from zero on $[0,1]$ of a polynomial $P_n \in H_n^*$ as

$$r_n^{-1} = \max_{0 \leqslant x \leqslant 1} |P_n(x)|^{1/n}. \tag{3}$$

*Supported by the University of the Basque Country and DGICYT PS 88-0075

It is easy to prove that

$$\rho \geqslant r_n \tag{4}$$

holds for any polynomial $P_n \in H_n^*$.
In fact, it is evident that

$$\rho_{nm}^{-nm} \leqslant r_n^{-nm}$$

for any natural numbers n and m, so $\rho_{nm} \geqslant r_n$. Finding the limit when $m \to \infty$ we obtain (4).
This inequality shows that if we find a polynomial of the class H_n^* with a very small relative deviation from zero (3), we will be able to get a good lower bound for ρ. So, for instance, using the polynomial of fifth degree proposed by L. G. Schnirelmann in 1934, the bound

$$\rho \geqslant \sqrt{5} \approx 2,236$$

was obtained, see (Aparicio Bernardo, 1954, 1967). This polynomial will occur in the text later.
Even a better lower bound is found in (Aparicio Bernardo, 1954, 1967).
I. N. Sanov affirms that considering a polynomial of seventeenth degree, which is not given in the paper, the asymptotic bound

$$\rho_n^{-n} = O(2,343^{-n})$$

was obtained (see (Sanov, 1949)).
Subsequently, A. D. Gorshkov pointed out a polynomial of fourteenth degree, with use of which he got the bound

$$\rho \geqslant 2,3176$$

(see (Aparicio Bernardo, 1978)).
Finally applying a polynomial of thirty fourth degree, proposed by C. Simó, it was found in (Aparicio Bernardo, 1981) that

$$\rho \geqslant 2,3307.$$

Now, we are going to construct a sequence of polynomials $\{T_m\}$ with small diophantic relative deviations from zero.
We shall employ the notation

$$t_n^{-1} = \lambda_n^{-\tau_n} = \max_{0 \leqslant x \leqslant 1} |T_n(x)|, \tag{5}$$

where τ_n is the degree of $T_n(x)$.
It is known that $T_1(x) = x$, $T_2(x) = x(1-x)$ are the polynomials of first and second degree, respectively, with integer coefficients, and with minimal uniform deviation from zero on $[0,1]$. Such polynomials can not be determined uniquely. For these polynomials we have $\lambda_1 = t_1 = 1$, $\lambda_2^{-1} = 1/2$. The maximum value of the last polynomial is t_2^{-1}, where

$$t_2 = (1+t_1)^{(1+t_1)} = 2^2. \tag{6}$$

We shall write the polynomial T_3 as

$$T_3 = T_2^k(1 - t_2 T_2),$$

and we shall find the natural number k in such a manner that the relative deviation from zero of T_3, as a polynomial with respect to T_2, will be minimal in the interval $[0, t_2^{-1}]$.
The nonzero root of its derivative

$$T_3' = T_2^{k-1}\left[k - (k+1)t_2 T_2\right]$$

is

$$\frac{k}{(k+1)t_2},$$

and the relative deviation from zero is attained in this point, its value being equal to

$$\left(\frac{k^k}{(k+1)^{k+1}t_2^k}\right)^{1/(k+1)} \tag{7}$$

To calculate the minimum of this expression as the function of k, we first observe that its logarithmic derivative is equal to

$$\frac{1}{(k+1)^2}\log\frac{k}{t_2},$$

so the minimum of it is reached for $k = t_2$, the value of the minimum being $(1+t_2)^{-1}$.
The polynomial we have been looking for is

$$T_3 = T_2^{t_2}(1 - t_2 T_2). \tag{8}$$

According to (7), its maximum value is t_3^{-1} where

$$t_3 = (1 + t_2)^{(1+t_2)}.$$

The expression of T_3 in terms of x is

$$T_3 = \left\{\left[x(1-x)\right]^2(2x-1)\right\}^2. \tag{9}$$

Curiously, the polynomial of fifth degree which appears between braces is Schnirelmann's polynomial of minimal deviation from zero, which had been mentioned before.
The other polynomials of the sequence we are looking for will be built through recurrency. Once we have the polynomials $T_1, \ldots, T_m$, we construct the polynomial $T_{m+1} = T_m^k(1 - t_m T_m)$, where t_m^{-1} is the maximum value of the modulus of T_m. We shall obtain the value $k = k_m$ in such a way that the deviation from zero of this polynomial, as a polynomial with respect to T_m, in the interval $[0, t_m^{-1}]$ will be minimal. To this end we see that nonzero root of its derivative

$$T'_{m+1} = T_m^{k-1}\left[k - (k+1)t_m T_m\right]$$

is

$$\frac{k}{(k+1)t_m}.$$

Thus, the minimal relative deviation from zero of the polynomial T_{m+1} in $[0, t_m^{-1}]$ is

$$\left(\frac{k^k}{(k+1)^{k+1}t_m^k}\right)^{1/(k+1)}$$

By the same way as before we find that the minimum of this expression is reached for $k = t_m$, the value of this minimum being equal to $(1+t_m)^{-1}$.
Hence the polynomial we are searching for is

$$T_{m+1} = T_m^{t_m}\left(1 - t_m T_m\right) \tag{10}$$

and its maximum modulus in $[0, t_m^{-1}]$ is equal to $(t_{m+1})^{-1}$ where

$$t_{m+1} = \left(1+t_m\right)^{1+t_m}. \tag{11}$$

The formulae (10) and (11) are valid for $m = 1$ too.
Beginning with $T_1 = x$, $t_1 = 1$, and applying (10), we construct the sequence of polynomials

$$T_1, T_2, \ldots, T_m, \ldots . \tag{12}$$

It is interesting to point out that the irreducible divisors x, $1-x$, $2x-1$, and $5x^2-5x-1$ of T_4 as the polynomial in x satisfy the conditions of the fundamental theorem of (Aparicio Bernardo, 1988).
Next, we shall apply the polynomials (12) to find a lower boundary of ρ.
The degree of T_{m+1}, as of a polynomial in x, is equal to

$$(1+t_1)\ldots(1+t_m), \tag{13}$$

and its relative deviation from zero in the interval $[0,1]$ is equal to

$$\left(\lambda_{m+1}\right)^{-1} = \left(t_{m+1}\right)^{-1/(1+t_1)\ldots(1+t_m)}.$$

Taking into account (11), we get the following expression for λ_{m+1}:

$$\lambda_{m+1} = \prod_{k=1}^{m}\left(1+t_k^{-1}\right)^{1/(1+t_1)\ldots(1+t_{k-1})} \tag{14}$$

where $t_0 = 0$.
This shows that the sequence $\{\lambda_m\}$ increases monotonically.
Let

$$\lambda = \lim_{m\to\infty} \lambda_m, \tag{15}$$

then from (4), we get that

$$\rho \geqslant \lambda.$$

Taking logarithms, and taking into account (14) and (15), we obtain finally

$$\log \rho \geqslant \sum_{k=1}^{\infty} \frac{1}{(1+t_1)\dots(1+t_{k-1})} \log\left(1+\frac{1}{t_k}\right). \tag{16}$$

Here the values t_k are calculated through the recurrent expression (11).
It can be easily seen that the terms of the series (16) approach to zero very rapidly. So we can use only few terms verified for to find the approximate value of the sum. But already first two terms of the series imply the Schnirelmann's bound, consequently, the polynomials with integer coefficients (12) are of small deviation from zero.
The expression

$$\rho \geqslant \lambda_m = \left(t_m\right)^{1/(1+t_1)\cdots(1+t_{m-1})},$$

which holds for every m, can also be used. In particular, if $m = 3$, Schnirelmann's bound is obtained.
It would be interesting to improve the method used in the present paper for to obtain another series of polynomials with small relative diophantic deviation from zero, which would allow us to obtain the inferior bounds of ρ found by the authors quoted previously. All this would also let us to delve deeper into the problem of the structure of the polynomials with minimal uniform diophantic deviation from zero on the interval [0, 1]. This extremely difficult problem is far from being solved. Some investigations by the author related to the asymptotic structure of such polynomials can be found in (Aparicio Bernardo, 1988).
We recommend to the reader to consult the extensive work (Ferguson, 1980).

REFERENCES

Aparicio Bernardo, E. (1954). *On Extremal Properties of Polynomials with Integral Coefficients and Approximation Functions and their Polynomials.* Author's abstract of dissertation. Moscow State University, Moscow.

Aparicio Bernardo, E. (1967). Sobre la aproximación de las funciones mediante polinomios de coeficientes enteros. In: *Actas de la Octava Reunión Anual de Mathemáticos Españoles.* Madrid, pp. 21–33.

Aparicio Bernardo, E. (1978). Métodos para el cálculo aproximado de la desviación diofántica uniforme mínima a cero en un segmento. *Rev. Mat. Hispano-Amer., Serie* 4, **38**, 6, 259–270.

Aparicio Bernardo, E. (1981). Sobre unos sistemas de números enteros algebraicos de Gorshkov y sus aplicaciones al cálculo. *Rev. Mat. Hispano-Amer., Serie* 4, **41**, 1–2, 3–17.

Aparicio Bernardo, E. (1984). On some results in the problem of diophantine aproximation of functions by polynomials. *Proc. Steklov Inst. Math.* **163**, 6–9.

Aparicio Bernardo, E. (1987). Generalization of the theorem of L. G. Schnirelmann. *J. d'Analyse Mathém.* **48**, 217–224.

Aparicio Bernardo, E. (1988). On the asymptotic structure of the polynomials of minimal diophantic deviation from zero. *J. Approxim. Theory* **55**, **3**, 270–278.

Ferguson, L. B. O. (1980). *Approximation by Polynomials with Integral Coefficients.* American Mathem. Soc., Providence, R.I.

Sanov, I. N. (1949). Functions with integral parameters, deviating the least from zero. *Uchen. Zap. Leningrad Gos. Univ., Ser. Mat. Nauk* **16**, 32–46 (in Russian).

New Trends in Probab. and Statist., pp. 9–15
F. Schweiger and E. Manstavičius (Eds.)
1992 VSP/TEV

A THEOREM OF GELFOND VIA SCHNEIDER'S METHOD

P. BUNDSCHUH
Mathematisches Institut der Universität zu Köln,
Weyertal 86–90, D-5000 Köln 41, Germany

ABSTRACT

In 1933 Gelfond proved that if an entire function f of sufficiently slow growth takes on integral values at all points q^n, $n = 1, 2, \ldots$, where $q > 1$ is some rational integer, then f must be a polynomial. In the present note we show that one can get "almost" the same result using Schneider's transcendence method. We study some related questions by this method.

THE RESULTS

It was proved in (Gelfond, 1933) that an entire function f satisfying $f(q^n) \in \mathbf{Z}$ for all $n \in \mathbf{N}$, and

$$\log \mu(r, f) \leqslant \frac{\log^2 r}{4 \log q} - \frac{1}{2} \log r - \omega(r),$$

ω being an increasing and unbounded function, must be a polynomial. Here the number $q \in \mathbf{N} \setminus \{1\}$ is fixed, and $\mu(r, f)$ denotes the maximum of $|f(z)|$ on the disc $|z| \leqslant r$. There are examples which show that Gelfond's result is the best possible up to the unspecified function ω, e.g.

$$f(z) := \sum_{k \geqslant 0} \frac{(z-q)\ldots(z-q^k)}{(q-1)\ldots(q^k-1)} q^{-k(k+1)/2}. \tag{1}$$

Whereas Gelfond used the method of Newton's interpolation series, some few years ago R. Wallisser got the same result using an analogue of Laplace transform (see Wallisser, 1985)).
In the present note we want to investigate mainly what Schneider's transcendence method can give when we try to apply it to prove Gelfond's above-mentioned result. The analogous question concerning Pólya's theorem on entire integer-valued functions was considered in (Waldschmidt, 1978). Our result is as follows.

THEOREM 1. *There exists an effectively computable absolute constant $c \in \mathbf{R}_+$ with the following property. Let $q \in \mathbf{N} \setminus \{1\}$ and let f: $\mathbf{C} \to \mathbf{C}$ be an entire function such that $f(q^n) \in \mathbf{Z}$ for all $n = 0, 1, \ldots$, and that*

$$\log \mu(r, f) \leqslant \frac{\log^2 r}{c \log q} \tag{2}$$

holds for all sufficiently large $r \in \mathbf{R}_+$. Then f is a polynomial.

Remark. In virtue of our remark on Gelfond's result, 4 is the best possible value for the constant c, but an easy computation at the end of the proof of Theorem 1 shows, that by Schneider's method we can get only 13.657... for that constant (at least at this moment).
In another situation we can come down to the constant 4, namely we have

THEOREM 2. *Suppose that the hypotheses of Theorem 1 on q and f are satisfied with some $c > 4$ in the inequality (2). Then f satisfies a certain functional equation of the type*

$$\sum_{i=0}^{I-1} \sum_{j=0}^{J-1} a_{ij} z^j f(zq^j) = 0 \tag{3}$$

with $a_{ij} \in \mathbf{Z}$, not all equal to zero.

Remark. It seems to be rather unlikely that one can show that a function f satisfying the conditions of Theorem 2 (and therefore also a functional equation of type (3)) must be a polynomial, which would settle Gelfond's theorem almost entirely. It should be noticed that the entire function f from (1) satisfies all conditions of Theorem 2, with $c = 4$ in (2), and furthermore the functional equation

$$f(q^2 z) - 2f(qz) + (1 - z)f(z) = 0$$

being exactly of the type (3). Nevertheless f is a transcendental function.
In the present context we have another application of Schneider's method which, in some sense, is a generalization of Theorem 1.

THEOREM 3. *Let $\mathbf{K}$ be an algebraic number field, and $\alpha \in \mathbf{R}_+$. Then there exists an effectively computable constant $c \in \mathbf{R}_+$, depending at most on α and on $d := [\mathbf{K} : \mathbf{Q}]$, with the following property.*
Suppose that $q \in \mathbf{N} \setminus \{1\}$, and let S be a subset of $\{1, q, q^2, \ldots\}$ with

$$\operatorname{card}(S \cap D_r(0)) \geqslant \alpha \frac{\log r}{\log q}$$

for all large $r \in \mathbf{R}_+$ where $D_r(0)$ denotes the closed disc of radius r around the origin. Let $f, g\colon \mathbf{C} \to \mathbf{C}$ be entire functions satisfying the following three conditions:

(a) $g(q^n) \neq 0$ and $\frac{f}{g}(q^n) \in \mathbf{K}$ *for all $n \in \mathbf{N}_0$ with $q^n \in S$,*

(b)

$$\sup_{r>1} \left(\frac{\log q}{\log^2 r} \max_{q^n \in S \cap D_r(0)} \left\{ -\log |g(q^n)|, \operatorname{size}\left(\frac{f}{g}(q^n)\right) \right\} \right)$$

is finite and independent on q,

(c) $\log \max(\mu(r, f), \mu(r, g)) \leqslant \dfrac{\log^2 r}{c \log q}$.

Then $\frac{f}{g}$ *is a rational function.*

Remark. For an algebraic number γ one usually defines

$$\operatorname{size}(\gamma) := \log \max(\overline{|\gamma|}, \operatorname{den} \gamma)$$

where den γ is the denominator of γ and $\overline{|\gamma|}$, the house of γ, is the maximum of the absolute values of γ and all its conjugates. Of course, this definition implies $\operatorname{size}(\gamma) \geqslant 0$. By condition (b) of our Theorem 3, there exists some $\beta \in \mathbf{R}_+$, independent on q, such that

$$\max_{q^n \in S \cap D_r(0)} \left\{ -\log|g(q^n)|, \quad \operatorname{size}\left(\frac{f}{g}(q^n)\right)\right\} \leqslant \beta \frac{\log^2 r}{\log q} \tag{4}$$

for all $r > 1$, and thus

$$\log \overline{\left|\frac{f}{g}(q^n)\right|} \leqslant \beta n^2 \log q \qquad \text{for} \quad q^n \in S. \tag{5}$$

It should be pointed out that A. O. Gelfond himself extended his above-mentioned results to the case when not only $f(q^n) \in \mathbf{Z}$, but $f^{(\sigma)}(q^n) \in \mathbf{Z}$ for all $n = 0, 1, \ldots$, and $\sigma = 0, \ldots, s-1$, where $s \in \mathbf{N}$ is fixed (see (Gelfond, 1967)). Again for this proof, A. O. Gelfond used the method of Newton's interpolation series. After our proof of Theorem 1 it will be clear that we could get a nontrivial result in the direction of (Gelfond, 1967) using his transcendence method instead of Schneider's one.

To conclude this section it should be reminded that different analogues of Gelfond's last result (including derivatives) for functions of several variables were proved in (Bézivin, 1984; Bundschuh, 1980). Obviously Gelfond's transcendence method in several variables could also be applied in this context.

THE PROOFS

Proof of Theorem 1. We suppose that $h, k \in \mathbf{R}$ and both are greater than 1. They will be fixed at the end of this proof. We proceed in several steps.

Step 1. Suppose $N \in \mathbf{N}$. Then by Siegel's lemma, see e.g. (Waldschmidt, 1979), there exist $a_0, \ldots, a_{[hN]-1} \in \mathbf{Z}$, not all equal to zero, such that the system of equations

$$\sum_{j=0}^{[hN]-1} a_j f(q^{n+j}) = 0, \qquad n = 0, \ldots, N-1,$$

is satisfied with

$$\log|a_j| \leqslant \frac{(h+1)^2}{c(h-1)} N^2 \log q + \gamma_1 N. \tag{6}$$

Here γ_1, and later $\gamma_2, \ldots$, denote positive real numbers which are independent of N (and M).
Now we consider the function

$$F(z) := \sum_{j=0}^{[hN]-1} a_j f(zq^j) \tag{7}$$

for which, using (6), we get the estimate

$$\log \mu(r, F) \leqslant \frac{(h+1)^2}{c(h-1)} N^2 \log q + \gamma_2 N + \frac{(\log r + hN \log q)^2}{c \log q} \tag{8}$$

as soon as r is large enough.

Step 2. The aim of this step is to prove that for every rational integer $M \geqslant N$ we have

(i)$_M$: $$F(q^n) = 0 \quad \text{for} \quad n = 0, \ldots, M-1$$

and

(ii)$_M$: $$\mu(q^M, F) < 1.$$

Of course, (i)$_N$ holds by the construction in Step 1. We conclude now as follows:

$$(\mathrm{i})_N \Longrightarrow (\mathrm{ii})_N \Longrightarrow (\mathrm{i})_{N+1} \Longrightarrow (\mathrm{ii})_{N+1} \Longrightarrow \ldots . \tag{9}$$

Here all implications (ii)$_M \Rightarrow$ (i)$_{M+1}$ are trivial, since

$$F(q^n) = \sum_j a_j f(q^{n+j})$$

is a rational integer less than 1 in absolute value, by (ii)$_M$, and therefore $F(q^n)$ vanishes.
To show (i)$_M \Rightarrow$ (ii)$_M$ for all $M \geqslant N$ we may suppose $F \neq 0$. Taking (i)$_M$ into account we find by Schwarz's lemma (see again (Waldschmidt, 1979))

$$\begin{aligned} \log \mu(q^M, F) &\leqslant \log \mu(q^{kM}, F) - M \log \frac{q^{2kM} + q^{2M}}{2q^{(k+1)M}} \\ &\leqslant \left(\frac{(h+1)^2}{h-1} + (h+k)^2\right) \frac{M^2 \log q}{c} - (k-1)M^2 \log q + \gamma_3 M \end{aligned} \tag{10}$$

if we use (8) and $M \geqslant N$. From the last inequality (ii)$_M$ follows (as soon as N, and so M, is large enough) if we choose c such that the inequality

$$\frac{1}{k-1}\left(\frac{(h+1)^2}{h-1} + (h+k)^2\right) < c \tag{11}$$

is satisfied under the above hypotheses $h, k > 1$. Using the usual necessary conditions for the left-hand side of (11) to be a minimum we get

$$h = 1.5224\ldots, \qquad k = 5.3062\ldots$$

which leads to $c > 13.657\ldots$

Step 3. Since we have $(\mathrm{ii})_M$ for all integers $M \geqslant N$, we know from Liouville's theorem that F is constant, and indeed it vanishes identically in virtue of $(\mathrm{i})_M$. Differentiating (7) we get

$$f^{(s)}(0)\sum_j a_j q^{js} = 0 \qquad (s = 0, 1, \ldots).$$

Since the polynomial $\sum_j a_j X^j \in \mathbf{Z}[X] \setminus \{0\}$ has only finitely many zeros, and a fortiori those of the form q^s, we see that $f^{(s)}(0) = 0$ for all $s = 0, 1, \ldots$ up to at most finitely many exceptions. Therefore f must be a polynomial, and Theorem 1 is proved.

Sketch of the proof of Theorem 2. Let again N be a positive integer to be fixed later. Put $I := [N^{2/3}]$ and $J := [N^{2/3}]$. Then, by Siegel's lemma, we can find a_{ij} (as is asserted in Theorem 2) such that the equations

$$\sum_{i=0}^{I-1}\sum_{j=0}^{J-1} a_{ij} q^{in} f\left(q^{n+j}\right) = 0, \qquad n = 0, \ldots, N-1,$$

and the inequalities

$$\log|a_{ij}| \leqslant \gamma_1 N^{5/3}, \qquad i = 0, \ldots, I-1, \quad j = 0, \ldots, J-1,$$

are satisfied with the same conventions on $\gamma_1, \ldots$ as above.
Now let us define the function F to be the left-hand side of (3) having zeros at the points q^n, $n = 0, \ldots, N-1$, by construction. For this new F we can prove again $(\mathrm{i})_M$ and $(\mathrm{ii})_M$ for all $M \geqslant N$ along the same lines as in our earlier Step 2. The only modification is to use now the first line of (10) with $k = 2$ which leads to the estimate

$$\log \mu(q^M, F) \leqslant \left(\frac{4}{c} - 1\right) M^2 \log q + \gamma_2 M^{5/3},$$

and thus finally to $F = 0$, if we fix N large enough.

Proof of Theorem 3. Choose $\delta \in \mathbf{R}_+$ with $\delta \leqslant \alpha$, and $J \in \mathbf{N}$ such that $\delta J > d$. Since $\alpha \leqslant 1$ we certainly have $J > d \geqslant 1$.

Step 1. Let us define

$$F(z) := \sum_{0 \leqslant i < \delta N} \sum_{0 \leqslant j < J} a_{ij} z^i \left(\frac{f}{g}(z)\right)^j.$$

To get

$$F(q^n) = 0 \qquad \text{for} \quad n = 0, \dots, N-1 \quad \text{with} \quad q^n \in S \tag{12}$$

we have to multiply the linear forms

$$\sum_i \sum_j a_{ij} q^{in} \left(\frac{f}{g}(q^n)\right)^j$$

by

$$\left(\operatorname{den} \frac{f}{g}(q^n)\right)^{J-1},$$

and then the new linear forms in the a_{ij} have integral coefficients in $\mathbf{K}$ whose houses are bounded by

$$q^{in} \overline{\left|\frac{f}{g}(q^n)\right|}^j \left(\operatorname{den} \frac{f}{g}(q^n)\right)^{J-1} \leqslant \exp\left(\left(\delta N n + 2\beta(J-1)n^2\right) \log q\right),$$

due to (4) and (5). Here we have at most N conditions for not less than δJN (greater than dN) unknowns. Using Siegel's lemma (in the case of the number field) we get a nontrivial set of $a_{ij} \in \mathbf{Z}$ such that (12) and

$$\log |a_{ij}| \leqslant \frac{d}{\delta J - d}\left(\frac{\log 2}{d} + \log J(\delta N + 1) + (\delta + 2\beta(J-1))N^2 \log q\right) \leqslant \gamma_1 N^2 \log q$$

are satisfied where γ_1 depends at most on α, β, and d.

Step 2. Here we assert the existence of some constant $\gamma_2 \in \mathbf{R}_+$, subject to the same conditions as γ_1, such that for every $n = 0, 1, \dots$ with $q^n \in S$ we have either

$$F(q^n) = 0$$

or

$$\log |F(q^n)| \geqslant -\gamma_2 n^2 \log q.$$

$F(q^n) \in \mathbf{K}$ for all n with $q^n \in S$ (compare with the hypothesis (a)) implies by the so-called fundamental inequality, using (4) and (5), that

$$\begin{aligned}
\log |F(q^n)| &\geqslant -(d-1) \log \overline{|F(q^n)|} - d \log \operatorname{den} F(q^n) \\
&\geqslant -d\left(\log J(\delta N + 1) + \gamma_1 N^2 \log q + \left(\delta N n + 2\beta(J-1)n^2\right) \log q\right) \\
&\geqslant -\gamma_2 n^2 \log q
\end{aligned}$$

since $n \geqslant N$ as soon as $F(q^n) \neq 0$, see Step 1.

Step 3. Now we define the entire function

$$G(z) := g(z)^{J-1} F(z),$$

and we assert for $M = N, N+1, \ldots$

$$(\mathrm{i})_M: \qquad F(q^n) = 0 \qquad \text{for all} \quad n = 0, \ldots, M-1 \quad \text{with} \quad q^n \in S,$$

$$(\mathrm{ii})_M: \qquad \log \mu(q^n, G) < -\gamma_3 M^2 \log q \qquad \text{where} \quad \gamma_3 := \gamma_2 + \beta(J-1).$$

The proof of this, again, follows the pattern (9), and $(\mathrm{i})_N$ is true by construction in Step 1.
First let us show $(\mathrm{ii})_M \Rightarrow (\mathrm{i})_{M+1}$. If $F(q^M) \neq 0$ then Step 2 implies the left-hand side of the following inequality

$$\begin{aligned} -\gamma_2 M^2 \log q \leqslant \log |F(q^M)| &= \log |G(q^M)| - (J-1) \log |g(q^M)| \\ &< -\gamma_3 M^2 \log q + \beta(J-1) M^2 \log q \end{aligned}$$

where we have used $(\mathrm{ii})_M$ and (4). By the definition of γ_3 this is a contradiction, and therefore we have $F(q^M) = 0$, and thus $(\mathrm{i})_{M+1}$.
Now let us prove $(\mathrm{i})_M \Rightarrow (\mathrm{ii})_M$. By Schwarz's lemma we have for every real $k > 1$

$$\begin{aligned} \log \mu(q^M, G) &\leqslant \log \mu(q^{kM}, G) - \alpha k(k-1) M^2 \log q + \gamma_4 M \\ &\leqslant \Big(\gamma_1 + \delta k + \frac{J-1}{c} k^2 - \alpha k(k-1)\Big) M^2 \log q + \gamma_5 M \end{aligned}$$

taking hypothesis (c) into consideration. Fix $c > (J-1)/\alpha$, where the right-hand side depends only on α and on d, and then choose k so large that

$$\gamma_1 + (\alpha + \delta)k - \Big(\alpha - \frac{J-1}{c}\Big)k^2 < -\gamma_3.$$

(Of course, k depends on α, β, and d).
By $(\mathrm{i})_M$ and $(\mathrm{ii})_M$ the function G vanishes identically, and therefore so does F. Thus $\frac{f}{g}$ is an algebraic function which is meromorphic, and therefore it must be rational as asserted in Theorem 3.

REFERENCES

Bézivin, J.-P. (1984). Une généralisation à plusieurs variables d'un résultat de Gel'fond. *Analysis* **4**, 125–141.

Bundschuh, P. (1980). Arithmetische Eigenschaften ganzer Funktionen mehrerer Variablen. *J. reine angew. Math.* **313**, 116–132.

Gelfond, A. O. (1933). Sur les fonctions entières, qui prennent des valeurs entières dans les points β^n, β est un nombre entier positif et $n = 1, 2, \ldots$. *Mat. Sb.* **40**, 42–47 (in Russian).

Gelfond, A. O. (1967). Functions which take on integral values. *Math. Notes* **1**, 337–339.

Waldschmidt, M. (1978). Pólya's theorem by Schneider's method. *Acta Math. Acad. Sci. Hungar.* **31**, 21–25.

Waldschmidt, M. (1979). *Transcendence Methods.* Queens University, Kingston/Ont., Canada.

Wallisser, R. (1985). Über ganze Funktionen, die in einer geometrischen Folge ganze Werte annehmen. *Monatsh. Math.* **100**, 329–335.

New Trends in Probab. and Statist., pp. 17–21
F. Schweiger and E. Manstavičius (Eds.)
1992 VSP/TEV

AN ESTIMATION OF THE DIFFERENCE BETWEEN TWO ZEROS OF A POLYNOMIAL

A. DUBICKAS
Department of Mathematics, Vilnius University,
Naugarduko 24, 2006 Vilnius, Lithuania

ABSTRACT

The lower bound for the modulus of the difference between two zeros of a polynomial is established.

Let $d \in \mathbf{N}$, $d \geqslant 2$,

$$P(x) = a_d x^d + a_{d-1} x^{d-1} + \ldots + a_1 x + a_0 \in \mathbf{C}[x], \qquad a_d a_0 \neq 0.$$

We denote the zeros of the polynomial $P(x)$ by $\alpha_1, \alpha_2, \ldots, \alpha_d$, where $|\alpha_1| \geqslant |\alpha_2| \geqslant \ldots \geqslant |\alpha_d|$. Let also $\Delta = \Delta(P)$ be the discriminant of $P(x)$,

$$M(P) = |a_d| \prod_{t=1}^{d} \max(1, |\alpha_t|),$$
$$T(P) = |a_d|^{d-1} |\alpha_1|^{d-1} |\alpha_2|^{d-2} \ldots |\alpha_{d-1}|.$$

Various estimates for $|\alpha_i - \alpha_j|$ have been obtained (Guting, 1967; Mignotte and Payafar, 1979). The most precise inequalities are

$$|\alpha_i - \alpha_j| \geqslant |\Delta|^{\frac{1}{2}} \max(1, |\alpha_i|, |\alpha_j|) d^{-\frac{d}{2}-1} M(P)^{1-d}, \tag{1}$$
$$|\alpha_i - \alpha_j| \geqslant |\Delta|^{\frac{1}{2}} \min(|\alpha_i|, |\alpha_j|) 2^{1-\frac{d(d-1)}{2}} T(P)^{-1}. \tag{2}$$

M. Mignotte and M. Payafar obtained some more precise inequalities than (2) for $P(x) \in \mathbf{R}[x]$, $\alpha_i \in \mathbf{C} \setminus \mathbf{R}$.
It is natural to compare the estimates (1) and (2). For large d we have

$$d^{-\frac{d}{2}-1} \approx \exp\Big(-\frac{d \log d}{2}\Big),$$
$$2^{1-\frac{d(d-1)}{2}} \approx \exp\Big(-\frac{\log 2}{2} d^2\Big).$$

Therefore

$$d^{-\frac{d}{2}-1} \gg 2^{1-\frac{d(d-1)}{2}}.$$

On the other hand $M(P)^{d-1} \geqslant T(P)$, and for some polynomials $T(P)^{-1} \gg M(P)^{1-d}$. In this paper we prove the inequality the right side of which includes both the "large parts" of (1), (2), i.e. $\exp(-cd \log d)$ and $T(P)^{-1}$.

THEOREM. *For any two zeros α_i and α_j of the polynomial $P(x) \in \mathbf{C}[x]$ the following inequality holds*

$$|\alpha_i - \alpha_j| > 2|\Delta|^{\frac{1}{2}} \max\left(|\alpha_i|, |\alpha_j|\right) d^{-\frac{d}{2}-\frac{3}{2}} T(P)^{-1}. \tag{3}$$

Furthermore, if

$$P(x) \in \mathbf{R}[x], \qquad \alpha_i, \alpha_j \in \mathbf{C} \setminus \mathbf{R}, \qquad \alpha_i \neq \overline{\alpha}_j$$

then

$$|\alpha_i - \alpha_j| > \sqrt{\frac{3}{2}} |\Delta|^{\frac{1}{4}} \max\left(|\alpha_i|, |\alpha_j|\right) d^{-\frac{d}{4}-1} T(P)^{-\frac{1}{2}}. \tag{4}$$

If

$$P(x) \in \mathbf{R}[x], \qquad \alpha_i \in \mathbf{C} \setminus \mathbf{R}, \qquad \alpha_j \in \mathbf{R},$$

then

$$|\alpha_i - \alpha_j| > \sqrt[3]{2} |\Delta|^{\frac{1}{6}} \max\left(|\alpha_i|, |\alpha_j|\right) d^{-\frac{d}{6}-1} T(P)^{-\frac{1}{3}}. \tag{5}$$

For example, let

$$P(x) = x^d - 1, \qquad \alpha_1 = e^{2\pi i/d}, \qquad \alpha_2 = e^{4\pi i/d}.$$

It is well known that $\Delta = \Delta(P) = d^d$. Since $T(P) = 1$ inequality (4) gives

$$|\alpha_1 - \alpha_2| = 2 \sin \frac{\pi}{d} > \sqrt{\frac{3}{2}} d^{-1}.$$

In order to prove this theorem we shall need the following lemma.

LEMMA. *Let*

$$\begin{array}{lll} u_i \in \mathbf{N}, & i = 1, 2, \ldots, r-1, & u_1 < u_2 < \ldots < u_{r-1}, \\ z_j \in \mathbf{C}, & j = 1, 2, \ldots, r, & |z_1| \geqslant |z_2| \geqslant \ldots \geqslant |z_r|, \end{array}$$

$$A = \begin{vmatrix} 1 & 1 & \ldots & 1 \\ z_1^{u_1} & z_2^{u_1} & \ldots & z_r^{u_1} \\ z_1^{u_2} & z_2^{u_2} & & z_r^{u_2} \\ \vdots & \vdots & & \vdots \\ z_1^{u_{r-1}} & z_2^{u_{r-1}} & \ldots & z_r^{u_{r-1}} \end{vmatrix}.$$

Then

$$|A| \leqslant r^{\frac{r}{2}} |z_1|^{u_{r-1}} |z_2|^{u_{r-2}} \ldots |z_{r-1}|^{u_1}. \tag{6}$$

For $u_i = i$, $i = 1, 2, \ldots, r-1$, this lemma has been proved in (Langevin, 1989).

Proof. We put

$$\frac{z_2}{z_1} = t_1,\ \frac{z_3}{z_2} = t_2,\ \ldots,\ \frac{z_r}{z_{r-1}} = t_{r-1}.$$

Then

$$|t_1| \leqslant 1,\ |t_2| \leqslant 1,\ \ldots,\ |t_{r-1}| \leqslant 1.$$

It is clear that

$$A = z_1^{u_{r-1}} z_2^{u_{r-2}} \ldots z_{r-1}^{u_1} p(t_1, t_2, \ldots, t_{r-1}),$$

where $p(t_1, t_2, \ldots, t_{r-1})$ is a polynomial. According to the maximum modulus principle we have

$$\begin{aligned} |A| &= |z_1|^{u_{r-1}} |z_2|^{u_{r-2}} \ldots |z_{r-1}|^{u_1} |p(t_1, t_2, \ldots, t_{r-1})| \\ &\leqslant |z_1|^{u_{r-1}} |z_2|^{u_{r-2}} \ldots |z_{r-1}|^{u_1} \max_{|t_1| \leqslant 1, \ldots, |t_{r-1}| \leqslant 1} |p(t_1, \ldots, t_{r-1})| \\ &= |z_1|^{u_{r-1}} |z_2|^{u_{r-2}} \ldots |z_{r-1}|^{u_1} |p(t_1^0, t_2^0, \ldots, t_{r-1}^0)|, \end{aligned} \tag{7}$$

where

$$|t_1^0| = |t_2^0| = \ldots = |t_{r-1}^0| = 1.$$

On the other hand

$$A = z_1^{u_{r-1}} \ldots z_{r-1}^{u_1} D \tag{8}$$

where

$$D = \begin{vmatrix} 1 & 1 & \ldots & 1 & 1 \\ (z_1/z_{r-1})^{u_1} & (z_2/z_{r-1})^{u_1} & & 1 & (z_r/z_{r-1})^{u_1} \\ \vdots & \vdots & & \vdots & \vdots \\ 1 & (z_2/z_1)^{u_{r-1}} & \ldots & (z_{r-1}/z_1)^{u_{r-1}} & (z_r/z_1)^{u_{r-1}} \end{vmatrix}$$

$$= \begin{vmatrix} 1 & 1 & \ldots & 1 \\ (t_1 t_2 \ldots t_{r-2})^{-u_1} & (t_2 \ldots t_{r-2})^{-u_1} & & t_{r-1}^{u_1} \\ \vdots & \vdots & & \vdots \\ 1 & t_1^{u_{r-1}} & \ldots & (t_1 t_2 \ldots t_{r-1})^{u_{r-1}} \end{vmatrix}$$

The determinant (8) is equal to $p(t_1, t_2, \ldots, t_{r-1})$. Then application of the Hadamard's inequality

$$\left|\det \|a_{ij}\|_{i,j=1,2,\ldots,r}\right| \leqslant \left(\prod_{j=1}^{r} \sum_{i=1}^{r} |a_{ij}|^2\right)^{1/2}$$

gives

$$|p(t_1^0, t_2^0, \ldots, t_{r-1}^0)| \leqslant r^{\frac{r}{2}}.$$

Now from (7) we obtain the inequality (6) and the lemma follows.

Proof of the theorem. Without loss of generality we assume that $i = 1$, $j = 2$. It is well known that

$$\Delta = a_d^{2d-2} \begin{vmatrix} 1 & 1 & \ldots & 1 \\ \alpha_1 & \alpha_2 & & \alpha_d \\ \vdots & \vdots & & \vdots \\ \alpha_1^{d-1} & \alpha_2^{d-1} & \ldots & \alpha_d^{d-1} \end{vmatrix}^2 \tag{9}$$

Subtracting the second column of the determinant (9) from the first one we find that

$$\frac{\Delta}{a_d^{2d-2}} = (\alpha_1-\alpha_2)^2 \begin{vmatrix} 0 & 1 & \dots & 1 \\ 1 & \alpha_2 & \dots & \alpha_d \\ \alpha_1+\alpha_2 & \alpha_2^2 & & \alpha_d^2 \\ \vdots & \vdots & & \vdots \\ \alpha_1^{d-1}+\alpha_1^{d-2}\alpha_2+\ldots+\alpha_2^{d-2} & \alpha_2^{d-1} & \dots & \alpha_d^{d-1} \end{vmatrix}^2 \tag{10}$$
$$= (\alpha_1-\alpha_2)^2\Big(\sum_{u=1}^{d}(-1)^{u-1}a_uM_u\Big)^2,$$

where a_u is the u-th element of the first column, and M_u is the proper minor. It is evident that

$$|a_u| \leqslant (u-1)|\alpha_1|^{u-2}. \tag{11}$$

Applying the Lemma to the determinant M_u, $u \geqslant 2$, we have

$$|M_u| \leqslant (d-1)^{\frac{d-1}{2}}|\alpha_2|^{d-1}|\alpha_3|^{d-2}\ldots|\alpha_{d-u+1}|^{u}|\alpha_{d-u+2}|^{u-2}\ldots|\alpha_{d-1}|.$$

Then by virtue of (11)

$$\begin{aligned} |a_uM_u| &\leqslant (d-1)^{\frac{d-1}{2}}(u-1)|\alpha_1|^{u-2}|\alpha_2|^{d-1}\ldots|\alpha_{d-u+1}|^{u}|\alpha_{d-u+2}|^{u-2}\ldots|\alpha_{d-1}| \\ &= (d-1)^{\frac{d-1}{2}}(u-1)|\alpha_1|^{u-2}|\alpha_2||\alpha_3|\ldots|\alpha_{d-u+1}|\frac{T(P)}{|\alpha_1|^{d-1}|a_d|^{d-1}} \\ &\leqslant (d-1)^{\frac{d-1}{2}}(u-1)\frac{T(P)}{|\alpha_1||a_d|^{d-1}}. \end{aligned}$$

Hence

$$\Big|\sum_{u=1}^{d}(-1)^{u-1}a_uM_u\Big| \leqslant (d-1)^{\frac{d-1}{2}}\frac{d(d-1)}{2}\frac{T(P)}{|\alpha_1||a_d|^{d-1}} < \frac{d^{\frac{d+3}{2}}T(P)}{2|\alpha_1||a_d|^{d-1}}.$$

Therefore from (10) we deduce

$$\frac{|\Delta|}{|a_d|^{2d-2}} < |\alpha_1-\alpha_2|^2\frac{d^{d+3}T(P)^2}{4|\alpha_1|^2|a_d|^{2d-2}},$$
$$|\alpha_1-\alpha_2| > 2|\Delta|^{\frac{1}{2}}|\alpha_1|d^{-\frac{d+3}{2}}T(P)^{-1},$$

and the inequality (3) takes place.
Now let $P(x) \in \mathbf{R}[x]$, $i=1$, $j=3$,

$$\alpha_1,\alpha_3 \in \mathbf{C}\setminus\mathbf{R}, \qquad \alpha_1=\overline{\alpha}_2, \qquad \alpha_3=\overline{\alpha}_4.$$

Then $|\alpha_1 - \alpha_3| = |\alpha_2 - \alpha_4|$. Subtracting the third column of the determinant (9) from the first, the fourth from the second, and applying the Laplace theorem to the first two columns, we find

$$\frac{|\Delta|}{|a_d^{2d-2}|} < |\alpha_1 - \alpha_3|^4 \Big((d-2)^{\frac{d-2}{2}} \frac{2d^3}{3} \frac{T(P)}{|\alpha_1|^2 |a_d|^{d-1}} \Big)^2 .$$

Hence

$$|\alpha_1 - \alpha_3| > \sqrt{\frac{3}{2}} |\Delta|^{\frac{1}{4}} |\alpha_1| d^{-\frac{d}{4}-1} T(P)^{-\frac{1}{2}},$$

and the inequality (4) holds.
If $P(x) \in \mathbf{R}[x]$, $i = 1$, $j = 3$,

$$\alpha_1 \in \mathbf{C} \setminus \mathbf{R}, \qquad \alpha_3 \in \mathbf{R}, \qquad \alpha_1 = \overline{\alpha}_2,$$

then $|\alpha_1 - \alpha_3| = |\alpha_2 - \alpha_3|$. Note that

$$|\alpha_1 - \alpha_2| \leqslant |\alpha_1 - \alpha_3| + |\alpha_2 - \alpha_3| = 2|\alpha_1 - \alpha_3|.$$

In this case after subtraction of the third column of the determinant (9) from the first and the second ones, and after subtraction of the second column from the first one we have

$$\frac{|\Delta|}{|a_d|^{2d-2}} < |\alpha_1 - \alpha_3|^2 |\alpha_2 - \alpha_3|^2 |\alpha_1 - \alpha_3|^2 \Big((d-2)^{\frac{d-2}{2}} \frac{d^4}{4} \frac{T(P)}{|\alpha_1|^3 |a_d|^{d-1}} \Big)^2$$
$$< \frac{1}{4} |\alpha_1 - \alpha_3|^6 d^{d+6} T(P)^2 |\alpha_1|^{-6} |a_d|^{2-2d}.$$

Hence we obtain the required estimate (5). This completes the proof of our theorem.

Remark. For large d the constants 2, $\sqrt{3/2}$, $\sqrt[3]{2}$ in the inequalities (3), (4), and (5) may be replaced by $2\sqrt{e}$, $\sqrt{3e/2}$, $\sqrt[3]{2e}$, respectively.

REFERENCES

Guting, R. (1967). Polynomials with multiple zeros. *Mathematika* **14**, 181–196.

Langevin, M. (1989). Systèmes complèts de conjugues sur un corps quadratique imaginaire et ensembles de largeur constante. *Number Theory and Applications, NATO ASI Series, Ser. C*, **265**, 445–457.

Mignotte, M. and Payafar, M. (1979). Distance entre les racines d'un polynome. *RAIRO Analyse Numerique* **13**, 181–192.

New Trends in Probab. and Statist., pp. 23–29
F. Schweiger and E. Manstavičius (Eds.)
1992 VSP/TEV

ON THE ARITHMETIC PROPERTIES OF THE VALUES OF E-FUNCTIONS

A. B. SHIDLOVSKII
Department of Mathematics, Moscow State University,
119899 Moscow, Russia

ABSTRACT

The paper presents the proof of a theorem which gives a necessary and sufficient condition for the algebraical independence of the values at algebraic points of a subset of E-functions in the case when the original set of E-functions is algebraically dependent over $\mathbf{C}(z)$. This result is a final step in a series of theorems, establishing conditions under which the values of the considered E-functions at algebraic points are algebraically independent.

C. L. Siegel extended the classical results on the arithmetic properties of the values of the exponential function at algebraic points after Hermite and Lindemann to the values of Bessel's functions (see (Siegel, 1929)). Siegel's method is applicable to a sufficiently wide class of entire functions, which he called the E-functions. The coefficients of the Taylor's expansions of these functions belong to an algebraic field of a finite degree over $\mathbf{Q}$ and satisfy some arithmetic properties.

Later C. L. Siegel presented his method in a form of a general theorem on the algebraical independence of the values of a set of E-functions constituting a solution of a system of linear differential equations of the first order with the coefficients from $\mathbf{C}(z)$ (see (Siegel, 1949)). In view of great difficulties of verifying the conditions of his theorem for concrete functions the book (Siegel, 1949)) contained no new arithmetical results compared to his original paper.

In the second half of the fifties Siegel's method was essentially developed and generalised. Many general theorems on the algebraical independence of the values of E-functions were proved, and they were applied to concrete functions. The possibilities of the applications of the method to E-functions were extended to their natural limits.

The detailed history of the considered problem is presented in (Shidlovskii, 1989a). It contains the definition of an E-function, another necessary definitions and the proofs of all the theorems formulated below, except from Theorems 1 and 3.

Let $\mathbf{A}$ denote the field of all algebraic numbers over $\mathbf{Q}$, $\mathbf{K}$ denote algebraic number field over $\mathbf{Q}$, and $\mathbf{I}$ denote some imaginary quadratic field over $\mathbf{Q}$. The E-functions with the coefficients of their Taylor's expansions from the field $\mathbf{K}$ (from $\mathbf{I}$) will be called $\mathbf{K}E$-functions ($\mathbf{I}E$-functions). We say that some assertion holds for almost

all $\xi \in \mathbf{A}$, if it is true for all $\xi \in \mathbf{A}$ with a possible exception of a finite set of points.
Consider now the set of E-functions

$$f_1(z), \dots, f_m(z) \tag{1}$$

constituting a solution of a system of linear differential equations

$$y'_k = Q_{k0} + \sum_{i=1}^{m} Q_{ki} y_i, \qquad k = 1, \dots, m, \quad Q_{ki} \in \mathbf{C}(z), \tag{2}$$

(which can be homogeneous as well) or, for $m \geqslant 2$, of a system of linear homogeneous differential equations

$$y'_k = \sum_{i=1}^{m} Q_{ki} y_i, \qquad k = 1, \dots, m, \quad Q_{ki} \in \mathbf{C}(z). \tag{3}$$

Let $T(z) \in \mathbf{C}[z]$ be the least common denominator of all the functions Q_{ki} in (2) (in (3)).
Further let $\xi \in \mathbf{A}$ and $\xi T(\xi) \neq 0$. It is proved that:
1°. The numbers

$$f_1(\xi), \dots, f_m(\xi) \tag{4}$$

are algebraically (homogeneously algebraically) independent if and only if the functions (1) are algebraically (homogeneously algebraically) independent over $\mathbf{C}(z)$.
This result enabled to prove the transcendence and the algebraical independence of the values at algebraic points of many subclasses of E-functions. The methods of proving the algebraical independence of functions over $\mathbf{C}(z)$ were developed for this purpose.
2°. The transcendence degree (homogeneous transcendence degree) of the numbers (4) is equal to l, $0 \leqslant l \leqslant m$, if and only if the transcendence degree (homogeneous transcendence degree) over $\mathbf{C}(z)$ of the functions (1) is equal to l.
3°. If $m \geqslant 2$ and the transcendence degree (homogeneous transcendence degree) of the functions (1) over $\mathbf{C}(z)$ is equal to l, $1 \leqslant l \leqslant m-1$, and the functions

$$f_1(z), \dots, f_l(z) \tag{5}$$

are algebraically (homogeneously algebraically) independent over $\mathbf{C}(z)$, then for almost all $\xi \in \mathbf{A}$ the numbers

$$f_1(\xi), \dots, f_l(\xi) \tag{6}$$

are algebraically (homogeneously algebraically) independent.
In particular, if an E-function $f(z)$ is transcendental and satisfies a linear differential equation with coefficients from $\mathbf{C}[z]$, then for almost all $\xi \in \mathbf{A}$ the number $f(\xi)$ is transcendental.
The set of points $\xi \in \mathbf{A}$ at which the numbers (6) are algebraically (homogeneously algebraically) dependent contains the number 0, possibly the zeroes of $T(z)$ and

several other points, depending on the structure of the algebraical equations, connecting the functions (1) over $\mathbf{C}(z)$, and the algebraic field $\mathbf{K}$, which contains the coefficients of the Taylor's expansions of the functions (1), and the number ξ.
The problems, connected with the description of the set of all the exceptional points of the functions (5) in the general case, or for some subclasses of E-functions arose in a natural way. In a series of cases such problems were solved. For example, if the transcendence degrees (the homogeneous transcendence degrees) of the functions (1) over $\mathbf{C}(z)$ and over $\mathbf{C}$ are equal to each other, then there are no additional exceptional points except from $z = 0$, and possibly from the zeroes $T(z)$ of the functions (5). In some cases for the functions (1), connected by algebraical equations over $\mathbf{C}(z)$ with a special structure, the finite sets were found, which contain all the exceptional points of the functions (5) (see (Shidlovskii, 1987)).
The papers (Shidlovskii, 1987, 1989b) present a complete solution of the problem of describing, the set of the exceptional points for the $\mathbf{I}E$-functions and $\xi \in \mathbf{I}$.
Consider the irreducible and primitive polynomials

$$P = P(z_1, \ldots, z_m) \in \mathbf{I}[z][z_1, \ldots, z_m]. \tag{7}$$

We shall denote by $D_{P,l}(z) \in \mathbf{I}[z]$, $1 \leqslant l \leqslant m-1$, the greatest common divisor of all the coefficients of the polynomial (7) at the terms, containing at least one of the variables $z_{l+1}, \ldots, z_m$.

THEOREM 1. *Let the set of* $\mathbf{I}E$*-functions* (1), $m \geqslant 2$, *constitute a solution of the system* (2) (*system* (3)), *and let the transcendence degree* (*homogeneous transcendence degree*) *of these functions over* $\mathbf{C}(z)$ *be equal to* l, $1 \leqslant l \leqslant m-1$. *Let the functions* (5) *be algebraically* (*homogeneously algebraically*) *independent over* $\mathbf{C}(z)$, *and* $\xi \in \mathbf{I}$, $\xi T(\xi) \neq 0$.
Then the numbers (6) *are algebraically* (*homogeneously algebraically*) *dependent if and only if there exists an irreducible and primitive* (*homogeneous*) *polynomial* (7) *such that* $P(f_1(z), \ldots, f_m(z)) = 0$ *and* $D_{P,l}(\xi) = 0$.

The case $l = m-1$ when the functions (1) are connected by only one algebraical equation (up to a constant factor) over $\mathbf{C}(z)$ is of special interest. To find the set of the exceptional points ξ, $\xi T(\xi) \neq 0$, one needs only to find the greatest common multiple of all the coefficients at the terms, containing $f_m(z)$ in the polynomial connecting the functions (1).
Note that a necessary and sufficient condition describing the properties of the exceptional points for the $\mathbf{I}E$-functions in the general case was published in (Shidlovskii, 1962). We prove this result in a form convenient for what follows.

THEOREM 2. *Let the set of* E*-functions* (1), $m \geqslant 2$, *constitute solution of the system* (2) (*the system* (3)), *and let the transcendence degree* (*the homogeneous transcendence degree*) *of these functions over* $\mathbf{C}(z)$ *be equal to* $l, 1 \leqslant l \leqslant m-1$. *Let the functions* (5) *be algebraically* (*homogeneously algebraically*) *independent over* $\mathbf{C}(z)$, *and let the functions* (1) *be connected by the algebraical* (*homogeneously algebraical*) *equations*

$$\begin{gathered} P_\nu(f_1(z), \ldots, f_l(z), f_\nu(z)) = 0, \qquad \nu = l+1, \ldots, m, \\ P_\nu = P_\nu(z_1, \ldots, z_l, z_m) \in \mathbf{A}[z][z_1, \ldots, z_l, z_\nu], \end{gathered} \tag{8}$$

here P_ν is an irreducible and a primitive (homogeneous) polynomial. Assume further $\xi \in \mathbf{A}$, $\xi T(\xi) \neq 0$.
Then the numbers (5) are algebraically independent if and only if at $z = \xi$ all the coefficients at various powers of $f_\nu(z)$, being the polynomials from $\mathbf{A}[z]\,[f_1(z), \ldots, f_l(z)]$, vanish in at least one of the equations (8).

Note that the polynomials (8) are determined uniquely up to constant factors.

Proof. We consider the inhomogeneous case. The homogeneous case is treated in analogy. According to the assertion 2° formulated above (the third fundamental theorem in (Shidlovskii, 1989a)

$$\deg\operatorname{tr}\{f_1(\xi), \ldots, f_m(\xi)\} = l. \tag{9}$$

Let $\Omega = \mathbf{A}(f_1(\xi), \ldots, f_l(\xi))$. If the numbers (6) are algebraically dependent, then $\deg\operatorname{tr}\Omega = r < l$. If for $z = \xi$ in every equation (8) there exist a non-vanishing coefficient at some powers of $f_\nu(z)$, then all the numbers

$$f_\nu(\xi), \qquad \nu = l+1, \ldots, m,$$

are algebraic over Ω, and therefore

$$\deg\operatorname{tr}\left(f_1(\xi), \ldots, f_m(\xi)\right) = r < l,$$

which contradicts to (9). Thus the numbers (6) are algebraically independent.
Assume now that in one of the equations (8), all the coefficients at various powers of $f_\nu(\xi)$ vanish at $z = \xi$. In view of the primitivity of the polynomial P this evidently implies that at least one of such coefficients presents a non-trivial algebraical relation among the numbers (6) over $\mathbf{A}$, and therefore these numbers are algebraically dependent.
Theorem 2 is useful in some cases as an auxiliary assertion, but, generally speaking, it does not allow to find the exceptional points of the functions (5) immediately.
This paper presents another necessary and sufficient condition for the algebraical independence of the numbers (6) for the $\mathbf{K}E$-functions, which demonstrates the reason for existing of the additional exceptional points for the functions (5). But this condition turns to be more complicated than in the case of Theorem 2 and Theorem 1 for the $\mathbf{I}E$-functions.

THEOREM 3. *Let the set of E-functions* (1), $m \geqslant 2$, *constitute a solution of the system of linear differential equations* (2) (*homogeneous differential equations* (3)), *and the transcendence degree* (*the homogeneous transcendence degree*) *of these functions over* $\mathbf{C}(z)$ *be equal to* l, $1 \leqslant l \leqslant m-1$, *let the functions* (5) *be algebraically* (*homogeneously algebraically*) *independent over* $\mathbf{C}(z)$. *Assume further* $\xi \in \mathbf{A}$, $\xi T(\xi) \neq 0$.
Then the numbers (6) *are algebraically* (*homogeneously algebraically*) *dependent if and only if there exists an irreducible polynomial* (*homogeneous polynomial*)

$$P(z, z_1, \ldots, z_m) \in \mathbf{A}[z, z_1, \ldots, z_m],$$

such that $P(z, f_1(z), \ldots, f_m(z)) = 0$, *and*

$$P(\xi, z_1, \ldots, z_m) = P_1(z_1, \ldots, z_l)P_2(z_1, \ldots, z_m),$$

$$P_1 = P_1(z_1, \ldots, z_l) \in \mathbf{A}[z_1, \ldots, z_l], \qquad P_2(z_1, \ldots, z_m) \in \mathbf{A}[z_1, \ldots, z_m],$$

with P_1 *irreducible and* $P_1(f_1(\xi), \ldots, f_l(\xi)) = 0$.

We can restrict ourselves to the inhomogeneous case of the theorem. To show the relation between Theorem 3 and Theorem 1, we first prove the simplest case, when $l = m - 1$.

The functions $f_1(z)$, ..., $f_{m-1}(z)$ are algebraically independent over $\mathbf{C}(z)$. Therefore there exist a unique (up to a constant factor) irreducible and containing z_m polynomial

$$P = P(z, z_1, \ldots, z_m) \in \mathbf{A}[z, z_1, \ldots, z_m]$$

such that

$$P(z, f_1(z), \ldots, f_m(z)) = 0.$$

Suppose that the numbers $f_1(\xi)$, ..., $f_{m-1}(\xi)$ are algebraically dependent. Then there exists a unique (up to a constant factor) polynomial

$$P_1(z_1, \ldots, z_{m-1}) \in \mathbf{A}[z_1, \ldots, z_{m-1}]$$

such that

$$P_1(f_1(\xi), \ldots, f_{m-1}(\xi)) = 0. \tag{10}$$

By Theorem 2 all the coefficients of the polynomial P at the powers of z_m, which are polynomials from $\mathbf{A}[z, z_1, \ldots, z_{m-1}]$. vanish while substituting

$$z = \xi, \; z_1 = f_1(\xi), \; \ldots, \; z_{m-1} = f_{m-1}(\xi).$$

Thus all the coefficients of the polynomial $P(\xi, z_1, \ldots, z_m)$ at the powers of z_m are divisible by an irreducible polynomial P_1, and

$$P(\xi, z_1, \ldots, z_m) = P_1(z_1, \ldots, z_{m-1})P_2(z_1, \ldots, z_m). \tag{11}$$

The assertion of the theorem follows now from (10) and (11).

Proof of Theorem 3. Let the numbers (6) be algebraically dependent. Denote

$$\Omega = \mathbf{A}(f_1(\xi), \ldots, f_l(\xi)).$$

Then

$$\deg \operatorname{tr} \Omega = r, \qquad \max(0, 2l - m) \leqslant r \leqslant l - 1.$$

Let the numbers $f_1(\xi)$, ..., $f_r(\xi)$ be algebraically independent. By the assertion 2°

$$\deg \operatorname{tr}(f_1(\xi), \ldots, f_m(\xi)) = l.$$

We may suppose that the l numbers

$$f_1(\xi), \; \ldots, \; f_r(\xi), \; f_{l+1}(\xi), \; \ldots, \; f_{2l-r}(\xi)$$

are algebraically independent. Then l functions

$$f_1(z),\ \dots,\ f_r(z),\ f_{l+1}(z),\ \dots,\ f_{2l-r}(z) \tag{12}$$

corresponding to them are algebraically independent over $\mathbf{C}(z)$.
But the functions (12) together with the functions $f_j(z)$, $r+1 \leqslant j \leqslant l$, are algebraically dependent over $\mathbf{C}(z)$, and they are connected by the unique (up to a constant factor) equation

$$P\big(z, f_1(z), \dots, f_r(z), f_j(z), f_{l+1}(z), \dots, f_{2l-r}(z)\big) = 0, \tag{13}$$

$$P\big(z, z_1, \dots, z_r, z_j, z_{l+1}, \dots, z_{2l-r}\big) \in \mathbf{A}\big[z, z_1, \dots, z_r, z_j, z_{l+1}, \dots, z_{2l-r}\big],$$

where P is an irreducible polynomial.
We consider P as a polynomial from the ring

$$\mathbf{A}\big[z, z_1, \dots, z_r, z_j\big]\big[z_{l+1}, \dots, z_{2l-r}\big].$$

Then all the coefficients of P, while substituting into them

$$z = \xi,\ z_1(\xi) = f_1(\xi),\ \dots,\ z_r(\xi) = f_r(\xi),\ z_j(\xi) = f_j(\xi),$$

belong to Ω. But the numbers $f_{l+1}(\xi)$, $\dots$, $f_{2l-r}(\xi)$ are algebraically independent over Ω. Therefore it follows from (13) that the coefficients of P vanish after the above substitution.
But the numbers $f_1(\xi)$, $\dots$, $f_r(\xi)$, $f_j(\xi)$ are connected by the unique algebraic equation with the coefficients from $\mathbf{A}$

$$P_1\big(f_1(\xi), \dots, f_r(\xi), f_j(\xi)\big) = 0, \tag{14}$$

$$P_1 = P_1\big(z_1, \dots, z_r, z_j\big) \in \mathbf{A}\big[z_1, \dots, z_r, z_j\big],$$

where P_1 is an irreducible polynomial. Thus all the coefficients of

$$P\big(\xi, z_1, \dots, z_r, z_j, z_{l+1}, \dots, z_{2l-r}\big)$$

as polynomials from

$$\mathbf{A}\big[z_1, \dots, z_r, z_j\big]\big[z_{l+1}, \dots, z_{2l-r}\big]$$

are divisible by $P_1(z_1, \dots, z_r, z_j)$. Now

$$\begin{aligned} &P\big(\xi, z_1, \dots, z_r, z_j, z_{l+1}, \dots, z_{2l-r}\big) \\ &\quad = P_1\big(z_1, \dots, z_r, z_j\big) P_2\big(z_1, \dots, z_r, z_j, z_{l+1}, \dots, z_{2l-r}\big). \end{aligned}$$

In view of (14) implies the Theorem.
The inverse assertion is evident.
Theorem 3 can be refined if one describes the structure of the polynomial P.

In some cases the assertion of Theorem 3 becomes analogous to the assertion of Theorem 1. For example, this occurs in the case of a linear polynomial in $z_1, \ldots, z_m$. It is unlikely that a similar situation can take place in the general case. Therefore it is of interest to construct an example of a concrete set of E-functions for which the assertion, analogous to Theorem 1, doesn't hold under the assumptions of Theorem 3.

Under the hypotheses of Theorem 3 denote by Λ the set of all the exceptional points ξ of the functions (5) such that $\xi T(\xi) \neq 0$, and by Λ^* denote some finite subset of points from $\mathbf{A}$ such that $\Lambda \subset \Lambda^*$.

In the case $l = m - 1$, $\mathbf{K} = \mathbf{I}$, $\xi \in \mathbf{I}$, by Theorem 1 the polynomial P is unique and Λ coincides with the set of the zeroes of the polynomial $D_{P,m-1}(z)$.

In the general case neither Theorem 3 nor Theorem 1 do not allow, generally speaking, to obtain the set Λ or some set Λ^* since they establish only the existence of the polynomial P. If the polynomial P is known, then, depending on its structure, sometimes it is possible, to determine either Λ or Λ^*. But the assertions of Theorems 1–3 make clear the reasons of existing of additional exceptional points of the functions (5).

Note, that with the help of the equations (8) one can present an algorithm, which enables to find some set Λ^* as a result of a great number of computations.

The formulation of Theorem 3 was published in (Shidlovskii, 1991).

REFERENCES

Siegel, C. L. (1929–1930). Über einige Anwendungen diophantischer Approximationen. *Abh. Preuss. Akad. Wiss. Berlin. Phys.-math. Kl.* 1, 1–70.

Shidlovskii, A. B. (1962). On transcendence and algebraic independence of the values of E-functions which are connected by an arbitrary number of algebraic equations over the field of rational functions. *Izv. Akad. Nauk SSSR, Ser. Mat.* **26**, 877–910 (in Russian).

Shidlovskii, A. B. (1987). On algebraic independence of the values of E-functions at algebraic points. *Vestnik Moscow Univ., Ser.* 1, *Mat. Mech.* 1, 30–33. (in Russian).

Siegel, C. L. (1949). *Transcendental Numbers.* Princeton University Press, Princeton.

Shidlovskii, A. B. (1989a). *Transcendental Numbers.* Walter de Gruyter, Berlin, New York.

Shidlovskii, A. B. (1989b). On a criterion of the algebraical independence of the values of E-functions at algebraic points. *Vestnik Moscow Univ., Ser.* 1, *Mat. Mech.* 6, 17-21 (in Russian).

Shidlovskii, A. B. (1991). On algebraic independence of the values of E-functions. *Uspekhi Mat. Nauk* **46**, 4, 175–176 (in Russian).

New Trends in Probab. and Statist., pp. 31–36
F. Schweiger and E. Manstavičius (Eds.)
1992 VSP/TEV

APPROXIMATION OF IRRATIONALS BY m/n, SATISFYING f(m/n)=a

V. STAKĖNAS
Department of Mathematics, Vilnius University,
Naugarduko 24, 2006 Vilnius, Lithuania

ABSTRACT

The diophantine approximations to an irrational number are being specified by the condition $f(m/n) = a$. Here f denotes a multiplicative function assuming its values in a commutative semigroup with unity e. An estimation of the discrepancy $|\xi - m/n|$ is given in connection with the behaviour of $\lambda(u) = \#\{p^\alpha \leqslant u\colon\ f(p^\alpha) \neq e\}$.

CONCEPTS AND RESULT

Let $\mathbf{Q}_+$ be the set of positive rational numbers, A denote a commutative semigroup with unity e. A function $f\colon\ \mathbf{Q}_+ \longrightarrow A$ is called multiplicative if

$$f\Big(\frac{m_1}{n_1}\cdot\frac{m_2}{n_2}\Big) = f\Big(\frac{m_1}{n_1}\Big)\cdot f\Big(\frac{m_2}{n_2}\Big)$$

holds for all integers m_1, n_1, m_2, n_2 which are pair-wise coprime.
We use the condition $f(m/n) = a$ to specify the rational approximations to the given irrational number $\xi > 0$. Indeed, some well-known requirements such as

$$\begin{array}{lr} m \equiv u \pmod s), \qquad n \equiv v \pmod s) & \text{(Hartman, 1949)},\\ (m, M) = 1 \quad \text{or} \quad (n, M) = 1 & \text{(Negoescu, 1968)},\\ m \text{ and } n \text{ are square-free integers} & \text{(Heath-Brown, 1984)} \end{array}$$

can be expressed in this form with an appropriate choice of f and a.
We consider the multiplicative function f satisfying the conditions

$$f(p^\alpha) = f\Big(\frac{1}{p^\alpha}\Big), \qquad \sum_{f(p)\neq e} \frac{1}{p} < \infty, \qquad f \not\equiv e, \tag{1}$$

where p always denote primes, $\alpha = 1,\ 2,\ \ldots$.

THEOREM. *Let $\xi > 0$ be an irrational number, the multiplicative function f satisfy* (1), *and*

$$\#\{p^\alpha \leqslant u\colon\ f(p^\alpha) \neq e\} = o\Big(\frac{u}{\varphi(u)}\Big) \qquad (u \to +\infty) \tag{2}$$

with an increasing function $\varphi(u)$.
Then for any value a of f there exist infinitely many irreducible fractions m/n satisfying

$$\left|\xi - \frac{m}{n}\right| < \frac{1}{\varphi(n)^2}, \qquad f\left(\frac{m}{n}\right) = a. \tag{3}$$

COROLLARY. *Let the multiplicative function f satisfy* (1) *and $f(p^\alpha) = e$ for $\alpha < u$, $u \geqslant 2$, let $\varepsilon(n) \to 0$ as $n \to \infty$. Then the statement of the Theorem holds for f with $\varphi(n) = \varepsilon(n)n^{1-1/u}\log n$.*

This is an improvement of a result announced by the author in (Stakėnas, 1989).

PROOF OF THE THEOREM.

Let the irrational number $\xi > 0$ be represented as a simple continued fraction $[a_0; a_1, \ldots, a_n, \ldots]$. Consider the set of its convergents

$$C_k = [a_0; a_1, \ldots, a_k] = p_k/q_k \qquad (k \geqslant 0).$$

For fixed k $(k \geqslant 1)$ define

$$s_n = p_k n + p_{k-1}, \qquad v_n = q_k n + q_{k-1} \qquad (n \geqslant 1).$$

It follows from the well-known equality

$$q_k p_{k-1} - p_k q_{k-1} = (-1)^k \qquad (k \geqslant 1) \tag{4}$$

that

$$v_{n+1}s_n - s_{n+1}v_n = (-1)^k \qquad (n \geqslant 1).$$

Then $(s_n, v_n) = 1$ for all integers $n \geqslant 1$. The condition $(s_1, v_1) = 1$ implies that $2 \mid p_k q_k$ or $2 \mid p_{k-1}q_{k-1}$. As a consequence we obtain that the subsequence of C_k with $2 \mid p_k q_k$ is infinite.
Define $t_n = s_n/v_n$ $(n \geqslant 1)$ and consider the discrepancy $|\xi - t_n|$. The sequence of t_n is lying between C_{k-1} and C_k. It approaches monotonically C_k as $n \to +\infty$. Then

$$|\xi - t_n| < |C_k - C_{k+1}| + |t_1 - C_{k+1}| \qquad (n \geqslant 1).$$

Using (4) and

$$p_{k+1} = a_{k+1}p_k + p_{k-1}, \qquad q_{k+1} = a_{k+1}q_k + q_{k-1}$$

we get

$$|C_k - C_{k+1}| = \frac{1}{q_k q_{k+1}} < \frac{1}{q_k^2},$$

$$|t_1 - C_{k+1}| = \left|(q_{k+1}p_{k-1} - p_{k+1}q_{k-1}) + (q_{k+1}p_k - p_{k+1}q_k)\right|\frac{1}{v_1 q_{k+1}}$$
$$\leqslant \frac{a_{k+1}-1}{q_k q_{k+1}} \leqslant \frac{a_{k+1}-1}{a_{k+1}} \cdot \frac{1}{q_k^2} < \frac{1}{q_k^2}.$$

Then it follows that

$$\left|\xi - t_n\right| < \frac{2}{q_k^2} \tag{5}$$

holds for all n $(n \geqslant 1)$.
For k satisfying $2 \mid p_k q_k$ let $Q_k = \{t_n\colon n \leqslant x\}$, where $x = x(k)$ becomes unbounded with $k \to +\infty$. The choice of x will be described later.
We are going to prove that the diophantine approximations satisfying (3) can be taken from the sets Q_k.
Let $r = r(x)$ $(r \geqslant 1)$ be some non-decreasing, unbounded, integer-valued function,

$$P = \prod_{p \leqslant r} p^r.$$

For any $l \mid P$ let

$$E(l) = \left\{t_n\colon\ n \leqslant x,\ l \mid s_n v_n,\ \left(\frac{s_n v_n}{l}, \frac{P}{l}\right) = 1\right\}.$$

For different values of l the sets $E(l)$ are disjoint.

LEMMA. *The function $r(x)$ can be chosen in a way that*

$$\#E(l) \geqslant \frac{1}{2}\frac{x}{l} g(l) \prod_{p \mid (P/l)} \left(1 - \frac{1}{p} g\left(\frac{p}{(p,l)}\right)\right) \tag{6}$$

holds for all $l \mid P$. Here $g(m)$ is a strongly multiplicative function defined on the set of natural numbers by

$$g(p) = \begin{cases} 1 & \text{if } \ p \mid p_k q_k, \\ 2 & \text{if } \ p \nmid p_k q_k. \end{cases}$$

Remark. The product on the right hand side of (6) is positive because $g(2) = 1$.
Proof of the Lemma. Without further references we shall use the following conditions:

$$2 \mid p_k q_k, \qquad (s_n, v_n) = (p_k, p_{k-1}) = (q_k, q_{k-1}) = (p_k, q_k) = 1.$$

Introducing the Möbius function we get

$$\#E(l) = \sum_{n \leqslant x, l \mid s_n v_n} \ \sum_{d \mid \left(\frac{s_n v_n}{l}, \frac{P}{l}\right)} \mu(d) = \sum_{d \mid \frac{P}{l}} \mu(d) \#\{n \leqslant x\colon\ ld \mid s_n v_n\}. \tag{7}$$

It is obvious that for any integer t

$$\#\{n \leqslant x\colon\ t \mid s_n v_n\} = \sum_{\substack{t_1 \cdot t_2 = 1 \\ (t_1, t_2) = 1}} \#\{n \leqslant x\colon\ t_1 \mid s_n, t_2 \mid v_n\} \geqslant \frac{x}{t} g(t) - \tau(t), \tag{8}$$

where $\tau(t)$ denotes the number of all distinct divisors of t and $g(t)$ denotes the number of distinct divisors t_1 of t satisfying

$$\left(t_1, \frac{t}{t_1}\right) = (t_1, p_k) = \left(\frac{t}{t_1}, q_k\right) = 1.$$

Evidently, the function $g(m)$ is the same as it was defined in Lemma.
By setting (8) in (7) we obtain

$$\#E(l) \geqslant \frac{x}{l} \sum_{d \mid \frac{P}{l}} \frac{\mu(d)}{d} g(ld) - \sum_{q \mid P} \tau(q).$$

The first sum in this estimate equals to

$$g(l) \sum_{d \mid \frac{P}{l}} \frac{\mu(d)}{d} g\left(\frac{d}{(l,d)}\right) = g(l) \prod_{p \mid \frac{P}{l}} \left(1 - \frac{1}{p} g\left(\frac{p}{(p,l)}\right)\right).$$

As for the second, it depends on r only. It is obvious that $r(x)$ can be chosen in a way to insure (6). This completes the proof of the Lemma.
Proof of the Theorem. Let $f(l) = a$ for some integer l. For sufficiently large values of k $r(x) = r(x(k))$ satisfies all the inequalities $r(x) > \max(p, \alpha + 1)$, where $p^\alpha \mid l$. Then

$$l \mid P, \qquad p \mid \frac{P}{l} \qquad \text{for all} \quad p \leqslant r.$$

Let u_p denote the minimal integer u for which $f(p^u) \neq e$, therefore $f(p^v) = e$ if $v < u$. Let

$$P^* = \left(P, \prod_{p \mid P,\ p \nmid l} p^{u_p - 1}\right).$$

Define two subsets of Q_k:

$$E = \sum_{s \mid P^*} E(ls), \qquad F = \{t_n \in E\colon\ f(t_n) = a\}.$$

Clearly,

$$\#F = \#E - \#(E \setminus F).$$

We shall specify the function $x = x(l)$ to obtain

$$\#E \gg x, \qquad \#(E \setminus F) = o(x) \qquad \text{as} \quad k \to \infty.$$

It will imply that the set of rational numbers with

$$f\left(\frac{m}{n}\right) = a, \qquad \frac{m}{n} \in Q_k,$$

is infinite.

Indeed, with an appropriate choice of r

$$\#E \gg x\frac{g(l)}{l}\sum_{s|P^*}\frac{g(s)}{s}\prod_{p|\frac{P^*}{s}}\left(1-\frac{1}{p}g\left(\frac{p}{(p,l)}\right)\right)$$

$$\geqslant x\frac{g(l)}{l}\prod_{p|P}\left(1-\frac{g(p)}{p}\right)\prod_{p|P^*,\ u_p>1}\left(1+\frac{g(p)}{p}\right)$$

$$\gg x\prod_{u_p=1}\left(1-\frac{g(p)}{p}\right)\gg x.$$

In the last step we have used (1).
For $\#(E\setminus F)$ we have the estimate

$$\#(E\setminus F)\leqslant \sum_{p\leqslant r,u_p>r} N_p+\sum_{p>r} N_p.$$

where $N_p=\#\{n\leqslant x:\ p^{u_p}\mid s_nv_n\}$,
We write

$$s_x=p_kx+p_{k-1},\qquad v_x=q_kx+q_{k-1}$$

not requiring for x ($x\geqslant 1$) to be integer. Note that $s_x<(1+\xi)v_x$.
Evidently, if

$$p^{u_p}>(1+\xi)v_x>\max(s_x,v_x),$$

then $N_p=0$. For any prime p

$$N_p\leqslant \#\{n\leqslant x:\ p^{u_p}\mid s_n\}+\#\{n\leqslant x:\ p^{u_p}\mid v_n\}\ll\frac{x}{p^{u_p}}+1.$$

Now

$$\#(E\setminus F)\ll x\left(\sum_{p\leqslant r}\frac{1}{p^r}+\sum_{p>r}\frac{1}{p^{u_p}}\right)+\sum_{p^{u_p}\leqslant(1+\xi)v_x}1=o(x)+\lambda\big((1+\xi)v_x\big),$$

where $\lambda(u)=\#\{p^\alpha\leqslant u:\ f(p^\alpha)\neq e\}$.
According to (2),

$$\lambda\big((1+\xi)v_x\big)=o\big(v_x/\varphi((1+\xi)v_x)\big),\qquad k\to\infty.$$

Now if k is sufficiently large we take for $x=x(k)$ the solution of the equation

$$x=v_x/\big(2\sqrt{2}\varphi((1+\xi)v_x)\big).$$

Then $\#(E\setminus F)=o(x)$ as $k\to\infty$.
For the proof, it is enough to show that

$$|\xi-t_n|<\frac{1}{\varphi(v_n)^2}\qquad\text{for all}\quad t_n\in Q_k.$$

This follows by combining the estimate

$$q_k=\frac{q_kx}{x}>\frac{1}{2}\frac{v_x}{x}=\sqrt{2}\varphi\big((1+\xi)v_x\big)>\sqrt{2}\varphi(v_n)$$

with (5).
The proof of the Theorem is complete.

REFERENCES

Hartman, S. (1949). Sur une condition supplémentaire dans les approximations diophantiques. *Colloq. Math.* **2**, 48–51.

Heath-Brown, D. R. (1984). Diophantine approximation with square-free numbers. *Math. Z.* **187**, 335–344.

Negoescu, N. (1968). Approximations to certain classes of real irrational numbers by rational fractions x/y with $(x, y) = 1$ and $(x, m) = 1$ or $(y, m) = 1$. *An. Şti. Univ. "Al. I. Cuza" Iaşi Secţ. I. a Mat. (N. S.)* **14**, 277–286.

Stakėnas, V. (1989). One the rational approximations of special type. In: *Abstr. Comm. Fifth Intern. Vilnius Conf. on Probab. Theory and Math. Statist.* IMC, Vilnius, pp. 161–162.

Part Two

FORMS AND CONGRUENCES

New Trends in Probab. and Statist., pp. 39–56
F. Schweiger and E. Manstavičius (Eds.)
1992 VSP/TEV

DISCRIMINANTS OF DECOMPOSABLE FORMS

J. H. EVERTSE* AND K. GYŐRY**

Department of Mathematics and Computer Science, University of Leiden,
P.O. Box 9512, 2300 RA Leiden, The Netherlands

Mathematical Institute, Kossuth Lajos University
H-4010 Debrecen, Hungary

ABSTRACT

In 1974 Győry obtained a sharp upper bound for the degree of a binary form F belonging to $\mathbf{Z}[X,Y]$ in terms of the absolute value of the discriminant of F. In this paper this result is generalized to decomposable forms in $n \geqslant 2$ variables over $\mathbf{Z}$. Some extension to decomposable forms over S-integers of a number field is also established. Together with the main result of a forthcoming paper of Evertse and Győry, the results of this paper imply that for every $n \geqslant 2$ and $D > 0$, there are only finitely many equivalence classes of primitive square-free decomposable forms in $\mathbf{Z}[X_1,\dots,X_n]$ of maximal rank and discriminant D, and a full set of representatives of these classes can be effectively determined. Finally, a criterion is given for the primes p dividing the discriminant of a decomposable form.

1. INTRODUCTION

Let

$$F(X,Y) = a_0X^r + a_1X^{r-1}Y + \dots + a_rY^r$$

be a binary form with coefficients in $\mathbf{Z}$. The form F can be factored as

$$\prod_{i=1}^{r}(\alpha_iX - \beta_iY)$$

where α_i and β_i are algebraic numbers for $i = 1,\dots,r$. The *discriminant* of F is defined by

$$D(F) = \prod_{1\leqslant i<j\leqslant r} (\alpha_i\beta_j - \alpha_j\beta_i)^2.$$

This does not depend on the choice of $\alpha_1,\dots,\alpha_r$, $\beta_1,\dots,\beta_r$. Further, since $D(F)$ is symmetric in $\alpha_1,\dots,\alpha_r$, $\beta_1,\dots,\beta_r$, the discriminant $D(F)$ is an integer. Clearly,

---

*The research of the first author has been made possible by a fellowship of the Royal Netherlands Academy of Arts and Sciences.

**The research of the second author has been supported in part by Grant 1641 from the Hungarian National Foundation for Scientific Research.

$D(F) \neq 0$ if and only if F has no zeros (in the algebraic closure $\overline{\mathbf{Q}}$ of $\mathbf{Q}$) of order greater than or equal to 2. Further, $D(F)$ is divisible by the prime p if and only if the reduction of F modulo p does have zeros of order greater than or equal to 2 in the algebraic closure of $\mathbf{F}_p$. For any matrix

$$A = \begin{pmatrix} a & b \\ c & d \end{pmatrix},$$

put

$$F_A = F(aX + bY, cX + dY).$$

Two binary forms F and G are said to be *equivalent*, if $G = F_U$ for some matrix $U \in \mathrm{SL}_2(\mathbf{Z})$ (i.e. U has entries in $\mathbf{Z}$ and determinant equal to 1). Clearly, two equivalent binary forms have the same discriminant.
Discriminants of binary forms have been studied extensively. Lagrange showed that there are only finitely many equivalence classes of quadratic forms in $\mathbf{Z}[X,Y]$ of given discriminant and, moreover, he gave an effective algorithm to determine a full set of representatives for these equivalence classes (see (Lagrange, 1773)). Hermite proved a similar result for binary cubic forms (see (Hermite, 1851)). It appeared to be much more difficult to prove such results for binary forms of degree greater than or equal to 4. Birch and Merriman succeeded in proving that for every $D \neq 0$, $r \geqslant 4$, there are only finitely many equivalence classes of binary forms of degree r and discriminant D (see (Birch and Merriman, 1972)). Their proof was ineffective in the sense that it does not provide an algorithm to determine representatives for all equivalence classes. Independently of Birch and Merriman, Győry proved in an effective way that the *monic* binary forms F (i.e. with $F(1,0) = 1$) of degree r and discriminant D are contained in only finitely many equivalence classes (see (Győry, 1973)). Further, he proved (see (Győry, 1974)) that independently of the fact that F is monic or not, the degree r can be bounded in terms of $D(F)$:

$$r \leqslant 3 + 2\frac{\log |D(F)|}{\log 3}, \tag{1.1}$$

and if F is not divisible by a linear form with coefficients in $\mathbf{Q}$, then

$$r \leqslant 2\frac{\log |D(F)|}{\log 3}. \tag{1.2}$$

These estimates cannot be improved in general. The inequalities (1.1) and (1.2) imply that, both in the theorem of Birch and Merriman and in the result of Győry, the finiteness statements remain true if r is not fixed. Recently, Evertse and Győry proved an effective result without the condition that F is monic (see (Evertse and Győry, 1991)). They proved that for all $D \neq 0$, there are only finitely many equivalence classes of binary forms in $\mathbf{Z}[X,Y]$ of degree greater than or equal to 2 and discriminant D, and that a full set of representatives of these equivalence classes can be determined effectively.
In a forthcoming paper of Evertse and Győry, the notion of discriminant of a binary form was generalised to decomposable forms (see (Evertse and Győry, 1992)).

In general, a decomposable form $F \in \mathbf{K}[X_1, \dots, X_n]$ (over any field $\mathbf{K}$) is a polynomial that factors into homogeneous linear forms over some extension of $\mathbf{K}$. The smallest extension of $\mathbf{K}$ over which F factors into linear forms is called the *splitting field* of $\mathbf{K}$; it is a finite normal extension of $\mathbf{K}$. The best known types of decomposable forms are *norm forms*, i.e. polynomials of the type

$$F(\underline{X}) = \lambda N_{\mathbf{M}/\mathbf{K}}(X_1 + \alpha_2 X_2 + \dots + \alpha_n X_n) = \lambda \prod_{i=1}^{r} (X_1 + \alpha_2^{(i)} X_2 + \dots + \alpha_n^{(i)} X_n),$$

where $\lambda \in \mathbf{K}^*$, $\mathbf{M} = \mathbf{K}(\alpha_2, \dots, \alpha_n)$ is a finite extension of $\mathbf{K}$, and $\alpha^{(1)}, \dots, \alpha^{(r)}$ are the conjugates over $\mathbf{K}$ of $\alpha \in \mathbf{M}$, *discriminant forms* and *index forms* (see, e.g. (Győry and Papp, 1977)). If

$$F(\underline{X}) = F(X_1, \dots, X_n) = l_1(\underline{X}) \cdot \dots \cdot l_r(\underline{X}),$$

where $l_1, \dots, l_r$ are linear forms in $\mathbf{L}[X_1, \dots, X_n]$ for some extension $\mathbf{L}$ of $\mathbf{K}$, then we put

$$\operatorname{rank}(F) = \operatorname{rank}_{\mathbf{L}}\{l_1, \dots, l_r\}.$$

The form F is said to be of *maximal rank* if $\operatorname{rank}(F) = n$. Further, F is called *square-free* if it is not divisible by the square of a linear form over $\mathbf{L}$.
Let $F(\underline{X}) \in \mathbf{Z}[X_1, \dots, X_n]$ be a decomposable form with the splitting field $\mathbf{L}$. The form F can be factored as

$$F(\underline{X}) = \lambda l_1(\underline{X})^{k_1} \cdot \dots \cdot l_s(\underline{X})^{k_s}, \tag{1.3}$$

where $\lambda \in \mathbf{Q}^*$, $k_1, \dots, k_s \in \mathbf{N}$ and $l_1, \dots, l_s$ are pairwise nonproportional linear forms in $\mathbf{L}[X_1, \dots, X_n]$. Denote by $\mathcal{O}_{\mathbf{L}}$ the ring of integers of $\mathbf{L}$, and by (l_i) the (possibly fractional) $\mathcal{O}_{\mathbf{L}}$-ideal generated by the coefficients of l_i. For any ($\mathbf{L}$-) linearly independent subset $\mathcal{L} = \{l_{i_1}, \dots, l_{i_p}\}$ of $\{l_1, \dots, l_s\}$, denote by $l_{i_1} \wedge \dots \wedge l_{i_p}$ the exterior product of the coefficient vectors of $l_{i_1}, \dots, l_{i_p}$, and by $(l_{i_1} \wedge \dots \wedge l_{i_p})$ the $\mathcal{O}_{\mathbf{L}}$-ideal generated by the coordinates of this exterior product. Now define the $\mathcal{O}_{\mathbf{L}}$-ideal

$$\mathfrak{D}(\mathcal{L}) = \frac{(l_{i_1} \wedge \dots \wedge l_{i_p})}{(l_{i_1}) \cdot \dots \cdot (l_{i_p})} \tag{1.4}$$

which is obviously integral. Assume that F has rank m, and let $\mathfrak{I}(F)$ denote the collection of linearly independent subsets of $\{l_1, \dots, l_s\}$ of cardinality m. There is a positive rational integer D_F, called the *discriminant* of F, such that

$$(D_F) = \prod_{\mathcal{L} \in \mathfrak{I}(F)} \mathfrak{D}(\mathcal{L})^2. \tag{1.5}$$

Here (α) denotes the $\mathcal{O}_{\mathbf{L}}$-ideal generated by α. (For polynomials in several variables, there exists also another concept of discriminant; see, e.g. (Gelfand *et al.*, 1990)). Indeed, the ideal on the right-hand side of (1.5) is clearly integral. It is possible to choose $l_1, \dots, l_s$ in (1.3) such that each $\mathbf{Q}$-automorphism of $\mathbf{L}$ permutes the linear forms $l_1, \dots, l_s$, and also the sets in the collection $\mathfrak{I}(F)$. Using

this, one can show that the ideal on the right-hand side of (1.5) is generated by a rational number.
From the definition of D_F it follows that D_F does not change when $l_1, \ldots, l_s$ are multiplied by constants. Hence

$$D_{\lambda F} = D_F \qquad \text{for every} \quad \lambda \in \mathbf{Q}^*. \tag{1.6}$$

Further, if F is a *primitive* square-free binary form (i.e. its coefficients are relatively prime) then $D_F = |D(F)|$. For any $n \times n$-matrix $A = (a_{ij})$, put

$$F_A(\underline{X}) = F\Big(\sum_{j=1}^{n} a_{1j}X_j, \ldots, \sum_{j=1}^{n} a_{nj}X_j\Big).$$

Two decomposable forms F and G are said to be equivalent if $G = F_U$ for some $U \in \mathrm{SL}_n(\mathbf{Z})$. For equivalent decomposable forms F and G we have

$$D_F = D_G. \tag{1.7}$$

In this paper we shall generalize some of the properties of discriminants of binary forms mentioned above to decomposable forms. First, we generalize the property of the discriminant $D(F)$ of a binary form F that $p \mid D(F)$ if and only if the reduction of F mod p has zeros of order greater than or equal to 2. This generalization roughly states that the prime p divides the discriminant D_F of the decomposable form F if and only if F has a zero $\underline{x}$ (in $\overline{\mathbf{Q}}^n$) of order $t \geqslant 1$ such that the reduction of $\underline{x}$ mod p is a zero of order greater than or equal to $t+1$ of the reduction of F mod p. The precise result is stated in § 2 (Theorem 3). Further, we prove the following generalization of (1.1) and (1.2).

THEOREM 1. *Let $F \in \mathbf{Z}[X_1, \ldots, X_n]$ be a square-free decomposable form of rank m and degree r. Then*

$$r \leqslant 2^m - 1 + \frac{m}{\log 3} \log D_F,$$

and if F is not divisible by a linear form with coefficients in $\mathbf{Q}$ then

$$r \leqslant \frac{m}{\log 3} \log D_F.$$

In (Evertse and Győry, 1992), the following result was proved: for every $n \geqslant 2$, $r \geqslant 2$, and $D > 0$, there are only finitely many equivalence classes of primitive decomposable forms in $\mathbf{Z}[X_1, \ldots, X_n]$ of maximal rank n, degree r, and discriminant D, and a full set of representatives of these classes can be determined effectively. Theorem 1 implies that it is not necessary to fix r, provided that we restrict ourselves to square-free forms.

THEOREM 2. *For every $n \geqslant 2$ and $D > 0$, there are only finitely many equivalence classes of primitive square-free decomposable forms in $\mathbf{Z}[X_1, \ldots, X_n]$ of maximal rank and discriminant D, and a full set of representatives of these classes can be determined effectively.*

In § 2 we consider decomposable forms over Dedekind domains, and state and prove Theorem 3 and some other properties of discriminants of decomposable forms which we need in the proof of Theorem 1. In § 3 we state and prove some version of Theorem 1 for decomposable forms over S-integers in number fields, and in § 4 we prove Theorem 1.

2. DISCRIMINANTS OF DECOMPOSABLE FORMS OVER DEDEKIND DOMAINS

The discriminant of a decomposable form over a Dedekind domain can be defined in precisely the same way as over $\mathbf{Z}$. In this section, we give some properties of this discriminant. $\mathcal{R}$ will denote a Dedekind domain with quotient field $\mathbf{K}$ of characteristic 0. By an $\mathcal{R}$-ideal we mean a nontrivial finitely generated $\mathcal{R}$-submodule of $\mathbf{K}$. We say that $\mathcal{R}$-ideals contained in $\mathcal{R}$ are integral. The $\mathcal{R}$-ideal generated by α is denoted by (α). For any finite extension $\mathbf{L}$ of $\mathbf{K}$, denote by $\mathcal{R}_{\mathbf{L}}$ the integral closure of $\mathcal{R}$ in $\mathbf{L}$. The integral closure $\mathcal{R}_{\mathbf{L}}$ is also a Dedekind domain. Let $F(\underline{X}) \in \mathbf{K}[X_1, \dots, X_n]$ be a decomposable form with splitting field $\mathbf{L}$. Then F can be factored as

$$F(\underline{X}) = \lambda l_1(\underline{X})^{k_1} \cdot \ldots \cdot l_s(\underline{X})^{k_s}, \tag{2.1}$$

where $\lambda \in \mathbf{K}^*$, $l_1, \dots, l_s$ are pairwise nonproportional linear forms with coefficients in $\mathbf{L}$, and $k_1, \dots, k_s$ are positive integers. Denote by (l_i) the $\mathcal{R}_{\mathbf{L}}$-ideal generated by the coefficients of l_i. Further, for any linearly independent subset $\mathcal{L} = \{l_{i_1}, \dots, l_{i_p}\}$ of $\{l_1, \dots, l_s\}$, denote by $(l_{i_1} \wedge \ldots \wedge l_{i_p})$ the $\mathcal{R}_{\mathbf{L}}$-ideal generated by the coordinates of the exterior product of the coefficient vectors of $l_{i_1}, \dots, l_{i_p}$. Define the $\mathcal{R}_{\mathbf{L}}$-ideal

$$\mathfrak{D}(\mathcal{L}) = \frac{(l_{i_1} \wedge \ldots \wedge l_{i_p})}{(l_{i_1}) \cdot \ldots \cdot (l_{i_p})}.$$

Let $\mathfrak{I}(F)$ be the collection of linearly independent subsets of $\{l_1, \dots, l_s\}$ of cardinality $m = \operatorname{rank}(F)$. Then there is an integral $\mathcal{R}$-ideal $\mathfrak{D}(F)$, called the *discriminant* of F, such that

$$\mathfrak{D}(F)\mathcal{R}_{\mathbf{L}} = \prod_{\mathcal{L} \in \mathfrak{I}(F)} \mathfrak{D}(\mathcal{L})^2. \tag{2.2}$$

Two decomposable forms F, $F' \in \mathbf{K}[X_1, \dots, X_n]$ are said to be *weakly equivalent* if there are $\lambda \in \mathbf{K}^*$ and a matrix $U \in \mathrm{GL}_n(\mathcal{R})$ such that $F' = \lambda F_U$. In that case we have

$$\mathfrak{D}(F') = \mathfrak{D}(F).$$

We now give some properties of the discriminant defined above. First we show a relationship between the prime ideals $\mathfrak{p}$ dividing the discriminant of the decomposable form F and the behaviour of F under reduction modulo $\mathfrak{p}$.

Let $\mathfrak{p}$ be a fixed prime ideal of $\mathcal{R}$, and v: $\mathbf{K} \longrightarrow \mathbf{Z} \cup \{\infty\}$ the corresponding discrete valuation, satisfying $\mathfrak{p} = \{x \in \mathcal{R}\colon\ v(x) > 0\}$. A polynomial $P \in \mathbf{K}[X_1, \dots, X_n]$ is said to be $\mathfrak{p}$-*primitive* if $v(a) \geqslant 0$ for every coefficient a of P, and $v(a) = 0$ for at least one coefficient a of P. For every nonzero $P \in \mathbf{K}[X_1, \dots, X_n]$ there is a $\lambda \in \mathbf{K}^*$ such that λP is $\mathfrak{p}$-primitive. For decomposable forms F we have $\mathfrak{D}(\lambda F) = \mathfrak{D}(F)$. Hence in order to investigate when $\mathfrak{p} \mid \mathfrak{D}(F)$ it suffices to consider $\mathfrak{p}$-primitive decomposable forms.

Denote by $\mathbf{K}_{\mathfrak{p}}$ the completion of $\mathbf{K}$ at v and by Ω the algebraic closure of $\mathbf{K}_{\mathfrak{p}}$. The valuation v can be extended uniquely to a (not necessarily discrete) valuation on Ω. Denote by $\mathcal{O}$ the ring $\{x \in \Omega\colon\ v(x) \geqslant 0\}$ and by $\mathfrak{P}$ the maximal ideal $\{x \in \Omega\colon\ v(x) > 0\}$ of $\mathcal{O}$. The canonical mapping $\mathcal{O} \longrightarrow \mathcal{O}/\mathfrak{P}$ induces in an obvious way mappings in $\mathcal{O}^n$, $\mathcal{O}[X_1,\dots,X_n]$ etc., and the images of $x \in \mathcal{O}$, $\underline{x} \in \mathcal{O}^n$, $P \in \mathcal{O}[X_1,\dots,X_n]$ are denoted by $x_{\mathfrak{P}}$, $\underline{x}_{\mathfrak{P}}$, $P_{\mathfrak{P}}$. We write

$$x \equiv y \bmod \mathfrak{P} \quad \text{if} \quad x - y \in \mathfrak{P},$$
$$\underline{x} \equiv \underline{y} \bmod \mathfrak{P} \quad \text{if} \quad \underline{x} - \underline{y} \in \mathfrak{P}^n,$$

$$P_1 \equiv P_2 \bmod \mathfrak{P} \quad \text{for} \quad P_1, P_2 \in \mathcal{O}[X_1,\dots,X_n] \quad \text{if} \quad P_1 - P_2 \in \mathfrak{P}[X_1,\dots,X_n].$$

If $\mathbf{A}$ is any algebraically closed field,

$$P \in \mathbf{A}[X_1,\dots,X_n] \quad \text{and} \quad \underline{x} = (x_1,\dots,x_n) \in \mathbf{A}^n,$$

then $\underline{x}$ is a zero of P of order t if t is the largest integer such that P can be expressed as

$$P(\underline{X}) = \sum_{\substack{i_1 \geqslant 0,\dots,i_n \geqslant 0 \\ i_1+\dots+i_n \geqslant t}} a(i_1,\dots,i_n)(X_1 - x_1)^{i_1} \cdot \ldots \cdot (X_n - x_n)^{i_n}$$

for certain $a(i_1,\dots,i_n) \in \mathbf{A}$.
Theorem 3 below states roughly that $\mathfrak{p}$ divides $\mathfrak{D}(F)$ if and only if F has a zero whose order increases after reduction mod $\mathfrak{P}$. More precisely, we have

THEOREM 3. *Let $\mathfrak{p}$ be a prime ideal of $\mathcal{R}$ and $F(\underline{X}) \in \mathbf{K}[X_1,\dots,X_n]$ be a $\mathfrak{p}$-primitive decomposable form. Then the following two statements are equivalent:*

(i) $\mathfrak{p} \mid \mathfrak{D}(F)$,

(ii) *F has a zero $\underline{x} \in \mathcal{O}^n$ of order $t \geqslant 1$ such that:*
every $\underline{y} \in \mathcal{O}^n$ with $\underline{y} \equiv \underline{x} \bmod \mathfrak{P}$ is a zero of F of order $\leqslant t$;
$\underline{x}_{\mathfrak{P}}$ is a zero of $F_{\mathfrak{P}}$ of order $> t$.

Remark. It is easy to check that independently of whether $\mathfrak{p} \mid \mathfrak{D}(F)$, if $\underline{x} \in \mathcal{O}^n$ is a zero of F of order t, then $\underline{x}_{\mathfrak{P}}$ is a zero of $F_{\mathfrak{P}}$ of order greater than or equal to t. This proposition can be considered as a generalization of the well-known fact that the discriminant of a binary form F is divisible by $\mathfrak{p}$ if and only if $F_{\mathfrak{P}}$ has zeros of order greater than or equal to 2.

Proof. We can define the discriminant of a decomposable form in $\Omega[X_1,\dots,X_n]$ with respect to $\mathcal{O}$ in precisely the same way as it was done for decomposable forms in $\mathbf{K}[X_1,\dots,X_n]$ with respect to $\mathcal{R}$. Let $\mathfrak{D}_{\mathfrak{P}}(F)$ denote the discriminant of F with respect to $\mathcal{O}$. Then $\mathfrak{D}_{\mathfrak{P}}(F)$ is the $\mathcal{O}$-ideal generated by $\mathfrak{D}(F)$, hence

$$\mathfrak{p} \mid \mathfrak{D}(F) \iff \mathfrak{D}_{\mathfrak{P}}(F) \subseteq \mathfrak{P}.$$

If $A \in \mathrm{GL}_n(\mathcal{O})$ then

$$\mathfrak{D}_{\mathfrak{P}}(F_A) = \mathfrak{D}_{\mathfrak{P}}(F)$$

and F_A satisfies (ii) if and only if F satisfies (ii). The form F can be factored over $\Omega[X_1,\dots,X_n]$ as

$$F(\underline{X}) = \lambda l_1(\underline{X})^{k_1} \cdot \ldots \cdot l_s(\underline{X})^{k_s}. \tag{2.3}$$

where $l_1, \ldots, l_s$ are pairwise nonproportional linear forms. The module $\mathcal{O}^n$ has a basis $\{\underline{a}_1, \ldots, \underline{a}_n\}$ such that $\{\underline{a}_{m+1}, \ldots, \underline{a}_n\}$, say, is a basis of

$$\{\underline{x} \in \mathcal{O}^n \colon\ l_1(\underline{x}) = \ldots = l_s(\underline{x}) = 0\};$$

here $m = \operatorname{rank}(F)$. Let A be the matrix of the $\mathcal{O}$-linear mapping

$$\underline{x} \longmapsto \underline{u} = (u_1, \ldots, u_n) \qquad \text{where} \quad \underline{x} = \sum_{i=1}^{n} u_i \underline{a}_i.$$

Then $A \in \mathrm{GL}_n(\mathcal{O})$ and

$$F_A(\underline{X}) = G(X_1, \ldots, X_m)$$

for some decomposable form $G \in \Omega[X_1, \ldots, X_m]$. Then

$$\mathfrak{D}_{\mathfrak{P}}(F) = \mathfrak{D}_{\mathfrak{P}}(G)$$

and G satisfies (ii) with n replaced by m if and only if F satisfies (ii). Further, G is of maximal rank n.

We say that a polynomial $P \in \Omega[X_1, \ldots, X_n]$ is $\mathfrak{P}$-primitive if all its coefficients belong to $\mathcal{O}$ but not all belong to $\mathfrak{P}$. Let $F \in \mathcal{O}[X_1, \ldots, X_n]$ be a $\mathfrak{P}$-primitive decomposable form of rank n. From what was mentioned above, it suffices to show that

$$\mathfrak{D}_{\mathfrak{P}}(F) \subseteq \mathfrak{P} \iff F \text{ satisfies (ii)}.$$

By Gauss' lemma, F can be expressed as

$$F(\underline{X}) = l_1(\underline{X})^{k_1} \cdot \ldots \cdot l_s(\underline{X})^{k_s},$$

where $l_1, \ldots, l_s \in \mathcal{O}[X_1, \ldots, X_n]$ are pairwise nonproportional $\mathfrak{P}$-primitive linear forms. Let $\mathfrak{I}(F)$ be the collection of linearly independent subsets of $\{l_1, \ldots, l_s\}$ of cardinality n. Then, for $\{l_{i_1}, \ldots, l_{i_n}\} \in \mathfrak{I}(F)$, the ideal

$$\frac{(l_{i_1} \wedge \ldots \wedge l_{i_n})}{(l_{i_1}) \cdot \ldots \cdot (l_{i_n})}$$

is equal to $(\det(l_{i_1}, \ldots, l_{i_n}))$, where (α) denotes the $\mathcal{O}$-ideal generated by α. Hence

$$\mathfrak{D}_{\mathfrak{P}}(F) = \prod_{\mathfrak{I}(F)} \left(\det(l_{i_1}, \ldots, l_{i_n})\right)^2,$$

where the product is taken over all sets $\{l_{i_1}, \ldots, l_{i_n}\}$ in $\mathfrak{I}(F)$. This implies that

$$\mathfrak{D}_{\mathfrak{P}}(F) \subseteq \mathfrak{P} \iff \det(l_{i_1}, \ldots, l_{i_n}) \in \mathfrak{P} \text{ for some } \{l_{i_1}, \ldots, l_{i_n}\} \in \mathfrak{I}(F).$$

In what follows it is more convenient to express F as

$$F(\underline{X}) = l_1(\underline{X}) \cdot \ldots \cdot l_r(\underline{X}), \tag{2.5}$$

where $l_1, \ldots, l_r$ are, possibly proportional, $\mathfrak{P}$-primitive linear forms. Let $\mathfrak{J}$ be the collection of sets of subscripts $\{i_1, \ldots, i_n\}$ such that the set $\{l_{i_1}, \ldots, l_{i_n}\}$ is linearly independent. Then

$$\mathfrak{D}_{\mathfrak{P}}(F) \subseteq \mathfrak{P} \iff \det(l_{i_1}, \ldots, l_{i_n}) \in \mathfrak{P} \text{ for some } \{l_{i_1}, \ldots, l_{i_n}\} \in \mathfrak{J}.$$

Hence it suffices to prove that

$$\det(l_{i_1}, \ldots, l_{i_n}) \in \mathfrak{P} \text{ for some } \{l_{i_1}, \ldots, l_{i_n}\} \in \mathfrak{J} \iff F \text{ satisfies (ii).} \tag{2.6}$$

($\Rightarrow$) Assume that $\{1, \ldots, n\} \in \mathfrak{J}$ and $\det(l_1, \ldots, l_n) \in \mathfrak{P}$. Then there are elements $c_1, \ldots, c_n \in \mathcal{O}$, not all in $\mathfrak{P}$, such that

$$c_1 l_1 + \ldots + c_n l_n \equiv 0 \bmod \mathfrak{P}$$

identically in $\underline{x}$. Assume that $c_n \notin \mathfrak{P}$. Then we have

$$l_n \equiv d_1 l_1 + \ldots + d_{n-1} l_{n-1} \bmod \mathfrak{P}$$

identically in $\underline{x}$. Let

$$I = \{i\colon\ l_i \text{ is a linear combination of } l_1, \ \ldots, \ l_{n-1}\}.$$

Then I has the cardinality t with $1 \leqslant t < r$. Note that $n \notin I$. Consider the vector space

$$V = \{\underline{x} \in \Omega^n\colon\ l_i(\underline{x}) = 0 \text{ for } i \in I\}.$$

Since $\operatorname{rank}\{l_i\colon\ i \in I\} = n - 1$, the space V is one-dimensional. For every $\underline{x} \in V$ with $\underline{x} \neq \underline{0}$ we have

$$\begin{aligned} l_i(\underline{x}) &= 0 \qquad \text{for} \quad i \in I, \\ l_i(\underline{x}) &\neq 0 \qquad \text{for} \quad i \in I^c \end{aligned}$$

(where for $J \subseteq \{1, \ldots, r\}$ we put $J^c := \{1, \ldots, r\} \setminus J$). Choose $\underline{x} \in V$ with $\underline{x} \in \mathcal{O}^n$, $\underline{x} \not\equiv 0 \bmod \mathfrak{P}$. Then

$$l_n(\underline{x}) = \sum_{i=1}^{n-1} d_i l_i(\underline{x}) \equiv 0 \bmod \mathfrak{P},$$

and

$$l_i(\underline{y}) \neq 0 \qquad \text{for} \quad i \in I^c, \quad \underline{y} \in V, \quad \underline{y} \equiv \underline{x} \bmod \mathfrak{P}.$$

So we have proved the existence of an integer t with $1 \leqslant t < r$ with the following properties:

$$\begin{aligned} &\text{there are a subset } I \text{ of } \{1, \ldots, r\} \text{ of cardinality } t, \\ &j \in I^c, \text{ and } \underline{x} \in \mathcal{O}^n, \text{ such that:} \\ &l_i(\underline{x}) = 0 \quad \text{for} \quad i \in I, \qquad l_i(\underline{x}) \neq 0 \quad \text{for} \quad i \in I^c,\ l_j(\underline{x}) \equiv 0 \bmod \mathfrak{P}, \\ &\text{for every } \underline{y} \text{ with } l_i(\underline{y}) = 0 \text{ for } i \in I,\ \underline{y} \equiv \underline{x} \bmod \mathfrak{P}, \\ &\text{one has } \ l_i(\underline{y}) \neq 0 \quad \text{for} \quad i \in I^c. \end{aligned} \tag{2.7}$$

Let $t < r$ be the largest integer with these properties, and $I \subset \{1, \ldots, r\}$ be the corresponding set of cardinality t, $j \in I^c$ be the corresponding subscript and $\underline{x} \in \mathcal{O}^n$ be the corresponding vector. Assume that there are

$$\underline{y} \in \mathcal{O}^n \qquad \text{with} \quad \underline{y} \equiv \underline{x} \bmod \mathfrak{P},$$

and a subset J of $\{1, \ldots, r\}$ of cardinality $q > t$, such that

$$l_i(\underline{y}) = 0 \qquad \text{for} \quad i \in J.$$

Assume that such an $\underline{y}$ is chosen with maximal q. Then

$$l_i(\underline{y}) \neq 0 \qquad \text{for} \quad i \in J^c$$

and if

$$\underline{z} \equiv \underline{y} \bmod \mathfrak{P}, \qquad l_i(\underline{z}) = 0 \qquad \text{for} \quad i \in J,$$

then

$$l_i(\underline{z}) \neq 0 \qquad \text{for} \quad i \in J^c.$$

Clearly,

$$l_i(\underline{y}) = 0 \bmod \mathfrak{P} \qquad \text{for} \quad i \in I.$$

So if $I \not\subset J$, then (2.7) is satisfied with q instead of t, contrary to the maximality of t. But then $I \subsetneqq J$ which is impossible in view of (2.7). We conclude that if

$$\underline{y} \in \mathcal{O}^n \quad \text{and} \quad \underline{y} \equiv \underline{x} \bmod \mathfrak{P}$$

then at most t of the linear forms $l_1, \ldots, l_r$ vanish at $\underline{y}$. Hence $\underline{x}$ has the following properties: $\underline{x}$ is a zero of F of order t; if $\underline{y} \equiv \underline{x} \bmod \mathfrak{P}$ then $\underline{y}$ is a zero of order less than or equal to t; and $\underline{x}_{\mathfrak{P}}$ is a zero of $F_{\mathfrak{P}}$ of order $> t$. This implies (ii).

($\Leftarrow$) From (ii) it follows that there are a subset I of $\{1, \ldots, r\}$ of cardinality t with $1 \leqslant t < r$, $j \in I^c$, and $\underline{x} \in \mathcal{O}^n$ satisfying (2.7). Then l_j is not a linear combination of the linear forms l_i ($i \in I$). Hence there is a set $\{i_1, \ldots, i_n\} \in \mathfrak{I}$, such that $i_n = j$, and each l_i with $i \in I$ is a linear combination of $l_{i_1}, \ldots, l_{i_{n-1}}$. For convenience, we assume that

$$i_1 = 1, \ \ldots, \ i_{n-1} = n-1, \ i_n = n = j.$$

Assume that $\det(l_1, \ldots, l_n) \notin \mathfrak{P}$. Then the linear mapping

$$\underline{x} \longmapsto (l_1(\underline{x}), \ldots, l_n(\underline{x}))$$

from $\mathcal{O}^n$ to itself is surjective. Choose $\underline{z} \in \mathcal{O}^n$ with

$$l_i(\underline{z}) = 0 \qquad \text{for} \quad i = 1, \ldots, n-1, \qquad l_n(\underline{z}) = 1$$

and put

$$\underline{y} = \underline{x} - l_n(\underline{x})\underline{z}.$$

Since $l_n(\underline{x}) \equiv 0 \bmod \mathfrak{P}$, we have $\underline{y} \equiv \underline{x} \bmod \mathfrak{P}$. Further,

$$l_i(\underline{y}) = l_i(\underline{x}) - l_n(\underline{x}) l_i(\underline{z}) = 0 \qquad \text{for} \quad i = 1, \dots, n.$$

This implies that

$$l_i(\underline{y}) = 0 \qquad \text{for} \quad i \in I.$$

But this contradicts (2.7). Hence our assumption that

$$\det(l_1, \dots, l_n) \notin \mathfrak{P}$$

was wrong, i.e.

$$\det(l_1, \dots, l_n) \in \mathfrak{P}.$$

This completes the proof of Theorem 3.

Now we give some other properties of the discriminant of a decomposable form which will be used in the proof of Theorem 1 in § 4.

LEMMA 1. *Let $F \in \mathbf{K}[X_1, \dots, X_n]$ be a square-free decomposable form such that $F = F_1 F_2$ where $F_1, F_2 \in \mathbf{K}[X_1, \dots, X_n]$. Then*

$$\mathfrak{D}(F) \subset \mathfrak{D}(F_1) \cdot \mathfrak{D}(F_2).$$

Proof. Let $\mathbf{L}$ denote the splitting field of F. We have

$$F_1(\underline{X}) = l_1(\underline{X}) \cdot \ldots \cdot l_s(\underline{X}), \qquad F_2(\underline{X}) = l_{s+1}(\underline{X}) \cdot \ldots \cdot l_r(\underline{X}),$$

where $l_1, \dots, l_r$ are pairwise nonproportional linear forms with coefficients in $\mathbf{L}$. We need the following auxiliary result.

Let $\mathfrak{L}$ be a linearly independent subset of $\{l_1, \dots, l_r\}$ with

$$\mathfrak{L} = \mathfrak{L}_1 \cup \mathfrak{L}_2 \qquad \text{where} \quad \mathfrak{L}_1 \neq \emptyset, \quad \mathfrak{L}_2 \neq \emptyset, \quad \mathfrak{L}_1 \cap \mathfrak{L}_2 = \emptyset.$$

Then

$$\mathfrak{D}(\mathfrak{L}) \subseteq \mathfrak{D}(\mathfrak{L}_1) \cdot \mathfrak{D}(\mathfrak{L}_2). \tag{2.8}$$

In order to prove this, we may assume after relabeling $l_1, \dots, l_r$, that

$$\mathfrak{L}_1 = \{l_1, \dots, l_v\}, \qquad \mathfrak{L}_2 = \{l_{v+1}, \dots, l_u\}.$$

Let A be one of the $u \times u$-submatrices of the $u \times n$-matrix of which the j-th row consist of the coefficients of l_j, for $j = 1, \dots, u$. We have the identity

$$\det A = \sum \pm \det B \cdot \det C,$$

where B runs through certain $v \times v$-submatrices from the first v rows of A, and C runs through certain $(u-v) \times (u-v)$-submatrices from the last $u - v$ rows of A. But

$$\det B \in (l_1 \wedge \ldots \wedge l_v), \qquad \det C \in (l_{v+1} \wedge \ldots \wedge l_u),$$

hence

$$\det A \in (l_1 \wedge \ldots \wedge l_v)(l_{v+1} \wedge \ldots \wedge l_u).$$

This holds for all matrices A as introduced above, hence for all coordinates of $(l_1 \wedge \ldots \wedge l_u)$. Therefore,

$$(l_1 \wedge \ldots \wedge l_u) \subseteq (l_1 \wedge \ldots \wedge l_v) \cdot (l_{v+1} \wedge \ldots \wedge l_u),$$

which implies (2.8).

Let $m = \operatorname{rank}(F)$ and $m_i = \operatorname{rank}(F_i)$ for $i = 1, 2$. We partition the collection $\mathfrak{I}(F)$ of $\mathbf{L}$-linearly independent subsets of $\{l_1, \ldots, l_r\}$ of cardinality m into four subcollections:

$\mathfrak{I}_1 = \{\mathcal{L} \in \mathfrak{I}(F)$: no subset of $\mathcal{L}$ belongs to $\mathfrak{I}(F_1)$ or $\mathfrak{I}(F_2)\}$;

$\mathfrak{I}_2 = \{\mathcal{L} \in \mathfrak{I}(F)$: some subset $\mathcal{L}_1$ of $\mathcal{L}$ belongs to $\mathfrak{I}(F_1)$ but no subset of $\mathcal{L}$ belongs to $\mathfrak{I}(F_2)\}$;

$\mathfrak{I}_3 = \{\mathcal{L} \in \mathfrak{I}(F)$: some subset $\mathcal{L}_2$ of $\mathcal{L}$ belongs to $\mathfrak{I}(F_2)$ but no subset of $\mathcal{L}$ belongs to $\mathfrak{I}(F_1)\}$;

$\mathfrak{I}_4 = \{\mathcal{L} \in \mathfrak{I}(F)$: some subset $\mathcal{L}_1$ of $\mathcal{L}$ belongs to $\mathfrak{I}(F_1)$ and some subset $\mathcal{L}_2$ of $\mathcal{L}$ belongs to $\mathfrak{I}(F_2)\}$.

Here $\mathfrak{I}(F_1)$ is the collection of $\mathbf{L}$-linearly independent subsets of $\{l_1, \ldots, l_s\}$ of cardinality $m_1 = \operatorname{rank}(F_1)$, and $\mathfrak{I}(F_2)$ is defined similarly.

For $\mathcal{L} \in \mathfrak{I}$ we have

$$\mathfrak{D}(\mathcal{L}) \subseteq \mathcal{R}_{\mathbf{L}} \tag{2.9a}$$

since $\mathfrak{D}(\mathcal{L})$ is integral. For $\mathcal{L} \in \mathfrak{I}_2$ we have

$$\mathfrak{D}(\mathcal{L}) \subseteq \mathfrak{D}(\mathcal{L}_1) \qquad \text{with} \quad \mathcal{L}_1 = \mathcal{L} \cap \{l_1, \ldots, l_s\} \tag{2.9b}$$

by (2.8) with $\mathcal{L}_2 = \mathcal{L} \setminus \mathcal{L}_1$ and the fact that $\mathfrak{D}(\mathcal{L}_2)$ is integral. Similarly, for $\mathcal{L} \in \mathfrak{I}_3$ we have

$$\mathfrak{D}(\mathcal{L}) \subseteq \mathfrak{D}(\mathcal{L}_2) \qquad \text{with} \quad \mathcal{L}_2 = \mathcal{L} \cap \{l_{s+1}, \ldots, l_u\}. \tag{2.9c}$$

Finally, if $\mathcal{L} \in \mathfrak{I}_4$ then $\mathcal{L} = \mathcal{L}_1 \cup \mathcal{L}_2$ with

$$\mathcal{L}_1 = \mathcal{L} \cap \{l_1, \ldots, l_s\}, \qquad \mathcal{L}_2 = \mathcal{L} \cap \{l_{s+1}, \ldots, l_u\}$$

and

$$\mathcal{L}_1 \cap \mathcal{L}_2 = \varnothing,$$

hence by (2.8),

$$\mathfrak{D}(\mathcal{L}) \subseteq \mathfrak{D}(\mathcal{L}_1)\mathfrak{D}(\mathcal{L}_2). \tag{2.9d}$$

The inclusions (2.9a–d) imply that

$$\begin{aligned} \mathfrak{D}(F) = \prod_{\mathcal{L} \in \mathfrak{I}(F)} \mathfrak{D}(\mathcal{L}^2) \subseteq \Big(\prod_{\mathcal{L}_1 \in \mathfrak{I}(F_1)} \mathfrak{D}(\mathcal{L}_1)^{2m(\mathcal{L}_1)} \Big) \\ \times \Big(\prod_{\mathcal{L}_2 \in \mathfrak{I}(F_2)} \mathfrak{D}(\mathcal{L}_2)^{2m(\mathcal{L}_2)} \Big) \end{aligned} \tag{2.10}$$

where $m(\mathfrak{L}_i)$ is the number of $\mathfrak{L} \in \mathfrak{I}(F)$ containing $\mathfrak{L}_i$ for $i = 1, 2$. It is easy to check that $m(\mathfrak{L}_1)$ and $m(\mathfrak{L}_2)$ are positive integers. Together with the fact that all ideals $\mathfrak{D}(\mathfrak{L}_1)$, $\mathfrak{D}(\mathfrak{L}_2)$ are integral this implies

$$\mathfrak{D}(F) \subseteq \mathfrak{D}(F_1)\mathfrak{D}(F_2),$$

i.e. Lemma 1.

The next lemma deals with norm forms

$$F(\underline{X}) = \lambda N_{\mathbf{M}/\mathbf{K}}(X_1 + \alpha_2 X_2 + \ldots + \alpha_n X_n), \tag{2.11}$$

where $\lambda \in \mathbf{K}^*$ and $\mathbf{M} = \mathbf{K}(\alpha_2, \ldots, \alpha_n)$ is a finite extension of $\mathbf{K}$. We denote by $\mathfrak{D}_{\mathbf{M}}$ the discriminant ideal of $\mathbf{M}$, i.e. the $\mathcal{R}$-ideal generated by all numbers $D_{\mathbf{M}/\mathbf{K}}(\omega_1, \ldots, \omega_r)$ where $r = [\mathbf{M} : \mathbf{K}]$ and $\omega_1, \ldots, \omega_r \in \mathcal{R}_{\mathbf{M}}$.

LEMMA 2. *Assume that F is given by* (2.11) *and that* $\operatorname{rank}(F) = m$. *Then*

$$\mathfrak{D}(F)^m \subseteq \mathfrak{D}_{\mathbf{M}}^2.$$

Proof. This can be proved in the same way as Lemma 3 of (Evertse and Győry, 1992).

In (Evertse and Győry, 1992), a different notion of decomposable forms was used. By an $\mathcal{R}$-lattice we mean a finitely generated $\mathcal{R}$-submodule of some $\mathbf{K}$-vector space. The dimension $\dim \mathfrak{M}$ of the $\mathcal{R}$-lattice $\mathfrak{M}$ is defined as the dimension of the $\mathbf{K}$-vector space $\mathbf{K}\mathfrak{M}$. An *$\mathcal{R}$-lattice-decomposable form pair* is a pair $(\mathfrak{M}, F)$, consisting of an $\mathcal{R}$-lattice $\mathfrak{M}$ and a mapping F: $\mathfrak{M} \longrightarrow \mathbf{K}$ with the following property: there are a set of generators $\{\underline{a}_1, \ldots, \underline{a}_m\}$ of $\mathfrak{M}$, and a decomposable form $\widetilde{F} \in \mathbf{K}[X_1, \ldots, X_m]$ with $\operatorname{rank}(\widetilde{F}) = \dim \mathfrak{M}$, such that

$$F\{x_1\underline{a}_1 + \ldots + x_m\underline{a}_m\} = \widetilde{F}(x_1, \ldots, x_m) \qquad \text{for} \quad x_1, \ldots, x_m \in \mathcal{R}.$$

The discriminant of $(\mathfrak{M}, F)$ is defined by

$$\mathfrak{D}(\mathfrak{M}, F) = \mathfrak{D}(\widetilde{F}); \tag{2.12}$$

it is easy to show that this is well defined, i.e. independent of the choice of $\widetilde{F}$. Two $\mathcal{R}$-lattice-decomposable form pairs $(\mathfrak{M}, F)$ and $(\mathfrak{M}', F')$ are said to be weakly equivalent if there are $\lambda \in \mathbf{K}^*$ and an $\mathcal{R}$-module isomorphism φ: $\mathfrak{M} \longrightarrow \mathfrak{M}'$, such that $F' \circ \varphi = \lambda F$. For two weakly equivalent $\mathcal{R}$-lattice-decomposable form pairs $(\mathfrak{M}, F)$ and $(\mathfrak{M}', F')$, we have

$$\mathfrak{D}(\mathfrak{M}, F) = \mathfrak{D}(\mathfrak{M}', F')$$

(cf. (Evertse and Győry, 1992, § 3)). Via (2.12), Lemmas 1 and 2 can be easily translated into similar properties for $\mathcal{R}$-lattice-decomposable form pairs.

3. DECOMPOSABLE FORMS OVER S-INTEGERS

Let $\mathbf{K}$ be an algebraic number field, and $\{| \ |_v \colon v \in M(\mathbf{K})\}$ be a maximal set of pairwise inequivalent absolute values on $\mathbf{K}$. Let $M_\infty(\mathbf{K})$ be the subset of $v \in M(\mathbf{K})$ for which $| \ |_v$ is archimedian. Let S be a finite subset of $M(\mathbf{K})$ containing $M_\infty(\mathbf{K})$. The ring of S-integers is defined by

$$\mathcal{O}_S = \{x \in \mathbf{K} \colon |x|_v \leqslant 1 \text{ for } v \in M(\mathbf{K}) \setminus S\}.$$

$\mathcal{O}_S$ is a Dedekind domain. In (Evertse and Győry, 1992), the following result was proved:

for every $r \geqslant 2$, $n \geqslant 2$, and integral $\mathcal{O}_S$-ideal $\mathfrak{D}$, there are only finitely many weak equivalence classes of $\mathcal{O}_S$-lattice-decomposable form pairs $(\mathfrak{M}, F)$ with

$$\operatorname{rank}(\mathfrak{M}) = n, \qquad \deg F = r, \qquad \mathfrak{D}(\mathfrak{M}, F) = \mathfrak{D},$$

and a full set of representatives of these classes can be determined effectively in terms of r, n, $\mathbf{K}$, S, and $\mathfrak{D}$.

In contrast with the case $\mathcal{O}_S = \mathbf{Z}$, it is not possible to bound the degree r from above in terms of only n, $\mathbf{K}$, S, and $\mathfrak{D}$, and so in the above result it is necessary to fix r. In this section, we shall show that r can be estimated from above in terms of n, $\mathbf{K}$, S, $\mathfrak{D}$, and the splitting field of F. In estimating r from above, we may and we shall consider discriminants of decomposable forms instead of discriminants of $\mathcal{O}_S$-lattice-decomposable form pairs.
Every $\mathcal{O}_S$-ideal $\mathfrak{a}$ can be decomposed uniquely as

$$\mathfrak{a} = \prod_{\mathfrak{p}} \mathfrak{p}^{w_\mathfrak{p}(\mathfrak{a})},$$

where $\mathfrak{p}$ runs through the $\mathcal{O}_S$-prime ideals and $w_\mathfrak{p}(\mathfrak{a}) \in \mathbf{Z}$. Put

$$\Omega(\mathfrak{a}) = \sum_{\mathfrak{p}} w_\mathfrak{p}(\mathfrak{a}).$$

Thus $\Omega(\mathfrak{a}) \geqslant 1$ if $\mathfrak{a}$ is integral and $\mathfrak{a} \neq (1) = \mathcal{O}_S$. Further,

$$\Omega(\mathfrak{a}\mathfrak{b}) = \Omega(\mathfrak{a}) + \Omega(\mathfrak{b})$$

for any $\mathcal{O}_S$-ideals $\mathfrak{a}$ and $\mathfrak{b}$. In what follows we use the following notation:

F is a square-free decomposable form in $\mathbf{K}[X_1, \ldots, X_n]$ of degree r, rank m, discriminant $\mathfrak{D}$ with respect to $\mathcal{O}_S$, and with splitting field $\mathbf{L}$ over $\mathbf{K}$;

s is the cardinality of S, and $g = [\mathbf{L} : \mathbf{K}]$, $d = [\mathbf{K} : \mathbf{Q}]$.

THEOREM 4. *We have*

$$r \leqslant \left(4 \times 7^{g(d+2s)}\right)^m + \frac{g}{2} \cdot \Omega(\mathfrak{D}).$$

It suffices to prove Theorem 4 in the case $\mathbf{L} = \mathbf{K}$. Indeed, let S' be the set of places on $\mathbf{L}$ lying above S, put $s' = \# S'$, let Ω' be defined in the same way as Ω but with respect to $\mathcal{O}_{S'}$, and put $d' = [\mathbf{L} : \mathbf{Q}]$. Then the discriminant of F with respect to $\mathcal{O}_{S'}$ is $\mathfrak{D}\mathcal{O}_{S'}$. Now it is obvious that

$$d' \leqslant gd, \qquad s' \leqslant gs, \qquad \Omega'(\mathfrak{D}\mathcal{O}_{S'}) \leqslant g\Omega(\mathfrak{D}).$$

So if Theorem 4 holds for $\mathbf{L} = \mathbf{K}$, then

$$r \leqslant \left(4 \times 7^{g(d+2s)}\right)^m + \frac{1}{2} \cdot \Omega'(\mathfrak{D}\mathcal{O}_{S'}),$$

and this implies Theorem 4 in full generality.

In the proof of Theorem 4 we need some lemmas. We can factor F as

$$F(\underline{X}) = l_1(\underline{X}) \cdot \ldots \cdot l_r(\underline{X}), \tag{3.1}$$

where $l_1, \ldots, l_r$ are pairwise nonproportional linear forms with coefficients in $\mathbf{K}$. Let $\mathfrak{I}(F)$ be the collection of linearly independent subsets of $\{l_1, \ldots, l_r\}$ of cardinality m. Then

$$\mathfrak{D} = \prod_{\mathcal{L} \in \mathfrak{I}(F)} \mathfrak{D}(\mathcal{L})^2,$$

where now each $\mathfrak{D}(\mathcal{L})$ is an integral $\mathcal{O}_S$-ideal. Denote by (α) the $\mathcal{O}_S$-ideal generated by α.

LEMMA 3. *Let*

$$\mathcal{L}_1 = \{l_{i_1}, \ldots, l_{i_m}\}, \ \mathcal{L}_2 = \{l_{j_1} \ldots, l_{j_m}\} \in \mathfrak{I}(F)$$

and suppose that

$$l_{j_k} = \sum_{p=1}^{m} c_{kp} l_{i_p} \qquad \textit{for} \quad k = 1, \ldots, m.$$

Then

$$\frac{\mathfrak{D}(\mathcal{L}_2)}{\mathfrak{D}(\mathcal{L}_1)} = \left(\det(c_{kp})\right) \frac{(l_{i_1}) \cdot \ldots \cdot (l_{i_m})}{(l_{j_1}) \cdot \ldots \cdot (l_{j_m})}. \tag{3.2}$$

Proof. By a well-known property of exterior products we have

$$l_{j_1} \wedge \ldots \wedge l_{j_m} = \det(c_{kp}) \cdot l_{i_1} \wedge \ldots \wedge l_{i_m}.$$

This implies (3.2) at once.

The following lemma is a special case of Theorem 1 of (Evertse, 1984). We denote by $\mathcal{O}_S^*$ the group of S-units $\{x \in \mathbf{K}\colon |x|_v = 1 \text{ for } v \in M(\mathbf{K}) \setminus S\}$.

LEMMA 4. *Let h denote the number of solutions of $x + y = 1$ in $x, y \in \mathcal{O}_S^*$. Then $h \leqslant 3 \cdot 7^{d+2s}$.*

Proof of Theorem 4. First assume that $\mathfrak{D}(\mathcal{L}) \subsetneqq (1)$ for each $\mathcal{L} \in \mathfrak{I}(F)$. We claim that $\mathfrak{I}(F)$ has cardinality greater than or equal to $r - m + 1$. Indeed, suppose that

$$\mathcal{L}_0 = \{l_1, \ldots, l_m\} \in \mathfrak{I}(F).$$

Every l_i with $m+1 \leqslant i \leqslant r$ can be expressed as

$$l_i = \sum_{j=1}^{m} c_{ij} l_j$$

for certain $c_{ij} \in \mathbf{K}$, at least one of which is different from zero. Let $j(i)$ be an index with $c_{i,\, j(i)} \neq 0$. Then

$$\mathfrak{L}_i := \mathfrak{L}_0 \cup \{l_i\} \setminus \{l_{j(i)}\} \in \mathfrak{I}(F).$$

Hence $\mathfrak{I}(F)$ contains the $r-m+1$ sets $\mathfrak{L}_0, \mathfrak{L}_{m+1}, \ldots, \mathfrak{L}_r$. This implies that

$$r - m + 1 \leqslant \Omega(\mathfrak{L}_0) + \Omega(\mathfrak{L}_{m+1}) + \ldots + \Omega(\mathfrak{L}_r) \leqslant \Omega(\mathfrak{D}),$$

i.e.

$$r \leqslant m - 1 + \Omega(\mathfrak{D}). \tag{3.3}$$

This certainly implies Theorem 2.
Now assume that there are $\mathfrak{L} \in \mathfrak{I}(F)$ with $\mathfrak{D}(\mathfrak{L}) = (1)$. Assume for instance that $\mathfrak{D}(\mathfrak{L}_0) = (1)$, where $\mathfrak{L}_0 = \{l_1, \ldots, l_m\}$. Every l_i with $m+1 \leqslant i \leqslant r$ can be expressed uniquely as

$$l_i = \sum_{j \in J_i} c_{ij} l_j \qquad \text{with} \quad J_i \subseteq \{1, \ldots, m\}, \quad c_{ij} \in \mathbf{K}^* \quad \text{for} \quad j \in J_i. \tag{3.4}$$

In fact, in (3.4) every J_i has cardinality greater than or equal to 2. Fix a subset J of $\{1, \ldots, m\}$ of cardinality greater than or equal to 2. Consider first the linear forms l_i with $J_i = J$ such that $(c_{ij}) \neq (l_i)/(l_j)$ for at least one $j \in J$. Put

$$\mathfrak{L}_i = \mathfrak{L}_0 \cup \{l_i\} \setminus \{l_j\}.$$

Then Lemma 3 implies that

$$\mathfrak{D}(\mathfrak{L}_i) = \frac{\mathfrak{D}(\mathfrak{L}_i)}{\mathfrak{D}(\mathfrak{L}_0)} = (c_{ij}) \frac{(l_j)}{(l_i)} \neq (1). \tag{3.5}$$

Assume that there are linear forms l_i with $J_i = J$ and $(c_{ij}) = (l_i)/(l_j)$ for each $j \in J$. Fix such a form l_{i_0}. Consider now the linear forms l_i with $J_i = J$ such that $(c_{ij}) = (l_i)/(l_j)$ for $j \in J$ and such that for at least one pair $p, q \in J$,

$$\left(c_{i_0 p} c_{iq} - c_{i_0 q} c_{ip}\right) \neq \frac{(l_i)(l_{i_0})}{(l_p)(l_q)} \qquad \text{and also} \qquad \neq (0).$$

Put

$$\mathfrak{L}_i = \mathfrak{L}_0 \cup \left\{l_{i_0}, l_i\right\} \setminus \left\{l_p, l_q\right\}.$$

Then $\mathfrak{L}_i \in \mathfrak{I}(F)$ and by Lemma 3,

$$\mathfrak{D}(\mathfrak{L}_i) = \frac{\mathfrak{D}(\mathfrak{L}_i)}{\mathfrak{D}(\mathfrak{L}_0)} = \left(c_{i_0 p} c_{iq} - c_{i_0 q} c_{ip}\right) \frac{(l_p)(l_q)}{(l_{i_0})(l_i)} \neq (1). \tag{3.6}$$

Let $\mathcal{L}_1(J)$ be the set of linear forms considered in (3.5) and (3.6). Let $\mathcal{L}_2(J)$ be the set of linear forms consisting of l_{i_0} and of the linear forms l_i with $J_i = J$ such that $(c_{ij}) = (l_i)/(l_j)$ for $j \in J$ and

$$\left(c_{i_0p}c_{iq} - c_{i_0q}c_{ip}\right) = \frac{(l_{i_0})(l_i)}{(l_p)(l_q)} \quad \text{or} \quad = (0) \qquad \text{for each pair} \quad \{p,q\} \subseteq J.$$

If

$$c_{i_0p}c_{iq} - c_{i_0q}c_{ip} = 0$$

then

$$c_{ip}/c_{iq} = c_{i_0p}/c_{i_0q}.$$

If

$$\left(c_{i_0p}c_{iq} - c_{i_0q}c_{ip}\right) = \frac{(l_{i_0})(l_i)}{(l_p)(l_q)},$$

then put

$$x = \frac{c_{i_0p}c_{iq}}{c_{i_0q}c_{ip}}, \qquad y = \frac{c_{i_0q}c_{ip} - c_{i_0p}c_{iq}}{c_{i_0q}c_{ip}}.$$

Then $x + y = 1$. Further,

$$(x) = \frac{(l_{i_0})}{(l_p)} \cdot \frac{(l_i)}{(l_q)} \cdot \left(\frac{(l_{i_0})}{(l_q)} \cdot \frac{(l_i)}{(l_p)}\right)^{-1} = (1)$$

and

$$(y) = \frac{(l_{i_0})(l_i)}{(l_p)(l_q)} \cdot \left(\frac{(l_{i_0})}{(l_q)} \cdot \frac{(l_i)}{(l_p)}\right)^{-1} = (1),$$

i.e. $x, y \in \mathcal{O}_S^*$. Now Lemma 4 implies that for x, i.e. for the quotient $c_{i_0p}c_{iq}/c_{i_0q}c_{ip}$, we have at most h possibilities. Including the possibility that $c_{ip}/c_{iq} = c_{i_0p}/c_{i_0q}$, we have at most $h+1$ possibilities for c_{ip}/c_{iq}. This implies that up to a constant factor, for the linear form

$$l_i = \sum_{j \in J} c_{ij} l_j$$

we have at most $(h+1)^{|J|-1}$ possibilities, where $|J|$ denotes the cardinality of J. Let $\mathfrak{I}(J)$ be the collection of sets $\mathcal{L}_i$ in (3.5), (3.6), and r_J be the cardinality of $\mathcal{L}_1(J) \cup \mathcal{L}_2(J)$. Then

$$r_J \leqslant (h+1)^{|J|-1} + \sum_{\mathcal{L} \in \mathfrak{I}(J)} \Omega(\mathfrak{D}(\mathcal{L})).$$

For $J = \{1\}, \ldots, \{m\}$, we put $r_J := 1$. But summing over all nonempty subsets J of $\{1, \ldots, m\}$ and using that the collection $\mathfrak{I}(J)$ are disjoint, we get

$$r = \sum_J r_J \leqslant \sum_{j=1}^{m} \binom{m}{j} (h+1)^{j-1} + \sum_{\mathcal{L} \in \mathfrak{I}(F)} \Omega(\mathfrak{D}(\mathcal{L})) \leqslant (h+2)^m - 1 + \frac{1}{2}\Omega(\mathfrak{D}). \quad (3.7)$$

Since $h \geqslant 0$, the inequality (3.7) supersedes (3.3). By inserting $h \leqslant 3 \times 7^{d+2s}$ in (3.7) we obtain Theorem 4 for $\mathbf{L} = \mathbf{K}$.

Remark. By using ideas from (Evertse and Győry, 1991), one can derive for $m = 2$ a better bound than that in Theorem 4, namely

$$r \leqslant 3\left(7^{g(d+2s)} + 1\right) + \sqrt{3g\Omega(\mathfrak{D})}.$$

4. PROOF OF THEOREM 1

Theorem 1 is a simple consequence of some of the results proved in §§ 2 and 3. Let $F(\underline{X}) \in \mathbf{Z}[X_1, \ldots, X_n]$ be a square-free decomposable form of rank m. We may assume without loss of generality that $F(1, 0, \ldots, 0) \neq 0$ (see, e.g. (Borevich and Shafarevich, 1967)). We can factor F as

$$F = F_0 F_1 \cdot \ldots \cdot F_t,$$

where F_0 is composed of linear forms with coefficients in $\mathbf{Z}$, and F_i is a norm form in $\mathbf{Z}[X_1, \ldots, X_n]$, i.e.

$$F_i(\underline{X}) = \lambda_i N_{\mathbf{K}_i/\mathbf{Q}}(X_1 + \alpha_{2i}X_2 + \ldots + \alpha_{ni}X_n),$$

where $\mathbf{K}_i = \mathbf{Q}(\alpha_{2i}, \ldots, \alpha_{ni})$ is a proper finite extension of $\mathbf{Q}$, and $\lambda_i \in \mathbf{Q}^*$. Let

$$r_i = \deg F_i, \qquad m_i = \operatorname{rank} F_i, \qquad D_i = D_{F_i} \qquad \text{for} \quad i = 0, \ldots, t.$$

In $\mathbf{Z}$ we have

$$\Omega(a) \leqslant \frac{\log |a|}{\log 2} \qquad \text{for every} \quad a \in \mathbf{Z} \quad \text{with} \quad a \neq 0.$$

Further the number h in Lemma 4 is equal to 0 for $\mathcal{O}_S^* = \mathbf{Z}^* = \{-1, 1\}$. Together with (3.7) this implies that

$$r_0 \leqslant 2^{m_0} - 1 + \frac{1}{2}\frac{\log D_0}{\log 2} \qquad \text{provided} \quad F_0 \neq 1. \tag{4.1}$$

By Lemma 2 we have

$$D^2_{\mathbf{K}_i/\mathbf{Q}} \mid D^{m_i}_{F_i} \qquad \text{for} \quad i = 1, \ldots, t.$$

Further, from a result of (Minkowski, 1891) it follows that

$$\left|D_{\mathbf{K}_i/\mathbf{Q}}\right| \geqslant 3^{[\mathbf{K}_i:\mathbf{Q}]/2}.$$

Since $[\mathbf{K}_i : \mathbf{Q}] = r_i$, this gives

$$r_i \leqslant \frac{m_i}{\log 3} \log D_i \qquad \text{for} \quad i = 1, \ldots, t. \tag{4.2}$$

Finally, by Lemma 1 we have $D_0 \cdot \ldots \cdot D_t \mid D_F$. Together with (4.1), (4.2), and $m_0, \ldots, m_t \leqslant m$ this implies

$$r \leqslant 2^m - 1 + \frac{m}{\log 3} \log D_F.$$

REFERENCES

Birch, B. J. and Merriman, J. R. (1972). Finiteness theorems for binary forms with given discriminant. *Proc. London Math. Soc.* **25**, 385–394.

Borevich, Z. I. and Shafarevich, I. R. (1967). *Number Theory.* Academic Press, New York.

Evertse, J. H. (1984). On equations in S-units and the Thue–Mahler equation. *Invent. Math.* **75**, 561–584.

Evertse, J. H. and Győry, K. (1991). Effective finiteness theorems for binary forms of given discriminant. *Compos. Math.* **79**, 169–204.

Evertse, J. H. and Győry, K. (1992). Effective finiteness theorems for decomposable forms of given discriminant. *Acta Arithm.* (to appear).

Gelfand, I. M., Zelevinsky, A. V., and Karpanov, M. M. (1990). On discriminants of polynomials of several variables. *Funct. Anal. and Appl.* **24**, 1–4 (in Russian).

Győry, K. (1973). Sur les polynômes à coefficients entiers et de discriminant donné. *Acta Arithm.* **23**, 419–426.

Győry, K. (1974). Sur les polynômes à coefficients entiers et de discriminant donné. II. *Publ. Math. Debrecen* **21**, 125–144.

Győry, K. and Papp, Z. Z. (1977). On discriminant form and index form equations. *Studia Sci. Math. Hung.* **12**, 47–60.

Hermite, C. (1851). Sur l'introduction des variables continues dans la théorie des nombres. *J. reine angew. Math.* **41**, 191–216.

Lagrange, J. L. (1773). *Recherches d'Arithmètique.* Nouv. Mém. Acad. Berlin.

Minkowski, H. (1891). Théorèmes arithmétiques. *Comptes Rendus Paris* **112**, 209–212.

New Trends in Probab. and Statist., pp. 57–67
F. Schweiger and E. Manstavičius (Eds.)
1992 VSP/TEV

ON SOME ENTIRE MODULAR FORMS OF HALF INTEGRAL WEIGHT ON THE GROUP $\Gamma_0(4N)$

G. LOMADZE
Department of Mechanics and Mathematics, Tbilisi University,
I. Chavchavadze 1, 380028 Tbilisi, Georgia

ABSTRACT

Three entire modular forms of weight $s/2$ ($s = 5, 7, 9$) and multiplier system $v(M)$ on the group $\Gamma_0(4N)$ are constructed.

In this paper N, a, k, q are positive integers; s and u are odd positive integers; c, g, h, l, m, n, x, α, β, γ, δ are integers; μ, ξ, η are real numbers; A, B, C, D, G, H are complex numbers; z and τ are complex variables. Let $\left(\frac{h}{u}\right)$ be the generalized Jacobi symbol, $e(z) = \exp(2\pi i z)$ (thus, $-1 = e(1/2)$, $i = e(1/4)$),

$$\eta(\gamma) = \begin{cases} 1 & \text{if } \gamma \geqslant 0, \\ -1 & \text{if } \gamma < 0. \end{cases}$$

Further, $\varphi(k)$ is Euler's function, $\mu(k)$ is the Möbius function, $S(h,q)$ is the Gauss sum, $c(h,q)$ is the Ramanujan sum, $r(n, f_s)$ is the number of representations of a positive integer n by the primitive positive definite quadratic form

$$f_s = a_1 x_1^2 + \ldots + a_s x_s^2 \qquad \text{with determinant} \quad \Delta,$$

a is the least common multiple of the coefficients a_k, $k = 1, \ldots, s$.
Let for $\operatorname{Im} \tau > 0$

$$\vartheta_{gh}(z \mid \tau; c, N) = \sum_{m \equiv c \,(\mathrm{mod}\ N)} e\Big(\frac{1}{2N} h(m-c)\Big) e\Big(\frac{1}{8N}(2Nm+g)^2 \tau\Big) e\Big(\frac{1}{2}(2m+g)z\Big).$$

Suppose that

$$\vartheta_{gh}(\tau; 0, N) = \vartheta_{gh}(0 \mid \tau; 0, N),$$
$$\vartheta'_{gh}(\tau; 0, N) = \frac{\partial}{\partial z} \vartheta_{gh}(z \mid \tau; 0, N)\Big|_{z=0}.$$

Then

$$\vartheta_{gh}(\tau;0,N)=\sum_{m=-\infty}^{\infty}(-1)^{hm}e\Big(\frac{1}{8N}(2Nm+g)^2\tau\Big), \tag{1}$$

$$\vartheta'_{gh}(\tau;0,N)=\pi i\sum_{m=-\infty}^{\infty}(-1)^{hm}(2Nm+g)e\Big(\frac{1}{8N}(2Nm+g)^2\tau\Big). \tag{2}$$

It follows from (1) that

$$\prod_{k=1}^{s}\vartheta_{00}(\tau;0,2a_k)=1+\sum_{n=1}^{\infty}r(n;f_s)e(n\tau). \tag{3}$$

Further, let

$$\Theta_s(\tau)=\Theta(\tau;f_s)=1+\sum_{n=1}^{\infty}\rho(n;f_s)e(n\tau), \tag{4}$$

where

$$\rho(n;f_s)=\frac{\pi^{s/2}}{\Gamma(s/2)\Delta^{1/2}}n^{(s/2)-1}\sum_{q=1}^{\infty}A_q(n), \tag{5}$$

$$A_q(n)=q^{-s}\sum_{h \bmod q,\ (h,q)=1}e\Big(-\frac{hn}{q}\Big)\prod_{k=1}^{s}S(a_kh,q). \tag{6}$$

Finally, let

$$\Gamma=\Big\{\frac{\alpha\tau+\beta}{\gamma\tau+\delta}:\ \alpha\delta-\beta\gamma=1\Big\},$$

$$\Gamma_0(4N)=\Big\{\frac{\alpha\tau+\beta}{\gamma\tau+\delta}\in\Gamma:\ \gamma=0\ (\mathrm{mod}\ 4N)\Big\}.$$

Throughout this paper $\mathcal{H}=\{\tau\in\mathbf{C}:\ \mathrm{Im}\,\tau>0\}$.
For $\tau\in\mathcal{H}$ put

$$(\gamma\tau+\delta)^{s/2}=\Big((\gamma\tau+\delta)^{s/2}\Big)^s,\qquad -\frac{\pi}{2}<\arg(\gamma\tau+\delta)^{1/2}\leqslant\frac{\pi}{2}.$$

DEFINITION 1. Let M be a matrix of an arbitrary substitution from $\Gamma_0(4N)$, and let $v(M)$ be a multiplier system for the group $\Gamma_0(4N)$ and of weight $s/2$. We shall say that a function F defined on $\mathcal{H}$ is an entire modular form of weight $s/2$ and multiplier system $v(M)$ on $\Gamma_0(4N)$ if

1) F is regular on $\mathcal{H}$,

2) for all matrices M of substitutions from $\Gamma_0(4N)$ and all $\tau\in\mathcal{H}$

$$F\Big(\frac{\alpha\tau+\beta}{\gamma\tau+\delta}\Big)=v(M)(\gamma\tau+\delta)^{s/2}F(\tau),$$

3) in the neighbourhood of the point $\tau = i\infty$

$$F(\tau) = \sum_{m=0}^{\infty} A_m e(m\tau),$$

4) for all substitutions from Γ, in the neighbourhood of each rational point $\tau = -\delta/\gamma$ ($\gamma \neq 0$, $(\gamma, \delta) = 1$),

$$(\gamma\tau + \delta)^{s/2} F(\tau) = \sum_{m=0}^{\infty} A'_m e\Big(\frac{m}{4N} \frac{\alpha\tau + \beta}{\gamma\tau + \delta}\Big).$$

DEFINITION 2. Let

$$L = \begin{pmatrix} \alpha & \beta \\ \gamma & \delta \end{pmatrix}$$

be a matrix of any substitution from Γ. For the function F defined on $\mathcal{H}$ and a fixed s we write after Hecke and Petersson

$$F(\tau) \mid L = (\gamma\tau + \delta)^{-s/2} F\Big(\frac{\alpha\tau + \beta}{\gamma\tau + \delta}\Big).$$

If

$$L_1 = \begin{pmatrix} \alpha_1 & \beta_1 \\ \gamma_1 & \delta_1 \end{pmatrix} \quad \text{and} \quad L_2 = \begin{pmatrix} \alpha_2 & \beta_2 \\ \gamma_2 & \delta_2 \end{pmatrix}$$

are matrices of two arbitrary substitutions from Γ, then for some root of unity ε

$$(F(\tau) \mid L_1) \mid L_2 = \varepsilon F(\tau) \mid L_1 L_2. \tag{7}$$

It was shown in (Lomadze, 1961, p. 244), that $\varepsilon = 1$ always in the following cases:

$$\begin{array}{lll} \text{a) } \gamma_1\gamma_2 \neq 0 & \text{and} & (\gamma_1\alpha_2 + \delta_1\gamma_2)\gamma_2 > 0, \\ \text{b) } \gamma_1 \neq 0,\ \gamma_2 = 0 & \text{and} & \delta_2 > 0, \\ \text{c) } \gamma_1 = \gamma_2 = 0 & \text{and} & \delta_1 > 0, \\ \text{d) } \gamma_1 = \gamma_2 = 0 & \text{and} & \delta_2 > 0. \end{array} \tag{8}$$

LEMMA 1 (Walfisz, 1957, p. 10, Lemma 1). *If* $(h, q) = 1$, *then*

$$S(kh, kq) = kS(h, q), \qquad |S(h, q)| \leqslant (2q)^{1/2}.$$

LEMMA 2 (Walfisz, 1957, p. 176, Lemma 7).

$$c(h, q) = \sum_{d \mid (h,q)} d\mu\Big(\frac{q}{d}\Big).$$

LEMMA 3 (Lomadze, 1971, Lemmas 9 and 10). *For even g and for all substitutions from $\Gamma_0(4N)$*

$$\begin{aligned}\vartheta_{gh}\Big(\frac{\alpha\tau+\beta}{\gamma\tau+\delta};0,2N\Big) &= e\Big(-\frac{\alpha\gamma\delta^2h^2}{16N}\Big)e\Big(\frac{\beta\delta g^2}{4}\frac{\delta^{2\varphi(2N)-2}}{4N}\Big)\\ &\times e\Big(\frac{1}{8}\eta(\gamma)(\operatorname{sgn}\delta-1)\Big)e\Big(\frac{1}{8}(1-|\delta|)\Big)\Big(\frac{2\beta N\operatorname{sgn}\delta}{|\delta|}\Big)\\ &\times(\gamma\tau+\delta)^{1/2}\vartheta_{\alpha g,h}(\tau;0,2N),\end{aligned}$$

$$\begin{aligned}\vartheta'_{gh}\Big(\frac{\alpha\beta+\beta}{\gamma\tau+\delta};0,2N\Big) &= \operatorname{sgn}\delta\, e\Big(-\frac{\alpha\gamma\delta^2h^2}{16N}\Big)e\Big(\frac{\beta\delta g^2}{4}\frac{\delta^{2\varphi(2N)-2}}{4N}\Big)\\ &\times e\Big(\frac{3}{8}\eta(\gamma)(\operatorname{sgn}\delta-1)\Big)e\Big(\frac{1}{8}(1-|\delta|)\Big)\Big(\frac{2\beta N\operatorname{sgn}\delta}{|\delta|}\Big)\\ &\times(\gamma\tau+\delta)^{3/2}\vartheta'_{\alpha g,h}(\tau;0,2N).\end{aligned}$$

LEMMA 4 (Lomadze, 1971, Lemma 11). *For even g and for all substitutions from Γ in the neighbourhood of each rational point*

$$\tau=-\delta/\gamma \qquad (\gamma\neq 0,\ (\gamma,\delta)=1),$$

$$(\gamma\tau+\delta)^{1/2}\vartheta_{gh}(\tau;0,2N)=\begin{cases}\displaystyle\sum_{m=0}^{\infty}B_m e\Big(\frac{m}{4N}\frac{\alpha\tau+\beta}{\gamma\tau+\delta}\Big) & \text{if } 2\mid\gamma,\\ \displaystyle e\Big(\frac{h}{16N}\frac{\alpha\tau+\beta}{\gamma\tau+\delta}\Big)\sum_{m=0}^{\infty}B'_m e\Big(\frac{m}{4N}\frac{\alpha\tau+\beta}{\gamma\tau+\delta}\Big) & \text{if } 2\nmid\gamma,\end{cases}$$

$$(\gamma\tau+\delta)^{3/2}\vartheta'_{gh}(\tau;0,2N)=\begin{cases}\displaystyle\sum_{m=0}^{\infty}D_m e\Big(\frac{m}{4N}\frac{\alpha\tau+\beta}{\gamma\tau+\delta}\Big) & \text{if } 2\mid\gamma,\\ \displaystyle e\Big(\frac{h}{16N}\frac{\alpha\tau+\beta}{\gamma\tau+\delta}\Big)\sum_{m=0}^{\infty}D'_m e\Big(\frac{m}{4N}\frac{\alpha\tau+\beta}{\gamma\tau+\delta}\Big) & \text{if } 2\nmid\gamma.\end{cases}$$

Here $h=0$ or 1.

LEMMA 5 (Lomadze, 1966, Lemmas 11 and 12). *The function $\Theta_s(\tau)$ is regular on $\mathcal{H}$ if $s>4$, and for all substitutions from $\Gamma_0(4a)$*

$$\Theta_s\Big(\frac{\alpha\tau+\beta}{\gamma\tau+\delta}\Big)=e\Big(\frac{s}{8}\eta(\gamma)(\operatorname{sgn}\delta-1)\Big)e\Big(\frac{s}{16}(|\delta|-1)^2\Big)\Big(\frac{\beta\Delta\operatorname{sgn}\delta}{|\delta|}\Big)(\gamma\tau+\delta)^{s/2}\Theta_s(\tau).$$

LEMMA 6. *If $s>4$, then for all substitutions from Γ in the neighbourhood of each rational point $\tau=-\delta/\gamma$ $(\gamma\neq 0,\ (\gamma,\delta)=1)$,*

$$(\gamma\tau+\delta)^{s/2}\Theta_s(\tau)=\sum_{n=0}^{\infty}A_n e\Big(\frac{n}{4a}\frac{\alpha\tau+\beta}{\gamma\tau+\delta}\Big).$$

Proof. For all matrices M of substitutions from $\Gamma_0(4a)$, by Lemma 5 and Definition 2,

$$\Theta_s(\tau) \mid M = v(M)\Theta_s(\tau), \tag{9}$$

where

$$v(M) = e\Big(\frac{s}{8}\eta(\gamma)(\operatorname{sgn}\delta - 1)\Big)e\Big(\frac{s}{16}(|\delta| - 1)^2\Big)\Big(\frac{\beta\Delta\operatorname{sgn}\delta}{|\delta|}\Big). \tag{10}$$

Let

$$L = \begin{pmatrix} \alpha & \beta \\ \gamma & \delta \end{pmatrix}$$

be a matrix of an arbitrary substitution from Γ and

$$U = \begin{pmatrix} 1 & 1 \\ 0 & 1 \end{pmatrix}.$$

Then

$$U^{4a} = \begin{pmatrix} 1 & 4a \\ 0 & 1 \end{pmatrix} \quad \text{and} \quad LU^{4a}L^{-1} = \begin{pmatrix} 1 - 4a\alpha\gamma & 4a\alpha^2 \\ -4a\gamma^2 & 4a\alpha\gamma + 1 \end{pmatrix}$$

are matrices of substitutions from $\Gamma_0(4a)$. Hence, by (9),

$$\Theta_s(\tau) \mid LU^{4a}L^{-1} = v\big(LU^{4a}L^{-1}\big)\Theta_s(\tau).$$

From this, by (10), it follows that

$$v\big(LU^{4a}L^{-1}\big) = e\Big(\frac{s}{8}\eta(-4a\gamma^2)(\operatorname{sgn}(4a\alpha\gamma + 1) - 1)\Big)e\Big(\frac{s}{16}(|4a\alpha\gamma + 1| - 1)^2\Big) \times \Big(\frac{4a\alpha^2\Delta\operatorname{sgn}(4a\alpha\gamma + 1)}{|4a\alpha\gamma + 1|}\Big),$$

where $a = 2^l u$ $(l \geqslant 0)$ and

$$\Delta = \prod_{k=1}^{s} a_k = \prod_{k=1}^{s} 2^{l_k}u_k, \qquad l_k \geqslant 0.$$

Hence, if $4a\alpha\gamma + 1 > 0$, then

$$\begin{aligned} v\big(LU^{4a}L^{-1}\big) &= \Big(\frac{a\Delta}{4a\alpha\gamma + 1}\Big) = \Big(\frac{2}{4a\alpha\gamma + 1}\Big)^{l+\sum_{k=1}^{s} l_k}\left(\frac{u\prod_{k=1}^{s} u_k}{4a\alpha\gamma + 1}\right) \\ &= \Big(\frac{4a\alpha\gamma + 1}{u}\Big)\prod_{k=1}^{s}\Big(\frac{4a\alpha\gamma + 1}{u_k}\Big) = 1, \end{aligned}$$

and if $4a\alpha\gamma + 1 < 0$, then

$$\begin{aligned} v\big(LU^{4a}L^{-1}\big) &= i^s e\Big(\frac{s}{4}(2a\alpha\gamma + 1)^2\Big)\Big(\frac{-a\Delta}{-4a\alpha\gamma - 1}\Big) = -i^{2s}\Big(\frac{a\Delta}{-4a\alpha\gamma - 1}\Big) \\ &= \Big(\frac{4a\alpha\gamma + 1}{u}\Big)\prod_{k=1}^{s}\Big(\frac{4a\alpha\gamma + 1}{u_k}\Big) = 1, \end{aligned}$$

since $2 \nmid s$. Thus

$$\Theta_s(\tau) \mid LU^{4a}L^{-1} = \Theta_s(\tau). \tag{11}$$

Let

$$\Theta_s(\tau) \mid L = \omega(\tau). \tag{12}$$

Then by (7), (8b), and (8d),

$$\Theta_s(\tau) \mid LU^{4a} = \omega(\tau) \mid U^{4a} = \omega(\tau + 4a).$$

On the other hand, by (7), (8a), (8c), (11), and (12)

$$\Theta_s(\tau) \mid LU^{4a} = (\Theta_s(\tau) \mid LU^{4a}L^{-1}) \mid L = \omega(\tau).$$

Hence

$$\omega(\tau + 4a) = \omega(\tau).$$

Since $\omega(\tau)$ is regular on $\mathcal{H}$ with the period $4a$, it has the Fourier expansion

$$\omega(\tau) = \sum_{n=-\infty}^{\infty} A_n e\Big(\frac{n\tau}{4a}\Big) \tag{13}$$

with the Fourier coefficients

$$A_n = \frac{1}{4a} \int_{\tau_0 - 2a}^{\tau_0 + 2a} \omega(\tau) e\Big(-\frac{n\tau}{4a}\Big)\, d\tau,$$

where τ_0 is an arbitrary point on $\mathcal{H}$.
It follows from (12) and (13) that for all substitutions from Γ

$$(\gamma\tau + \delta)^{-s/2} \Theta_s\Big(\frac{\alpha\tau + \beta}{\gamma\tau + \delta}\Big) = \sum_{n=-\infty}^{\infty} A_n e\Big(\frac{n\tau}{4a}\Big), \tag{14}$$

$$A_n = \frac{1}{4a} \int_{\tau_0 - 2a}^{\tau_0 + 2a} (\gamma\tau + \delta)^{-s/2} \Theta_s\Big(\frac{\alpha\tau + \beta}{\gamma\tau + \delta}\Big) e\Big(-\frac{n\tau}{4a}\Big)\, d\tau. \tag{15}$$

If $s > 4$, it follows from (4) and (5) that

$$\Theta_s(\tau) = 1 + \frac{\pi^{s/2}}{\Gamma(s/2)\Delta^{(1/2)}} \sum_{q=1}^{\infty} A_q(n) \sum_{n=1}^{\infty} n^{(s/2)-1} e(n\tau). \tag{16}$$

To estimate the Gauss sums in $A_q(n)$, let

$$q = (q, a_k) q_k, \qquad a_k = (q, a_k) a_k'.$$

Hence, $q_k = q/(q, a_k)$ and, by Lemma 1,

$$\left|\prod_{k=1}^{s} S(a_k h, q)\right| \leqslant \prod_{k=1}^{s}(q, a_k)(2q_k)^{1/2} = 2^{s/2} q^{s/2} \prod_{k=1}^{s}(q, a_k)^{1/2}.$$

Therefore, by (6) and Lemma 2,

$$|A_q(n)| \leqslant 2^{s/2} q^{-s/2} \prod_{k=1}^{s}(q, a_k)^{1/2} |c(n, q)| < K q^{-s/2} n^2,$$

where K does not depend on n.
Thus, if we put $\tau = \mu_1 + i\mu_2$ ($\mu_2 > 0$) in (16), then

$$|\Theta_s(\tau)| < 1 + K \sum_{n=1}^{\infty} n^{(s/2)+1} \exp(-2\pi n \mu_2).$$

From this, since $\exp(2\pi\mu_2 n) \geqslant K n^s \mu_2^s$ and $s > 4$, it follows that

$$|\Theta_s(\tau)| < 1 + K \sum_{n=1}^{\infty} n^{-(s/2)+1} \mu_2^{-s} < 1 + K\mu_2^{-s}. \tag{17}$$

Let

$$\tau_0 = -\frac{\delta}{\gamma} + i\eta_0, \qquad \eta_0 > 0.$$

Then in (15)

$$\tau = \xi + i\eta_0, \qquad -2a - \frac{\delta}{\gamma} \leqslant \xi \leqslant 2a - \frac{\delta}{\gamma},$$

hence

$$\gamma\tau + \delta = (\gamma\xi + \delta) + i\gamma\eta_0, \qquad -2a\gamma \leqslant \gamma\xi + \delta \leqslant 2a\gamma,$$

i.e.

$$(\gamma\xi + \delta)^2 \leqslant 4a^2\gamma^2.$$

Thus in (15)

$$\operatorname{Im}\Big(-\frac{1}{\gamma\tau + \delta}\Big) = \frac{\gamma\eta_0}{(\gamma\xi + \delta)^2 + \gamma^2\eta_0^2} \geqslant \frac{\eta_0}{\gamma(4a^2 + \eta_0^2)} \qquad \text{if} \quad \gamma > 0,$$

$$\operatorname{Im}\Big(\frac{1}{\gamma\tau + \delta}\Big) = \frac{-\gamma\eta_0}{(\gamma\xi + \delta)^2 + \gamma^2\eta_0^2} \geqslant \frac{\eta_0}{-\gamma(4a^2 + \eta_0^2)} \qquad \text{if} \quad \gamma < 0,$$

i.e.

$$\operatorname{Im}\Big(\frac{\alpha\tau + \beta}{\gamma\tau + \delta}\Big) = \operatorname{Im}\Big(\frac{\alpha}{\gamma} - \frac{1}{\gamma(\gamma\tau + \delta)}\Big) \geqslant \frac{\eta_0}{\gamma^2(4a^2 + \eta_0^2)}.$$

Hence, by (17), in (15) we have

$$\left|\Theta_s\Big(\frac{\alpha\tau + \beta}{\gamma\tau + \delta}\Big)\right| < 1 + K\big(\eta_0^{-1}\gamma^2(4a^2 + \eta_0^2)\big)^s. \tag{18}$$

Further,

$$|\gamma\tau+\delta|^{s/2} \geqslant |\gamma|^{s/2}\eta_0^{s/2}. \tag{19}$$

It follows from (18) and (19) that in (15)

$$\left|(\gamma\tau+\delta)^{-s/2}\Theta_s\Big(\frac{\alpha\tau+\beta}{\gamma\tau+\delta}\Big)\right| \leqslant \eta_0^{-s/2} + K\eta_0^{-3s/2}(4a^2+\eta_0^2)^s,$$

and therefore

$$|A_n| \leqslant \exp\Big(\frac{\pi\eta_0 n}{2a}\Big)\Big(\eta_0^{-s/2} + K\eta_0^{-3s/2}(4a^2+\eta_0^2)\Big).$$

From this it follows that $|A_n| \to 0$, when $\eta_0 \to \infty$ and $n < 0$. Thus (14) has the form

$$(\gamma\tau+\delta)^{-s/2}\Theta_s\Big(\frac{\alpha\tau+\beta}{\gamma\tau+\delta}\Big) = \sum_{n=0}^{\infty} A_n e\Big(\frac{n\tau}{4a}\Big).$$

The lemma follows from this equality if we write

$$\delta, \quad -\beta, \quad -\gamma, \quad \alpha, \quad \frac{\alpha\tau+\beta}{\gamma\tau+\delta}$$

instead of α, β, γ, δ, τ.

THEOREM. *Let $N \geqslant a$, and*

$$\Phi_{gh}(\tau) = \begin{cases} \vartheta'_{g_1h_1}(\tau;0,2N_1)\vartheta_{g_2h_2}(\tau;0,2N_2)\vartheta_{g_3h_3}(\tau;0,2N_3) & \text{if } s=5,\\ \vartheta'_{g_1h_1}(\tau;0,2N_1)\vartheta'_{g_2h_2}(\tau;0,2N_2)\vartheta_{g_3h_3}(\tau;0,2N_3) & \text{if } s=7,\\ \vartheta'_{g_1h_1}(\tau;0,2N_1)\vartheta'_{g_2h_2}(\tau;0,2N_2)\vartheta'_{g_3h_3}(\tau;0,2N_3) & \text{if } s=9. \end{cases}$$

The function

$$\psi_s(\tau) = \psi(\tau;f_s) = \prod_{k=1}^{s}\vartheta_{00}(\tau;0,2a_k) - \Theta(\tau;f_s) - \lambda\Phi_{gh}(\tau), \tag{20}$$

where λ is an arbitrary constant, is an entire modular form of weight $s/2$ and multiplier system

$$v(M) = e\Big(\frac{s}{8}\eta(\gamma)(\operatorname{sgn}\delta-1)\Big)e\Big(\frac{s}{16}(|\delta|-1)^2\Big)\Big(\frac{\beta\Delta\operatorname{sgn}\delta}{|\delta|}\Big)$$

($M = \begin{pmatrix}\alpha & \beta\\ \gamma & \delta\end{pmatrix}$ is a matrix of the substitution from $\Gamma_0(4N)$) on $\Gamma_0(4N)$ if the following conditions hold:

$$1)\quad 2\mid g_k, \qquad N_k\mid N \quad (k=1,2,3), \qquad a\mid N, \tag{21}$$

$$2)\quad 4\mid N\sum_{k=1}^{3}\frac{h_k^2}{N_k}, \qquad 4\mid\sum_{k=1}^{3}\frac{g_k^2}{4N_k}, \tag{22}$$

3) for all α and δ with $\alpha\delta \equiv 1 \pmod{4N}$

$$\begin{aligned} \operatorname{sgn}\delta\Big(\frac{N_1N_2N_3}{|\delta|}\Big)\Phi_{\alpha g,h}(\tau) &= \Big(\frac{-\Delta}{|\delta|}\Big)\Phi_{gh}(\tau) \qquad \text{when} \quad s=5 \text{ or } s=9,\\ \Big(\frac{N_1N_2N_3}{|\delta|}\Big)\Phi_{\alpha g,h}(\tau) &= \Big(\frac{\Delta}{|\delta|}\Big)\Phi_{gh}(\tau) \qquad \text{when} \quad s=7. \end{aligned} \tag{23}$$

Proof. I. It is well known that the theta-series (1)–(3) are regular on $\mathcal{H}$. By Lemma 5 the function $\Theta_s(\tau)$ is regular on $\mathcal{H}$, too. Hence, $\psi_s(\tau)$ satisfies the condition 1) of the Definition 1.
II. It may be easily verified that (22) implies

$$4 \mid \sum_{k=1}^{3} \delta^{2\varphi(2N_k)-2}\frac{g_k^2}{4N_k}, \tag{24}$$

when $2 \nmid \delta$, because $\alpha\delta \equiv 1 \pmod{4N}$.
From (21) it follows that each substitution from $\Gamma_0(4N)$ is also a substitution from $\Gamma_0(4a)$, $\Gamma_0(4a_k)$ $(k=1,2,\ldots,s)$, and $\Gamma_0(4N_k)$ $(k=1,2,3)$. Hence, by (22)–(24) and Lemma 3 (with a_k and N_k instead of N), for each substitution from $\Gamma_0(4N)$ we have

$$\begin{aligned} \prod_{k=1}^{s}\vartheta_{00}\Big(\frac{\alpha\tau+\beta}{\gamma\tau+\delta};0,2a_k\Big) &= e\Big(\frac{s}{8}\eta(\gamma)(\operatorname{sgn}\delta-1)\Big)e\Big(\frac{s}{8}(1-|\delta|)\Big)\\ &\quad\times\Big(\frac{2\beta\Delta\operatorname{sgn}\delta}{|\delta|}\Big)(\gamma\tau+\delta)^{s/2}\prod_{k=1}^{s}\vartheta_{00}(\tau;0,2a_k),\\ \Phi_{gh}\Big(\frac{\alpha\tau+\beta}{\gamma\tau+\delta}\Big) &= e\Big(\frac{s}{8}\eta(\gamma)(\operatorname{sgn}\delta-1)\Big)e\Big(\frac{3}{8}(1-|\delta|)\Big)\Big(\frac{-2\beta\Delta\operatorname{sgn}\delta}{|\delta|}\Big)\\ &\quad\times(\gamma\tau+\delta)^{s/2}\Phi_{gh}(\tau) \qquad \text{if} \quad s=5 \text{ or } s=9,\\ \Phi_{gh}\Big(\frac{\alpha\tau+\beta}{\gamma\tau+\delta}\Big) &= e\Big(\frac{s}{8}\eta(\gamma)(\operatorname{sgn}\delta-1)\Big)e\Big(\frac{3}{8}(1-|\delta|)\Big)\Big(\frac{2\beta\Delta\operatorname{sgn}\delta}{|\delta|}\Big)\\ &\quad\times(\gamma\tau+\delta)^{s/2}\Phi_{gh}(\tau) \qquad \text{if} \quad s=7. \end{aligned}$$

It is not difficult to verify that

$$\begin{aligned} e\Big(\frac{s}{8}(|1-|\delta|)\Big)\Big(\frac{2}{|\delta|}\Big) &= e\Big(\frac{3}{8}(1-|\delta|)\Big)\Big(\frac{-2}{|\delta|}\Big)\\ &= e\Big(\frac{s}{16}(|\delta|-1)^2\Big) \qquad \text{if} \quad s=5 \text{ or } s=9,\\ e\Big(\frac{s}{8}(1-|\delta|)\Big)\Big(\frac{2}{|\delta|}\Big) &= e\Big(\frac{3}{8}(1-|\delta|)\Big)\Big(\frac{2}{|\delta|}\Big)\\ &= e\Big(\frac{s}{16}(|\delta|-1)^2\Big) \qquad \text{if} \quad s=7. \end{aligned}$$

Hence, by (20) and Lemma 5, for each substitution from $\Gamma_0(4N)$ we have

$$\begin{aligned} \psi_s\Big(\frac{\alpha\tau+\beta}{\gamma\tau+\delta}\Big) &= e\Big(\frac{s}{8}\eta(\gamma)(\operatorname{sgn}\delta-1)\Big)e\Big(\frac{s}{16}(|\delta|-1)^2\Big)\Big(\frac{\beta\Delta\operatorname{sgn}\delta}{|\delta|}\Big)\\ &\quad\times(\gamma r+\delta)^{s/2}\psi_s(\tau). \end{aligned} \tag{25}$$

Thus $\psi_s(\tau)$ satisfies the condition 2) of the Definition 1.
III. By (3) and (4), the functions

$$\prod_{k=1}^{s} \vartheta_{00}(\tau; 0, 2a_k) \quad \text{and} \quad \theta(\tau; f_s)$$

in (20) satisfy the condition 3) of the Definition 1. By (1) and (2), we have

$$\Phi_{gh}(\tau) = \pi i \sum_{m_1,m_2,m_3=-\infty}^{\infty} e\left(\frac{1}{2}\sum_{k=1}^{3} h_k m_k\right)(4N_1 m_1 + g_1)e(\Lambda\tau) \qquad \text{if} \quad s = 5,$$

$$\Phi_{gh}(\tau) = -\pi^2 \sum_{m_1,m_2,m_3=-\infty}^{\infty} e\left(\frac{1}{2}\sum_{k=1}^{3} h_k m_k\right)$$
$$\times (4N_1 m_1 + g_1)(4N_2 m_2 + g_2)e(\Lambda\tau) \qquad \text{if} \quad s = 7,$$

$$\Phi_{gh}(\tau) = -\pi^3 i \sum_{m_1,m_2,m_3=-\infty}^{\infty} e\left(\frac{1}{2}\sum_{k=1}^{3} h_k m_k\right)$$
$$\times \prod_{k=1}^{3}(4N_k m_k + g_k)e(\Lambda\tau) \qquad \text{if} \quad s = 9,$$

where

$$\Lambda = \sum_{k=1}^{3} \frac{1}{16N_k}(4N_k m_k + g_k)^2 = \sum_{k=1}^{3}(N_k m_k^2 + \frac{1}{2}m_k g_k) + \frac{1}{4}\sum_{k=1}^{3}\frac{g_k^2}{4N_k},$$

and, by (22), Λ is an integer. Thus $\psi_s(\tau)$ satisfies also the condition 3) of the Definition 1.
IV. By Lemma 4 (with a_k and N_k instead of N), (21), and (22) for even g and for all substitutions from Γ, in the neighbourhood of each rational point

$$\tau = -\delta/\gamma \qquad (\gamma \neq 0,\ (\gamma, \delta) = 1),$$

the following propositions are true:
1)

$$(\gamma\tau + \delta)^{s/2} \prod_{k=1}^{s} \vartheta_{00}(\tau; 0, 2a_k) = \prod_{k=1}^{s} \sum_{m_k=0}^{\infty} B_{m_k} e\Big(\frac{m_k}{4a_k}\,\frac{\alpha\tau + \beta}{\gamma\tau + \delta}\Big)$$
$$= \prod_{k=1}^{s} \sum_{m_k=0}^{\infty} B_{m_k} e\Big(\frac{1}{4N}\,\frac{Nm_k}{a_k}\,\frac{\alpha\tau + \beta}{\gamma\tau + \delta}\Big) = \sum_{n=0}^{\infty} C_n e\Big(\frac{n}{4N}\,\frac{\alpha\tau + \beta}{\gamma\tau + \delta}\Big),$$

2) if $2 \mid \gamma$ or if $2 \nmid \gamma$ and $h_1 = h_2 = h_3 = 0$, then

$$(\gamma\tau + \delta)^{s/2}\Phi_{gh}(\tau) = \prod_{k=1}^{3} \sum_{m_k=0}^{\infty} D_{m_k} e\Big(\frac{m_k}{4N_k}\,\frac{\alpha\tau + \beta}{\gamma\tau + \delta}\Big)$$
$$= \prod_{k=1}^{3} \sum_{m_k=0}^{\infty} D_{m_k} e\Big(\frac{1}{4N}\,\frac{Nm_k}{N_k}\,\frac{\alpha\tau + \beta}{\gamma\tau + \delta}\Big) = \sum_{n=0}^{\infty} G_n e\Big(\frac{n}{4N}\,\frac{\alpha\tau + \beta}{\gamma\tau + \delta}\Big),$$

3) if $2 \nmid \gamma$ and at least one of h_1, h_2 and h_3 is 1,

$$(\gamma\tau+\delta)^{s/2}\Phi_{gh}(\tau) = \prod_{k=1}^{3}\left\{e\Big(\frac{h_k}{16N_k}\,\frac{\alpha\tau+\beta}{\gamma\tau+\delta}\Big)\sum_{m_k=0}^{\infty} D'_{m_k} e\Big(\frac{m_k}{4N_k}\,\frac{\alpha\tau+\beta}{\gamma\tau+\delta}\Big)\right\}$$

$$= e\Big(\frac{1}{4N}\sum_{k=1}^{3}\frac{Nh_k}{4N_k}\,\frac{\alpha\tau+\beta}{\gamma\tau+\delta}\Big)\prod_{k=1}^{3}\sum_{m_k=0}^{\infty} D'_{m_k} e\Big(\frac{1}{4N}\,\frac{Nm_k}{N_k}\,\frac{\alpha\tau+\beta}{\gamma\tau+\delta}\Big)$$

$$= e\Big(\frac{h'}{4N}\,\frac{\alpha\tau+\beta}{\gamma\tau+\delta}\Big)\sum_{n'=0}^{\infty} H_{n'} e\Big(\frac{n'}{4N}\,\frac{\alpha\tau+\beta}{\gamma\tau+\delta}\Big) = \sum_{n=0}^{\infty} H_n e\Big(\frac{n}{4N}\,\frac{\alpha\tau+\beta}{\gamma\tau+\delta}\Big).$$

Thus, by (20) and Lemma 6, the function $\psi_s(\tau)$ satisfies the condition 4) of the Definition 1. This completes the proof.

The proved theorem will be useful in revealing the arithmetical sense of the so-called additional terms in formulae for the number of representations of positive integers by positive quadratic forms in 5, 7, and 9 variables.

For example, it was noticed in (Mordell, 1919, p. 372), that the number of representations of a positive integer as a sum of 9 squares can be expressed by the help of the number of representations as a sum of 7 squares. But from our theorem it follows that the number of representations of a positive integer n as a sum of 9 squares is

$$r_9(n) = \rho_9(n) + \frac{32}{17}\sum_{\substack{x_1^2+x_2^2+x_3^2=3n \\ x_1\equiv x_2\equiv x_3\equiv 1 \pmod 3}} x_1x_2x_3,$$

where $\rho_9(n)$ is the sum of Hardy's singular series. The proof of this corollary will be published elsewhere.

REFERENCES

Lomadze, G. (1961). Über die Darstellung der Zahlen durch einige ternäre quadratishe Formen. *Acta Arithm.* **6**, 225–275.

Lomadze, G. A. (1966). Representation of numbers by certain quadratic forms with six variables. I. *Proc. Tbilisi Univ.* **117**, 7-43 (in Russian).

Lomadze, G. A. (1971). The number of representations of numbers by quadratic forms with four variables. *Proc. Tbilisi A. M. Razmadze Math. Inst.* **40**, 106–139 (in Russian).

Mordell, L. J. (1919). On the representations of a number as a sum of an odd number of squares. *Trans. Cambr. Philos. Soc.* **22**, 361–372.

Walfisz, A. (1957). *Gitterpunkte in mehrdimensionalen Kugeln.* PWN, Warszawa.

New Trends in Probab. and Statist., pp. 69–71
F. Schweiger and E. Manstavičius (Eds.)
1992 VSP/TEV

ON SOLUTIONS OF LINEAR CONGRUENCES

V. I. NECHAEV
Tcherniahovskij 5–166,
125319 Moscow, Russia

ABSTRACT

We study the distribution of solutions of a system of linear congruences to prime modulus with the condition that values of unknowns belong to intervals.

We consider the problem of distribution of solutions of a system of linear congruences to prime modulus with a condition that values of unknowns belong to certain intervals.
Let $\mathbf{Z}$ be the set of integers, $\mathbf{C}$ be the set of complex numbers, p be a prime, n and m be fixed positive integers.
Let

$$h_1, \dots, h_m \in \mathbf{Z}, \qquad 0 \leqslant h_j < p \qquad \text{for} \quad 1 \leqslant j \leqslant n,$$

$$e_p(t) = \exp\left(\frac{2\pi i t}{p}\right) \qquad \text{for real} \quad t, \qquad x_0 = 1.$$

Let

$$A = \begin{pmatrix} a_{10} & a_{11} & \dots & a_{1n} \\ . & . & \dots & . \\ a_{m0} & a_{m1} & \dots & a_{mn} \end{pmatrix} \tag{1}$$

be a $m \times (n+1)$ matrix with

$$a_{ij} \in \mathbf{Z}, \qquad 0 \leqslant a_{ij} < p, \qquad \text{for} \quad 1 \leqslant i \leqslant m, \quad 0 \leqslant j \leqslant n. \tag{2}$$

In what follows the sum taken over the set of matrices (1) satisfying (2) will be denoted by $\sum_A$.

DEFINITION. For a matrix A described in (1), (2) we denote by $R(A)$ the number of solutions of the system of congruences

$$\sum_{j=0}^{n} a_{ij} x_j \equiv 0 \pmod{p}, \qquad i = 1, \dots, m,$$

with the conditions

$$0 \leqslant x_j < h_j \qquad \text{for} \quad 1 \leqslant j \leqslant n.$$

THEOREM 1.
$$\frac{1}{p^{m(n+1)}}\sum_A R(A)=\frac{h_1\cdot\ldots\cdot h_n}{p^m}. \tag{3}$$

THEOREM 2.
$$\frac{1}{p^{m(n+1)}}\sum_A\Big(R(A)-\frac{h_1\cdot\ldots\cdot h_n}{p^m}\Big)^2=\frac{h_1\cdot\ldots\cdot h_n}{p^m}\Big(1-\frac{1}{p^m}\Big).$$

The following lemma holds.

LEMMA 1. *Let $a\in\mathbf{Z}$. Then*

$$\text{(i)}\qquad \frac{1}{p}\sum_{u=0}^{p-1}e_p(ua)=\begin{cases}1 & \text{if } p\mid a,\\ 0 & \text{if } p\nmid a,\end{cases}$$

$$\text{(ii)}\qquad \sum_{u=1}^{p-1}e_p(ua)=\begin{cases}p-1 & \text{if } p\mid a,\\ -1 & \text{if } p\nmid a.\end{cases}$$

Proof of Theorem 1. It is well known (see (Vinogradov, 1954)), that

$$R(A)=\frac{1}{p^m}\sum_{x_1=0}^{h_1-1}\cdots\sum_{x_n=0}^{h_n-1}\sum_{u_1=0}^{p-1}\cdots\sum_{u_m=0}^{p-1}e_p\Big(\sum_{i=1}^m u_i\sum_{j=0}^n a_{ij}x_j\Big). \tag{4}$$

We note that

$$\sum_{u=0}^{p-1}\sum_{a_0=0}^{p-1}\cdots\sum_{a_n=0}^{p-1}e_p\Big(u\sum_{j=0}^n a_jx_j\Big)$$
$$=\sum_{u=0}^{p-1}\Big(\prod_{j=1}^n\sum_{a_j=0}^{p-1}e_p(ua_jx_j)\Big)\sum_{a_0=0}^{p-1}e_p(ua_0)=p^{n+1}.$$

Hence equality (3) follows easily.
Proof of Theorem 2. From equality (4) we obtain

$$\frac{1}{p^{m(n+1)}}\sum_A|R(A)|^2=\frac{1}{p^{m(n+3)}}\sum\nolimits_1,$$

where $\sum_1$ is the multiple sum with $m(n+1)+2n+2m$ variables

$$a_{ij},\qquad 1\leqslant i\leqslant m,\quad 0\leqslant j\leqslant n,$$
$$x_j,x_j',\qquad 1\leqslant j\leqslant n,$$
$$u_i,u_i',\qquad 1\leqslant i\leqslant m.$$

Let
$$u_i,u_i'\in\mathbf{Z},\qquad 0\leqslant u_i\leqslant p-1,\qquad 0\leqslant u_i'\leqslant p-1.$$

Then

$$\sum_{a_{j0}=0}^{p-1} e_p\big(a_{j0}(u_i - u_i')\big) = \begin{cases} p & \text{if } u_i = u_i', \\ 0 & \text{if } u_i \neq u_i. \end{cases}$$

Therefore

$$\frac{1}{p^{m(n+1)}} \sum_A |R(A)|^2 = \frac{1}{p^{m(n+2)}} \sum\nolimits_2, \tag{5}$$

where $\sum_2$ is the multiple sum with $mn + 2n + m$ variables

$$\begin{aligned} &a_{ij}, && 1 \leqslant i \leqslant m, \quad 1 \leqslant j \leqslant n, \\ &x_j, x_j', && 1 \leqslant j \leqslant n, \\ &u_i, && 1 \leqslant i \leqslant m. \end{aligned}$$

If f is some function, $f\colon \mathbf{Z} \longrightarrow \mathbf{C}$, then

$$\sum_{u=0}^{p-1} f(u) = f(0) + \sum_{u=1}^{p-1} f(u). \tag{6}$$

From (5), applying the identity (6), Lemma 1, and the simple identity

$$e_p(a+b) = e_p(a)\, e_p(b),$$

we obtain

$$\begin{aligned} \frac{1}{p^{m(n+1)}} \sum_A |R(A)|^2 &= \left(\frac{h_1 \cdot \ldots \cdot h_n}{p^m}\right)^2 + \sum_{k=1}^{m} \binom{m}{k} \frac{h_1 \cdot \ldots \cdot h_n}{p^{2m}} (p-1)^k \\ &= \left(\frac{h_1 \cdot \ldots \cdot h_n}{p^m}\right)^2 + \frac{h_1 \cdot \ldots \cdot h_n}{p^{2m}} (p^m - 1). \end{aligned}$$

Theorem 2 follows from this equality and from Theorem 1.

REFERENCES

Vinogradov, I. M. (1954). *Elements of Number Theory.* Dover Publications, New York.

Part Three

METRIC NUMBER THEORY

New Trends in Probab. and Statist., pp. 75–81
F. Schweiger and E. Manstavičius (Eds.)
1992 VSP/TEV

THE METHOD OF TRIGONOMETRIC SUMS AND LOWER ESTIMATES OF HAUSDORFF DIMENSION

V. I. BERNIK and N.A. PEREVERSEVA
Institute of Mathematics, Academy of Sciences of Byelorussia,
Surganovo 11, 220604 Minsk, Byelorussia

ABSTRACT

Let $f_i(x_i)$ be the functions defined on some intervals I_i and satisfying $f_i(x_i) \in C^3(I_i)$, $|f_i''(x_i)| > 0$ for $x \in I_i$. Let $M(v)$ denote the set of $\bar{x}$, $\bar{x} \in \mathbf{R}^m$, for which the inequality $\max_{1\leqslant i\leqslant m}(\|x_i q\|, \|f_i(x_i)q\|) < q^{-v}$ has infinitely many integer solutions in q. The upper and lower bounds for the Hausdorff dimension of $M(v)$ are obtained.

Most of methods of the metric theory of Diophantine approximation using the estimates for trigonometric sums lead us to nontrivial estimates for Hausdorff dimension. This possibility occurs when closer approximations are considered and the Lebesgue measure of the sets involved in such approximations is equal to zero, as it is in (Kubilius, 1949; Cassels, 1950; Sprindžiuk, 1974). On this way you get the upper estimates. In this paper we shall show how to get lower estimates in some case as well.

Let $f_i(x_i)$, $1 \leqslant i \leqslant m$, be the collection of functions having three continuous derivatives on the interval J_i, and let $|f_i''(x_i)| > 0$ for $x_i \in J_i$, $1 \leqslant i \leqslant m$. Consider the inequality

$$\max_{1\leqslant i\leqslant m} (\|x_i q\|, \|f_i(x_i)q\|) < q^{-v}. \tag{1}$$

Note that an arbitrary system with separable variables can be reduced to (1). Without loss of generality we assume that

$$\prod_{i=1}^{m} J_i = [0,1]^m$$

and $f_i''(x_i) > 0$ for all $x_i \in [0,1]$. Let $v > 0$ and $M(v)$ be the set of $\bar{x}$, $\bar{x} \in \mathbf{R}^m$, for which inequality (1) has an infinite number of integer solutions q. It is clear that if $v \leqslant 1/2m$, then $M(v) = \mathbf{R}^m$, and this is why we take $v > 1/2m$ in what follows. For every integer q consider the inequality

$$\max_{1\leqslant i\leqslant m} \|x_i q\| < q^{-v} \tag{2}$$

assuming the points $\bar{x} \in [0,1]^m$ lie inside the cube with the sides of length $2q^{-v}$ and centre

$$\bar{a}_q = (a_1/q, \ldots, a_m/q), \qquad 0 \leqslant a_i \leqslant q.$$

Now we try to establish the restrictions on the points $\bar{a}_q$ arising by considering the inequality

$$\max_{1\leqslant i\leqslant m} \|qf_i(x_i)\| < q^{-v}. \tag{3}$$

Let

$$K_i = \max_{x\in[0,1]} \left|f_i'(x_i)\right|, \qquad K = \max_{1\leqslant i\leqslant m} K_i. \tag{4}$$

For each i the function $f_i(x_i)$ may be represented in the neighbourhood of the point a_i/q as

$$f_i(x_i) = f_i(a_i/q) + f_i'(\theta_i)(x_i - a_i/q), \qquad \theta_i \in [0,1]. \tag{5}$$

Relations (3)–(5) imply that if

$$\|qf_i(a_i/q)\| > (K+1)q^{-v}$$

for at least one i then the inequality (3) can not be satisfied for sufficiently large q, since for such i we have

$$\|qf_i(x_i)\| = \|qf_i(a_i/q) + qf_i'(\theta_i)(x - a_i/q)\| > \left|(K+1)q^{-v} - qKq^{-1-v}\right| > q^{-v}.$$

On the other hand if

$$\|f_i(a_i/q)\| < \frac{1}{2}q^{-1-v}$$

and

$$|x_i - a_i/q| < \frac{1}{2K}q^{-1-v},$$

then

$$\|qf_i(x_i)\| = \|qf_i(a_i/q) + qf_i'(\theta_i)(x - a_i/q)\| < \frac{1}{2}q^{-v} + \frac{1}{2}q^{-v} = q^{-v},$$

and the inequality (1) is satisfied. Thus the inequality (1) is satisfied only for those $\bar{x}$ which satisfy (2) and are such that

$$\max_{1\leqslant i\leqslant m} \|qf_i(a_i/q)\| < (K+1)q^{-v}. \tag{6}$$

Certainly there may exist $\bar{x}$ such that (2) and (5) are simultaneously satisfied, but the inequality (1) is not. Alternatively, (1) is always satisfied if the system of inequalities

$$\begin{cases} \max\limits_{1\leqslant i\leqslant m} |x_i - a_i/q| < \dfrac{1}{2K}q^{-v-1}, \\ \max\limits_{1\leqslant i\leqslant m} \|qf_i(a_i/q)\| < \dfrac{1}{2}q^{-v} \end{cases} \tag{7}$$

is satisfied. Upper estimates for dim $M(v)$ will be based on the inequalities (2) and (6) and the lower estimates for (7). Both the upper and lower estimates have in common the following. Let $N(q,v) = N_d(q,v)$ be the number of solutions of the inequality

$$\max_{1\leqslant i\leqslant m} \|qf(a_i/q)\| < dq^{-v}, \tag{8}$$

where $d > 0$ is fixed and

$$\bar{a}_q(a_1/q,\ldots,a_m/q), \qquad \bar{a}_q \in P_m,$$

$$P_m = \prod_{i=1}^{m}[\alpha_i,\beta_i), \qquad P_m \subset [0,1]^m.$$

Now we apply the following lemma due to I. M. Vinogradov (1952).

LEMMA 1. *Let τ be integer, α,β,δ be real numbers, $0<\delta<1/2$, and*

$$\delta \leqslant \beta-\alpha < 1-\delta.$$

Then there exists a periodic function $\psi(x)$ with the period 1, having the following properties:

1) $\psi(x) = 1$ *for* $\alpha + \frac{\delta}{2} \leqslant x \leqslant \beta - \frac{\delta}{2}$;
2) $0 \leqslant \psi(x) \leqslant 1$ *for* $x \in \left[\alpha-\frac{\delta}{2},\alpha+\frac{\delta}{2}\right] \cup \left[\beta-\frac{\delta}{2},\beta+\frac{\delta}{2}\right]$;
3) $\psi(x) = 0$ *for* $x \in \left[\beta+\frac{\delta}{2},\ 1+\alpha-\frac{\delta}{2}\right]$;
4) *the Fourier series for $\psi(x)$ is*

$$\psi(x) = \beta-\alpha+\sum_{\nu=1}^{\infty} c_\nu e^{2\pi i\nu x}$$

and for any natural r we have

$$|c_\nu| < l_1 \min\left(\beta-\alpha, |\nu|^{-1}, \delta^{-r}|\nu|^{r+1}\right). \tag{9}$$

For each $j = 1,2,\ldots$ let l_j denote a constant depending only on n. Now consider the function

$$\psi(x) = \begin{cases} 1 & \text{if } \|x\| < dq^{-v}, \\ 0 & \text{if } \|x\| \geqslant dq^{-v}. \end{cases}$$

Then the number of solutions of (8) may be written in the form

$$N(q,v) = \sum_{\bar{a}\in P_m} \prod_{i=1}^{m} \psi(qf(a_i/q)). \tag{10}$$

Beside the function $\psi(x)$ consider the functions $\psi_1(x)$ and $\psi_2(x)$ satisfying the conditions of Lemma 1:

$$\psi_1(x) = \begin{cases} 1 & \text{if} \quad \|x\| < dq^{-v}, \\ 0 & \text{if} \quad \|x\| > dq^{-v}, \end{cases}$$

$$0 \leqslant \psi_1(x) \leqslant 1 \qquad \text{if} \quad dq^{-v} \leqslant \|x\| \leqslant dq^{-v} + \delta,$$

$$\psi_2(x) = \begin{cases} 1 & \text{if} \quad \|x\| < dq^{-v} - \delta, \\ 0 & \text{if} \quad \|x\| > dq^{-v}, \end{cases}$$

$$0 \leqslant \psi_2(x) \leqslant 1 \qquad \text{if} \quad dq^{-v} \leqslant \|x\| \leqslant dq^{-v},$$

where the value of δ will be fixed later. Writing $M_i(q,v)$ for the number of the solutions of inequality

$$\|qf_i(a_i/q)\| \leqslant dq^{-v}, \qquad a_i \in [\alpha_i q, \beta_i q),$$

we note that

$$N_{P_m}(q,v) = M_1(q,v)\dots M_m(q,v). \tag{11}$$

Also writing

$$M_{1i}(q,v) = \sum_{q_i \in [\alpha_i q, \beta_i q)} \psi_1(qf_i(a_i/q)), \tag{12}$$

$$M_{2i}(q,v) = \sum_{q_i \in [\alpha_i q, \beta_i q)} \psi_2(qf_i(a_i/q)), \tag{13}$$

we obtain that

$$M_{1i}(q,v) \leqslant M_i(q,v) \leqslant M_{2i}(q,v)$$

and that there exists θ_i, $0 \leqslant \theta_i \leqslant 1$, such that

$$M_i(q,v) = M_{1i}(q,v) + \theta_i(M_{2i}(q,v) - M_{1i}(q,v)).$$

We find that the expression for $M_i(g,v)$ is the same for each i, and therefore we shall omit the index i in what follows. We now represent the function $\psi_1(x)$ in the terms of its Fourier series

$$\psi_1(qf(a/q)) = \sum_{\nu=-\infty}^{\infty} c_\nu e^{2\pi i\nu qf(a/q)},$$

where

$$c_0 = 2(dq^{-v} - \delta) + \mu\delta = 2dq^{-v} + \mu_1\delta$$

and $\max(|\mu|,\ |\mu_1|) \leqslant 2$.
Using the estimates (9), we obtain that

$$|c_\nu| < \begin{cases} l_2 q^{-\nu} & \text{if} \quad |\nu| < q^v, \\ l_3\delta^{-1}\nu^{-2} & \text{if} \quad |\nu| \geqslant q^v. \end{cases}$$

Then

$$M(q,v) = c_0 q(\beta-\alpha) + R(q) = 2dq^{1-v}(\beta-\alpha) + \mu_1\delta q(\beta-\alpha) + R_1(q), \qquad (14)$$

where

$$R(q) = \sum_{a\in[\alpha q,\beta q]} \sum_{\nu\neq 0} c_\nu e^{2\pi i\nu q f(a/q)}. \qquad (15)$$

The order of summation in (15) is interchangeable since $|c_\nu| < l_1\delta^{-1}|\nu|^{-2}$ and the series (15) is convergent. Let

$$S(q) = \sum_{a\in[\alpha q,\beta q)} e^{2\pi i q\nu f(a/b)} \qquad (16)$$

For estimation of the sum (16) we shall use the following lemma (see (Patterson, 1988)).

LEMMA 2. *Suppose that $f(x)\in C^2([a,b],\mathbf{R})$ where $b-a \geqslant 1$. Then we have*

$$\sum_{a<n\leqslant b} e^{2\pi i f(n)} < l_4\left[(b-a)\sup f''(x)+1\right]\left(\inf f''(x)\right)^{1/2} \qquad (17)$$

if $\inf f''(x) > 0$.

As

$$(\nu q f(x/q))'' = \nu q^{-1} f''(x/q) \qquad (18)$$

in view of the restrictions on the function $f(x)$, we may assume that

$$l_5 < f''(x/q) < l_6.$$

Hence from (16)–(18) we obtain that

$$|S(q)| < l_7\left(q^{|\nu|}|\nu|^{1/2}(\beta-\alpha) + q^{1/2}|\nu|^{-1/2}\right). \qquad (19)$$

Summing the inequalities (19) over all ν, $|\nu|\leqslant q^\nu$, and using the estimates (9) for c_ν, and $\beta-\alpha > q^{-v}$ we obtain that

$$|R_1(q) < l_8 q^{1/2+v/2}(\beta-\alpha) \qquad \text{for all} \quad \nu,\ |\nu|\leqslant q^v.$$

Now let

$$\delta = \frac{1}{2}\, d(\beta-\alpha)^{1/2} q^{-v/4-1/4}.$$

We estimate $R_1(q)$ for $|\nu| > q^v$ by the help of bound for $S(q)$ from Lemma 2 under $|\nu|\leqslant q$ and trivial estimate for $(\beta-\alpha)q$ under $|\nu|>q$. So we obtain

$$|R(q)| < l_9 q^{3/4-v/4}(\beta-\alpha)^{1/2} \qquad (21)$$

and hence

$$M(q,v) = 2dq^{1-v}(\beta-\alpha) + kl_{10}(\beta-\alpha)^{1/2}q^{3/4-v/4}, \qquad |k|\leqslant 1. \qquad (22)$$

The last equality implies that for sufficiently large q in every interval $I \in [0,1)$ with

$$\operatorname{mes} I > l_{11} q^{-1/2+3v/2}$$

there exists a rational point a/q with sufficiently large q. Now applying this to (10) we may confirm that in every cube $P_1, P_1 \subset [0,1)^m$, with the sides of length exceeding $l_{11}q^{-1/2+3v/2}$ there exists a point $(a_1/q, \ldots, a_m/q)$. As the functions $M(q,v)$ and $N(q,v)$ are nonincreasing in v, the following upper estimates for $N_{P_m}(q,v)$ follow from (10) and (22):

$$N_{P_m}(q,v) < l_{12} q^{(1-v)m} \qquad \text{if} \quad v < 1/3, \tag{23}$$

and

$$N_{P_m}(q,v) < l_{13} q^{2/3} \qquad \text{if} \quad v \geqslant 1/3.$$

For each point $\bar{a}_q = (a_1/q, \ldots, a_m/q)$ we define the function

$$R(\bar{a}_q) = q^{-1/2+3v/2}.$$

Let $\rho(a,b)$ denote the Euclidean distance between points $a, b \in \mathbf{R}^d$. The following definition and lemma can be found in (Melian and Pestana, 1991).

DEFINITION. Let W be a countable collection of Euclidean balls $B_i = B(a_i, R_i)$, $B_i \subset \mathbf{R}^d$. We shall say that W is a well distributed system of balls with constant l if for every ball B in $\mathbf{R}^d$ there exists a positive number $K(B)$ such that for every K with $K \geqslant K(B)$ we have a subcollection $W(K,B)$, $W(K,B) \subset W$, satisfying
1) $a_i \in B$ and $R_i \geqslant K^{-1}$ for all $B_i \in W(K,B)$,
2) $\rho(a_i, a_j) > \min(R_i, R_j)$ if $i \neq j$ and $B_i, B_j \in W(K,B)$,
3) the number of elements in $W(K,B)$ is not less than $lK^d \operatorname{mes} B$.

LEMMA 3. *Let $W = \{B_i\}$, $i = 1, 2, \ldots$, be a well distributed system in $\mathbf{R}^d$ with constant l, let B be a ball in $\mathbf{R}^n$. If $0 < \tau < 1$ then the Hausdorff dimension of the set of points $\xi \in B$ such that*

$$\rho(\xi, a_i) < c(\xi) R_i^{1/\tau} \qquad \textit{for infinitely many} \quad i$$

is at least τd.

From (10) and (22) it follows that the collection of balls $B_q(\bar{a}/q,\ l_{11}q^{-1/2+3v/2})$ is a well distributed system in $\mathbf{R}^m$ with constant l_{10}. The set of $\bar{x}$ satisfying the inequality (1) consists of those $\bar{x}$ for which the inequality (2) holds with rational points $\bar{a}_q$ satisfying the inequality (6). This implies that

$$\rho(\bar{x}, \bar{a}_q) < m^{1/2} q^{-v-1} < l_{14} \left(l_{11} q^{-1/2+3v/2}\right)^t, \tag{24}$$

where

$$t = \frac{2(v+1)}{1-3v}.$$

Lemma 3 and (24) imply that

$$\dim M(v) \geqslant \frac{1-3v}{2(v+1)}m. \tag{25}$$

The following upper estimates for $\dim M(v)$ are obtained from the estimates for $N_{P_m}(q,v)$ in (23) and (24):

$$\dim M(v) \leqslant \frac{1+(1-v)m}{1+v}, \qquad v < 1/3, \tag{26}$$

$$\dim M(v) \leqslant \frac{2m+3}{3(v+1)}, \qquad v \geqslant 1/3. \tag{27}$$

Combining (25)–(27) we obtain the following result.

THEOREM. *Let $f_i(x_i)$ be some functions defined on the intervals I_i, $1 \leqslant i \leqslant m$, where $m \geqslant 2$. Moreover, let $f_i(x_i) \in C^3(I_i)$ and $|f''(x_i)| > 0$ as $x_i \in I_i$. If $M(v)$ is the set of $\bar{x}$, $\bar{x} \in \mathbf{R}^m$, for which the system (1) has infinitely many integer solutions in q, then*

$$\frac{m(1-3v)}{2(v+1)} \leqslant \dim M(v) \leqslant \frac{1+(1-v)m}{1+v} \qquad \textit{if} \quad v < 1/3,$$

and

$$\dim M(v) \leqslant \frac{2m+3}{3(1+v)} \qquad \textit{if} \quad v \geqslant 1/3.$$

REFERENCES

Cassels, J. W. S. (1950). Some metrical theorems in Diophantine approximation. *Proc. Cambr. Phil. Soc.* **46**, 2, 209–218.

Kubilius, J. (1949). On an application of the method of I. M. Vinogradov to solution of a problem in the metric theory of numbers. *Dokl. Akad. Nauk SSSR* **67**, 5, 783–786 (in Russian).

Sprindžiuk, V. G. (1974). Metric theory of Diophantine approximation. In: *Actual Problems of Analytic Number Theory.* Nauka i Technika, Minsk, pp. 178–198 (in Russian).

Vinogradov, I. M. (1952). *Selected Works.* Izd. AN SSSR, Moscow (in Russian).

New Trends in Probab. and Statist., pp. 83-90
F. Schweiger and E. Manstavičius (Eds.)
1992 VSP/TEV

THE OPTIMAL ZONE FOR LARGE DEVIATIONS OF THE DENOMINATORS OF CONTINUED FRACTIONS

G. MISEVIČIUS
Department of Mathematics, Vilnius University,
Naugarduko 24, 2006 Vilnius, Lithuania

ABSTRACT

The limit theorem for large deviations of the denominators of continued fractions is proved in the paper.

1. INTRODUCTION AND RESULTS

Let $t = [a_1(t),\ a_2(t)\ldots]$ be the expansion of $t \in [0,1]$ into continued fraction, and let p_n/q_n be the n-th convergent of this fraction. The denominator may be expressed by the sum

$$\log q_n = \sum_{j=1}^{n} \log(q_j/q_{j-1}) = \sum_{j=1}^{n} \log\left[a_j, a_{j-1}, \ldots, a_1\right]. \tag{1}$$

The central limit theorem for $\log q_n(t)$ was proved by Ibragimov (1961). In this paper he also proved the equalities

$$\mathbf{E} \log q_n(t) = \int_0^1 q_n(t)\,dt = an + c_1 + o(1), \tag{2}$$

$$\mathbf{D} \log q_n(t) = \mathbf{E}\big(\log q_n(t) - an\big)^2 = \int_0^1 \big(\log q_n(t) - an\big)^2\,dt$$
$$= \sigma^2 n + c_2 + o(1). \tag{2*}$$

It was shown in (Gordin, 1971) that the constant in the relation (2*) is positive. Note that results similar to those of Ibragimov and Gordin were obtained in (Philipp and Stackelberg, 1969).

J. Kubilius was the first who had estimated the error term in the central limit theorem for $\log q_n(t)$ (see (Kubilius, 1965)).

We introduce the following notations: $\Phi(x)$ is the standard normal distribution function; mes $\{\ldots\}$ denotes the Lebesgue measure of the set consisting of those

$t \in [0,1]$ which satisfy the conditions indicated in the braces, or of the set indicated by its name;

$$F_n(x) = \text{mes}\left\{(\log q_n - an)/\sigma\sqrt{n} < x\right\}; \tag{3}$$

$$\Delta_n = H_1\sqrt{n}/\log n.$$

Our main result is the following limit theorem for large deviations.

THEOREM. *In the interval $0 < x < \Delta_n$, the following equalities are true:*

$$\begin{aligned} \frac{1-F_n(x)}{1-\Phi(x)} &= \exp\{L(x)\}\Big(1+\theta_1 f_1(x)\frac{x+1}{\Delta_n}\Big), \\ \frac{F_n(x)}{\Phi(-x)} &= \exp\{-L(x)\}\Big(1+\theta_2 f_2(x)\frac{x+1}{\Delta_n}\Big). \end{aligned} \tag{4}$$

Here

$$L(x) = \sum_{k=3}^{\infty} \lambda_k x^k$$

denotes the power series of Cramér, $f_j(x)$, $j = 1,2$, are bounded functions. Moreover,

$$|\lambda_k| \leqslant (2/k)(16/\Delta_n)^{k-2},$$

$$-x^3/3\Delta_n \leqslant L(\pm x) \leqslant (x^2/2)(x/(x+8\Delta_n)).$$

2. AUXILIARY RELATIONS

The digits of continued fractions a_j, $j = 0,1,\ldots$, are random variables with respect to the Lebesgue measure on the interval (0,1). Let $\mathfrak{F}_\alpha^\beta$ be the σ-algebra of Lebesgue measurable sets, generated by digits a_j, $\alpha \leqslant j \leqslant \beta$. We define the "conditional measure" mes $\{A \mid \mathfrak{F}\}$ by the relation

$$\int_B \text{mes}\{A \mid \mathfrak{F}\}\,dt = \text{mes}\{A \cap B\}, \qquad B \in \mathfrak{F}_\alpha^\beta.$$

The digits are weakly dependent, i.e. they satisfy the so-called condition of "very strong mixing"

$$\big|\text{mes}\{A \cap B\} - \text{mes}\{A\}\text{mes}\{B\}\big| \leqslant A_0 e^{-\lambda n}\text{mes}\{A\}\text{mes}\{B\}, \tag{5}$$

where A_0 and λ are constants and $A \in \mathfrak{F}_1^k$, $B \in \mathfrak{F}_{k+n}^\infty$; see (Iosifescu, 1980). The digits satisfy also a condition which is stronger than the Markov regularity condition (RMT); see (Saulis and Statulevičius, 1991). We present this condition in the following statement.

LEMMA 1. *We have*

$$\sup_{0\leqslant j\leqslant n}\ \sup_{C\in\mathfrak{F}_{j+k}^n} \Big|\text{mes}\big\{C \mid \mathfrak{F}_1^{j+k}\big\} - \text{mes}\big\{C \mid \mathfrak{F}_j^{j+k}\big\}\Big| \leqslant \text{mes}\{C\}\chi(k), \tag{6}$$

where $\chi(k) = B_1 \exp\{-\beta_1 k\}$.

Proof. Having in mind that the σ-algebras $\mathcal{F}_i^j$ are generated by a_j which take only integer values, it is sufficient to prove the relation

$$\left|\frac{\operatorname{mes}\{a_1 = i_1, \ldots, a_n = i_n\}}{\operatorname{mes}\{a_1 = i_1, \ldots, a_{j+k} = i_{j+k}\}} - \frac{\operatorname{mes}\{a_{j+1} = i_{j+1}, \ldots, a_n = i_n\}}{\operatorname{mes}\{a_{j+1} = i_{j+1}, \ldots, a_{j+k} = i_{j+k}\}}\right| \leqslant A_2 \rho^k \operatorname{mes}\left\{a_{j+k+1} = i_{j+k+1}, \ldots, a_n = i_n\right\} \tag{7}$$

with $|\rho| < 1$, the constants A_z and ρ being independent on j, k, and n. We prove the inequality

$$\left|\frac{\operatorname{mes}\{a_1 = i_1, \ldots, a_n = i_n\}}{\operatorname{mes}\{a_1 = i_1, \ldots, a_{j+k} = i_{j+k}\}} - \frac{\operatorname{mes}\{a_1 = l_1, \ldots, a_j = l_j, a_{j+1} = i_{j+1}, \ldots, a_n = i_n\}}{\operatorname{mes}\{a_1 = l_1, \ldots, a_j = l_j, a_{j+1} = i_{j+1}, \ldots, a_{j+k} = i_{j+k}\}}\right| \leqslant A_3 \rho^k \operatorname{mes}\left\{a_{j+k+1} = i_{j+k+1}, \ldots, a_n = i_n\right\}. \tag{8}$$

Denote by

$$E\begin{pmatrix} k_1 & k_2 & \ldots & k_s \\ i_1 & i_2 & \ldots & i_s \end{pmatrix}$$

the set of numbers $t \in [0, 1]$ for which

$$a_{k_1}(t) = i_1, \ \ldots, \ a_{k_s}(t) = i_s.$$

Let

$$I_1 = E\begin{pmatrix} 1 & \ldots & k+j \\ i_1 & \ldots & i_{k+j} \end{pmatrix} \quad \text{and} \quad I_2 = E\begin{pmatrix} 1 & \ldots & n \\ i_1 & \ldots & i_n \end{pmatrix}.$$

Then the end points of the interval I_2 can be written in the form

$$\frac{p_{k+j}s_1 + p_{k+j-1}}{q_{k+j}s_1 + q_{k+j-1}} \quad \text{and} \quad \frac{p_{k+j}s_2 + p_{k+j-1}}{q_{k+j}s_2 + p_{k+j-1}} \tag{9}$$

where $s_1 = [i_{j+k+1}, \ldots, i_n]$ and $s_2 = [i_{j+k+1}, \ldots, i_n + 1]$,

$$\frac{p_{k+j}}{q_{k+j}} = [i_1, \ldots, i_{k+j}] \quad \text{and} \quad \frac{p_{k+j-1}}{q_{k+j-1}} = [i_1, \ldots, i_{k+j-1}].$$

If

$$J_1 = E\begin{pmatrix} 1 & \ldots & j & j+1 & \ldots & j+k \\ l_1 & \ldots & l_j & l_{j+1} & \ldots & l_{j+k} \end{pmatrix}$$

and

$$J_2 = E\begin{pmatrix} 1 & \ldots & j & j+1 & \ldots & n \\ l_1 & \ldots & l_j & l_{j+1} & \ldots & l_{j+k} \end{pmatrix}$$

then a similar representation applies. The lengths of the intervals I_1 and I_2 are equal to

$$\frac{1}{q_{i+k}\bigl(q_{j+k}+q_{j+k-1}\bigr)} \quad \text{and} \quad \frac{|s_2-s_1|}{\bigl(q_{j+k}s_1+q_{j+k-1}\bigr)\bigl(q_{j+k}s_2+q_{j+k-1}\bigr)}.$$

The lengths of J_1 and J_2 may be expressed in a similar manner.
We shall demonstrate that

$$\frac{|s_1-s_2|}{3s_1} \leqslant \operatorname{mes}\left\{E\begin{pmatrix} j+k+1 & \dots & n \\ i_{j+k+1} & \dots & i_n \end{pmatrix}\right\} \leqslant \frac{2|s_1-s_2|}{s_1}. \tag{10}$$

Indeed, the ratio of the lengths of intervals may be represented in the form

$$\frac{|s_1-s_2|}{s_1^2}\cdot\frac{\left(1+\dfrac{q_{j+k-1}}{q_{j+k}}\right)}{\left(1+\dfrac{q_{j+k-1}}{q_{j+k}s_1}\right)\left(\dfrac{s_2}{s_1}+\dfrac{q_{j+k-1}}{q_{j+k}s_1}\right)}.$$

We get

$$\begin{aligned}\frac{|s_1-s_2|}{3s_1}\operatorname{mes}\left\{E\begin{pmatrix} 1 & \dots & j+k \\ i_1 & \dots & i_{j+k} \end{pmatrix}\right\} &\leqslant \operatorname{mes}\left\{E\begin{pmatrix} 1 & \dots & n \\ i_1 & \dots & i_n \end{pmatrix}\right\} \\ &\leqslant \frac{2|s_1-s_2|}{s_1}\operatorname{mes}\left\{E\begin{pmatrix} 1 & \dots & j+k \\ i_1 & \dots & i_{j+k} \end{pmatrix}\right\}.\end{aligned}$$

Adding these inequalities for intervals of rank $j+k$ we obtain the inequality (10).
Having in mind that

$$\frac{q_n}{q_{n-1}} = [a_n, a_{n-1}, \dots, a_1],$$
$$\left|\alpha-\frac{p_k}{q_k}\right| \leqslant \frac{1}{q_kq_{k-1}}, \qquad q_k \geqslant 2^{(k-1)/2},$$

we deduce the formulae

$$\left|\frac{\operatorname{mes} I_2}{\operatorname{mes} I_1}-\frac{\operatorname{mes} J_2}{\operatorname{mes} J_1}\right| = \frac{|s_1-s_2|}{s_1},$$

$$\begin{aligned}&\frac{\alpha_k(\alpha_k+1)(\alpha'_k+1/s)(\alpha'_ks_2/s_1+1/s_1)-\alpha'_k(\alpha'_k+1)(\alpha_k+1/s_1)(\alpha_ks_2/s_1+1/s_1)}{(\alpha_k+1/s_1)(\alpha_ks_2/s_1+1/s_1)(\alpha'_k+1/s_1)(\alpha'_ks_2/s_1+1/s_1)} \\ &\leqslant A_3\rho^k\operatorname{mes}\left\{E\begin{pmatrix} j+k+1 & \dots & n \\ i_{j+k+1} & \dots & i_n \end{pmatrix}\right\},\end{aligned}$$

where

$$\alpha_k = \frac{q_{j+k}}{q_{j+k-1}}, \qquad \alpha'_k = \frac{\widetilde{q}_{j+k}}{\widetilde{q}_{j+k-1}}, \qquad A_3 < 12, \qquad \rho = \frac{\sqrt{2}}{2}.$$

The inequality (8) is proved. After multiplying this inequality by

$$\text{mes}\left\{a_1 = l_1, \ldots, a_j = l_j,\ a_{j+1} = l_{j+1}, \ldots, a_n = i_n\right\},$$

and summing over the set of $l_1, \ldots, l_j$ we get the inequality (7).
The lemma is proved.
Now we introduce the new random variables by

$$\eta_j^{(k)} = \begin{cases} \log\left([a_j, a_{j-1}, \ldots, a_{j-k}]\right) & \text{if } j > k, \\ \log\left([a_j, a_{j-1}, \ldots a_1]\right) & \text{if } j \leqslant k, \end{cases} \tag{11}$$

where $k = k_0 \log n$, k_0 is a constant to be chosen, $\eta_j^{(k)}$ are mixing random variables with the mixing coefficient

$$\varphi_k(\tau) = \begin{cases} 1 & \text{if } \tau \leqslant k, \\ A_0 \exp\left\{-\lambda(\tau - k)\right\} & \text{if } \tau > k. \end{cases} \tag{12}$$

Putting

$$S_n^1 = \sum_{j=1}^{n} \eta_j^{(k)},$$

we obtain

$$\mathbf{E}(S_n - S_n^1)^2 \leqslant ne^{-\lambda k}. \tag{13}$$

LEMMA 2. *The random variables η_j, $j = 1, 2\ldots$, satisfy the inequality*

$$\mathbf{E}\left\{|\eta_j|^m \mid \mathfrak{F}_1^{j-1}\right\} \leqslant m! H_2^m. \tag{14}$$

Proof. We shall use the relation

$$\frac{1}{3s^2} \leqslant \text{mes}\left\{E\binom{n+1}{s}\right\} \leqslant \frac{2}{s^2}. \tag{15}$$

According to the definition, we write

$$\begin{aligned} \mathbf{E}\left\{|\eta_j|^m \mid \mathfrak{F}_1^{j-1}\right\} &= \int_0^1 \left|\log\left[a_k, a_{k-1}, \ldots, a_1\right]\right|^m \text{mes}\left\{dt \mid \mathfrak{F}_1^{k-1}\right\} \\ &= \frac{1}{\text{mes}\, \Xi} \int_{\Xi} \left|\log\left[a_k, a_{k-1}, \ldots, a_1\right]\right|^m \text{mes}\left\{dt\right\} \end{aligned} \tag{16}$$

where

$$\Xi = E\begin{pmatrix} 1 & \ldots & k-1 \\ i_1 & \ldots & i_{k-1} \end{pmatrix}.$$

Using the relations (5) and (15), from formula (16) we get the proof of Lemma.

3. PROOF OF THE THEOREM

From (11), (12), and (13) we get that

$$\mathbf{E}\big(S_n^1 - \mathbf{E}S_n^1\big)^2 = \sigma_k^2 n + c_3 + R, \tag{17}$$

$$|\sigma - \sigma_k| < \tilde{c}_3 2^{-\alpha k/4}, \qquad |R_k| < c_4 2^{-\alpha k/4},$$

where k is taken from the relation (11).
Further we shall apply the arguments from (Saulis and Statulevičius, 1991, Thm. 4.14 and Cor.). We conclude from (6) that the following relation is valid:

$$\sup_{0\leqslant j\leqslant n} \sup_{C\in\mathcal{F}_{j+k}^{n(k)}} \Big|\mathrm{mes}\,\{C \mid \mathcal{F}_1^{j+l}(k)\} - \mathrm{mes}\,\{C \mid \mathcal{F}_j^{j+l}(k)\}\Big| \leqslant \mathrm{mes}\,\{C\}\chi_k(l), \tag{18}$$

where $\mathcal{F}_1^{j+k}(k)$, $\mathcal{F}_j^{j+l}(k)$, and $\mathcal{F}_{j+k}^{n}(k)$ are the σ-algebras generated by $\eta_j^{(k)}$ from (11),

$$\chi_k(l) = \begin{cases} 1 & \text{if } \ l \leqslant k, \\ B_1 \exp\{-\beta(l-k)\} & \text{if } \ l > k, \end{cases} \tag{19}$$

and $k = k_1 \log n$. From (19) and (14) we deduce that

$$\big|\widehat{\mathbf{E}}\eta_{t_1}^{(k)} \ldots \eta_{t_m}^{(k)}\big| \leqslant 2^{m-1} \prod_{j=1}^{r-1} \chi_k(l_{j+1} - l_j) \prod_{j=1}^{r} \mathrm{ess\,sup}\,\big|\mathbf{E}\,(\eta_{l_j}^{(k)} \mid \mathcal{F}_1^{l_j-1}(k))\big|, \tag{20}$$

$r \in \{2, 3, \ldots\}$, and

$$\widehat{\mathbf{E}}\eta_{t_1}^{(k)} \ldots \eta_{t_m}^{(k)} = \mathbf{E}\eta_{t_1}^{(k)} \overbrace{\eta_{t_2}^{(k)} \ldots \overbrace{\eta_{t_{m-1}}^{(k)} \overbrace{\eta}_{t_m}^{(k)}}}$$

where $\frown$ means the centering, i.e. $\overset{\frown}{\eta} = \eta - \mathbf{E}\eta$. We put

$$Z_n' = S_n'/B_n', \qquad B_n'^2 = \mathbf{E}(S_n' - \mathbf{E}S_n')^2$$

and following the arguments in (Saulis and Statulevičius, 1991, Thm. 4.22), from (20) we get the estimates for cumulants:

$$\left|\Gamma\left(\eta_{t_1}^{(k)} \ldots \eta_{t_m}^{(k)}\right)\right| \leqslant m!c_5^{k-1} \prod_{j=1}^{r-1} \chi_k(l_{j+1} - l_j)$$

and

$$|\Gamma_m(Z_n')| \leqslant m! H_3^m \big(k_3 \log n / B_n\big)^{m-2}.$$

The last inequality implies the estimates for the large deviations of $S_n' - \mathbf{E}S_n'$ in the interval $0 \leqslant x < \widetilde{\Delta}_n$ (see (Saulis and Statulevičius, 1991)):

$$\frac{1 - \mathrm{mes}\,\{S_n' - \mathbf{E}S_n' < x\sigma_k\sqrt{n}\}}{1 - \Phi(x)} = \exp\{\tilde{L}(x)\}\Big(1 + \theta_1 f_1(x)\frac{x+1}{\widetilde{\Delta}_n}\Big), \tag{21}$$

$$\frac{\operatorname{mes}\{S_n' - \mathbf{E}S_n' < -x\sigma_k\sqrt{n}\}}{\Phi(-x)} = \exp\{-\tilde{L}(x)\}\Big(+ \theta_2 f_2(x)\frac{x+1}{\tilde{\Delta}_n}\Big),$$

where

$$\tilde{\Delta}_n = c_6\gamma\sqrt{n}/\ln n,$$

$f_i(x)$, $i = 1, 2$, are bounded functions,

$$\tilde{L}(x) = \sum_{j=3}^{\infty} \lambda_j^{(k)} x^j$$

is the power series of Cramér, whose coefficients depend on the moments of the sum S_n'. In order to pass from the distribution of S_n' to the distribution of $\log q_n$, first of all we shall use the inequality

$$(1 - \Phi(x))\exp\{x^2/2\} \geqslant \frac{3.1}{4\sqrt{2\pi}(x+1)}. \tag{22}$$

The inequalities (13) and (17) give that

$$\left|\operatorname{mes}\{S_n' - \mathbf{E}S_n' < x\sigma_k\sqrt{n}\} - F_n(x)\right| < \frac{c_7}{n_2}, \tag{23}$$

if in (11) we choose

$$k_0 > 12/\lambda. \tag{24}$$

It remains to examine the coefficients of the Cramér series, which are expressed by cumulants. According to the definition of S_n' we get

$$|\log q_n(t) - S_n'| \leqslant 2^{-k\log n}.$$

Using the Minkowsky inequality,

$$\left|\mathbf{E}^{1/r}|x|^r - \mathbf{E}^{1/r}|y|^r\right| \leqslant \mathbf{E}^{1/r}|x-y|^r,$$

we estimate

$$\left|\Big(\int_0^1 \log^r q_n(t)\,dt\Big)^{1/r} - \Big(\int_0^1 (S_n')^r\,dt\Big)^{1/r}\right|$$
$$\leqslant \Big(\int_0^1 |\log q_n(t) - S_n'|^r\,dt\Big)^{1/r} \leqslant \frac{c_8}{n^2}$$

if $k_0 = 12/\lambda$. Therefore we obtain that for any j, $1 \leqslant j \leqslant r$, the moments of S_n' tend to the moments of $\log q_n$. The theorem is proved.

REFERENCES

Billingsley, P. (1965). *Ergodic Theory and Information*. John Willey and Sons, New York.

Gordin, M. (1971). The behaviour of the dispersion of sums of random variables that generate a stationary process. *Theory Probab. Appl.* **16**, 474–484 (in Russian).

Heinrich, L. (1987). A method for the derivation of limit theorems for sums of weekly dependent random variables. *Optimization* **18**, 5 715–735.

Ibragimov, I. A. (1961). A theorem of metrical theory of continued fractions. *Vestnik Leningrad. Univ.* **16**, 13–24 (in Russian).

Iosifescu, M. (1990). A survey of the metrical theory of continued fractions. In: *Probab. Theory and Math. Statist. Proc. Fifth Vilnius Conf.*, I. VSP BV/Mokslas, Utrecht, Vilnius, pp. 550–572.

Jakimavičius, D. and Statulevičius, V. (1987). Estimates of cumulants and centered moments of mixing random processes. *Preprint* No 3. Inst. Mathem. Cybern. of Acad. Sci. Lithuanian SSR, Vilnius.

Kubilius, J. (1965). Limit theorems for sums of weakly dependent random variables in the theory of diophantine approximations. *Liet. Matem. Rink.* **5**, 2, 337–338 (in Russian).

Misevičius, G. (1971). Asymptotic expansions for distribution functions of sums of the form $\sum f(T^j t)$. *Ann. Univ. Sci. Budapest, Sect. Math.* **14**, 77–92 (in Russian).

Misevičius, G. (1990). On large deviations for a sum of type $\sum f(T^j t)$. *Acta et Comment. Univ. Tartuensis* **893**, 3–10.

Philipp, W. and Stackelberg, O. P. (1969). Zwei Grenzwertsätze für Kettenbrüche. *Math. Ann.* **181**, 152–156.

Saulis, L. and Statulevičius, V. (1991). *Limit Theorems for Large Deviations*. Kluwer Academic Publishers, Dordrecht, Boston, London.

New Trends in Probab. and Statist., pp. 91–100
F. Schweiger and E. Manstavičius (Eds.)
1992 VSP/TEV

ERGODIC PROPERTIES OF MULTI-DIMENSIONAL SUBTRACTIVE ALGORITHMS

F. SCHWEIGER
Universität Salzburg, Institut für Mathematik,
Hellbrunnerstr. 34, A 5020 Salzburg, Austria

ABSTRACT

In the present paper we investigate the ergodic properties of a class of multi-dimensional algorithms which contains the algorithms of V. Brun and E. Selmer as special cases.

0. In last years a renewed interest in multi-dimensional continued fractions was being seen. A good survey is given in (Brentjes, 1981). We mention further (Greiter, 1977) and (Toussaint, 1986). Both papers are concerned with the difficult questions of convergence and periodicity of such algorithms. The ergodic properties of these algorithms have been considered in (Schweiger, 1982).

1. Let

$$\Delta^{n+1} = \{b = (b_0, b_1, \dots, b_n)\colon\ 1 \geqslant b_0 \geqslant \dots \geqslant b_n \geqslant 0\}.$$

Then fix an index $p \in \{1, \dots, n\}$ and define $\sigma b := (b_0 - b_p, b_1, \dots, b_n)$. There is an index $i = i(b)$, $0 \leqslant i \leqslant n$, such that the permuted $(n+1)$-tuple

$$\pi\sigma b = (b_1, b_2, \dots, b_i, b_0 - b_p, b_{i+1}, \dots, b_n)$$

belongs to Δ^{n+1}. If we introduce the projection

$$p\colon\ \Delta^{n+1} \longrightarrow \Delta^n, \qquad p(b_0, b_1, \dots, b_n) = \left(\frac{b_1}{b_0}, \dots, \frac{b_n}{b_0}\right),$$

then there is a unique map $T\colon\ \Delta^n \longrightarrow \Delta^n$ such that the diagram

$$\begin{array}{ccc} \Delta^{n+1} & \xrightarrow{\pi\sigma} & \Delta^{n+1} \\ \downarrow p & & \downarrow p \\ \Delta^n & \xrightarrow{T} & \Delta^n \end{array}$$

is commutative.

We denote by $B(i)$, $0 \leqslant i \leqslant n$, the set of all points $x = (x_1, \ldots, x_n) \in \Delta^n$ which correspond to index $i = i(b)$. Then the following properties hold.

a) $B(i) = \{x\colon\ x_i + x_p \geqslant 1 \geqslant x_{i+1} + x_p\}$.
Here we put $x_0 = 1$ and $x_{n+1} = 0$.

b) T is given piecewise as follows:

$$Tx = \left(\frac{x_1}{1-x_p}, \frac{x_2}{1-x_p}, \ldots, \frac{x_n}{1-x_p}\right), \qquad x \in B(0),$$

$$Tx = \left(\frac{x_2}{x_1}, \ldots, \frac{x_i}{x_1}, \frac{1-x_p}{x_1}, \frac{x_{i+1}}{x_1}, \ldots, \frac{x_n}{x_1}\right), \qquad x \in B(i), \quad 1 \leqslant i \leqslant n.$$

Note that if $0 \leqslant i < j \leqslant n$ then

$$B(i) \cap B(j) = \{x\colon\ x_i + x_p \geqslant 1 \geqslant x_{j+1} + x_p,\ x_{i+1} = x_{i+2} = \ldots = x_j\},$$

and therefore T is continuous on Δ^n.

THEOREM 1. (1) *If* $0 \leqslant i \leqslant p-1$ *then*

$$TB(i) = \bigcup_{j \geqslant i} B(j) = \{x \in \Delta^n\colon\ x_i + x_p \geqslant 1\}.$$

(2) *If* $p \leqslant i \leqslant n$ *then*

$$TB(i) = \{x \in \Delta^n\colon\ x_i + x_{p-1} \geqslant 1\}.$$

Proof. (1) The assertion is obvious for $i = 0$.
Assume that $1 \leqslant i \leqslant p-1$ and put $y = Tx$. Then

$$y_i + y_p = \frac{1-x_p}{x_1} + \frac{x_p}{x_1} = \frac{1}{x_1} \geqslant 1.$$

(2) Again, put $y = Tx$. Then

$$y_i + y_{p-1} = \frac{1-x_p}{x_1} + \frac{x_p}{x_1} = \frac{1}{x_1} \geqslant 1.$$

THEOREM 2. *The set*

$$B := \bigcup_{j \geqslant p-1} B(j)$$

is invariant under T. *Furthermore,*

$$TB(p-1) = TB(p) = B,$$

$$TB(j) = B \cap \{x\colon\ x_j + x_{p-1} \geqslant 1\}, \qquad p+1 \leqslant j \leqslant n.$$

Proof. Note that

$$B = \{x \in \Delta^n\colon\ 1 \leqslant x_{p-1} + x_p\}.$$

Furthermore, if $x \in \Delta^n$ and $p \leqslant j \leqslant n$ then $x_j + x_{p-1} \geqslant 1$ implies $x_p + x_{p-1} \geqslant 1$. The next theorem shows that we may restrict our attention entirely to the ergodic behaviour of T restricted to B.

THEOREM 3. *If $i \leqslant p - 2$ then the orbit of almost every $x \in B(i)$ eventually leaves $B(i)$. If $p \geqslant 2$, (by Theorem 2) this orbit never returns to*

$$\bigcup_{j \leqslant i} B(j).$$

Proof. We consider the cases $i = 0$ and $i = 1$ only. We leave the other cases to the reader.
Let $x,\ Tx, \ldots, T^{k-1}x \in B(0)$. Then

$$T^k x = \Big(\frac{x_1}{1 - kx_p}, \ldots, \frac{x_n}{1 - kx_p}\Big).$$

If k satisfies the inequalities $1 - kx_p > x_1 > 1 - (k+1)x_p$ then

$$\frac{x_1}{1 - kx_p} + \frac{x_p}{1 - kx_p} > 1,$$

which means that $T^k x \in \Delta^n \setminus B(0)$.
Iteration of T on $B(1)$ is slightly more complicated. The pattern is the following. As long as the orbit stays in $B(1)$ we obtain

$$T^{2k}x = \Big(\frac{x_1 - kx_p}{1 - kx_p}, \frac{x_2}{1 - kx_p}, \ldots, \frac{x_p}{1 - kx_p}, \ldots\Big),$$

$$T^{2k+1}x = \Big(\frac{1 - (k+1)x_p}{x_1 - kx_p}, \frac{x_2}{x_1 - kx_p}, \ldots, \frac{x_p}{x_1 - kx_p, \ldots}\Big).$$

Then the orbit leaves $B(1)$ if

$$x_2 > 1 - (k+1)x_p \quad \text{or} \quad x_2 > x_1 - (k+1)x_p.$$

Remark. The case $p = 1$ corresponds to Brun's algorithm (see (Brun, 1957)). In this case ($x_{p-1} = x_0 = 1$) all cylinders are full (note that $B = \Delta^n$), and no transient part (Theorem 3) appears.
The case $p = n$ corresponds to Selmer's algorithm (see (Selmer, 1961)). Then $B = B(n-1) \cup B(n)$ and again $B(n-1)$ and $B(n)$ are full cylinders (if T is restricted to B).

2. In (Schweiger, 1982) it is shown:
(α) Brun's algorithm is ergodic and there exists an invariant measure with density

$$h(x_1, \ldots, x_n) = \int_0^\infty dy_1 \int_0^1 dy_2 \ldots \int_0^1 dy_n \, (1 + x_1 y_1 + \ldots + x_n y_n)^{-n-1}.$$

(β) Selmer's algorithm is ergodic (when restricted to B) and there exists an invariant measure with density

$$h(x_1,\ldots,x_n) = \int_0^\infty dy_1 \int_0^\infty dy_2 \ldots \int_0^\infty dy_n\,(1+x_1y_1+\ldots+x_ny_n)^{-n-1}.$$

A common background for these integral representations is given in (Schweiger, 1990).
The remaining cases ($n \geqslant 3$ and $2 \leqslant p \leqslant n-1$) seem to cause new problems.
It is sufficient to consider T restricted to B. We restate some properties of the cylinders of the time-1-partition which follow at once from Theorem 1 and Theorem 2:

(i) $B = B(p-1) \cup B(p) \cup \ldots \cup B(n)$,

(ii) $TB(p-1) = TB(p) = B$,

(iii) $TB(p+1) \supseteq \ldots \supseteq TB(n)$,

(iv) $TB(i) \supseteq \bigcup_{j \geqslant i} B(j), \qquad p+1 \leqslant i \leqslant n.$

The dynamics of the system is more interesting and complicated since points from $B(i)$, $i \geqslant p+1$, may also return to $B(j)$, $p-1 \leqslant j < i$, as is seen in the following result.

THEOREM 4. *For all pairs i,j, $p-1 \leqslant i,j \leqslant n$, the intersection $TB(i) \cap B(j)$ has the nonempty interior.*

Proof. Theorem 4 follows at once from a geometric lemma which gives a complete description of the cylinders $B(i)$, $p-1 \leqslant i \leqslant n$, as convex polytopes. Clearly it will be sufficient to show that for any j the intersection $TB(n) \cap B(j)$ contains a nondegenerate simplex (which depends on j).
We introduce points $E(s,t)$ as follows:

$$E(s,t) \quad \text{has coordinates} \quad x_1 = \ldots = x_s = 1,$$

$$x_{s+1} = \ldots = x_{s+t} = \frac{1}{2}, \qquad x_{s+t+1} = \ldots = x_n = 0.$$

In particular, $E(s,0)$ is given by

$$x_1 = \ldots = x_s = 1, \qquad x_{s+1} = \ldots = x_n = 0$$

and $E(0,t)$ is given by

$$x_1 = \ldots = x_t = \frac{1}{2}, \qquad x_{t+1} = \ldots = x_n = 0.$$

Clearly $0 \leqslant s,t \leqslant n$ and $s+t \leqslant n$.

LEMMA 1. (a) *$B(p-1)$ is a convex polytope with $1+p(n-p+1)$ vertices, namely,*

$$E(p-1,0) \quad \text{and} \quad E(k,p+m-k), \qquad 0 \leqslant k \leqslant p-1, \quad 0 \leqslant m \leqslant n-p.$$

(b) *$B(i)$, $p \leqslant 1 \leqslant n$, is a convex polytope with*

$$i-p+1+p(n+1-i) = i+1+p(n-i)$$

vertices, namely,

$$E(p,0), \ \ldots, \ E(i,0)$$

and

$$E(k,i+m-k), \qquad 0 \leqslant k \leqslant p-1, \quad 0 \leqslant m \leqslant n-i.$$

(c) *B is a convex polytope with $1+p(n-p+1)$ vertices, namely,*

$$E(p-1,0), \ \ldots, \ E(n,0),$$

and

$$E(k,p+m-k), \qquad 0 \leqslant k \leqslant p-2, \quad 0 \leqslant m \leqslant n-p.$$

Proof. Since

$$B(i) = \{x \in B: \ x_i + x_p \geqslant 1 \geqslant x_{i+1} + x_p\}, \qquad p-1 \leqslant i \leqslant n,$$

the lemma may be verified by elementary but long calculations.
Furthermore, $B(n)$ has just $n+1$ vertices, namely,

$$E(p,0), \ \ldots, \ E(n,0)$$

and

$$E(0,n), \ E(1,n-1), . E(p-1,n-p+1).$$

Then $TB(n)$ has the $n+1$ vertices

$$E(p-1,0), \ E(p,0), \ \ldots, \ E(n,0), \ E(0,n), \ E(1,n-1), \ \ldots, \ E(p-2,n-p+2).$$

Now let us look at $B(j)$, $p \leqslant j \leqslant n$. The polytopes $B(j)$ and $TB(n)$ have the vertices

$$E(p,0), \ \ldots, \ E(j,0), \ E(0,n), \ \ldots, \ E(p-2,n-p+2)$$

in common. The points

$$E(p-1,j+m-p+1), \qquad 0 \leqslant m \leqslant n-j,$$

are vertices of $B(j)$ which belong to $TB(n)$ (since they satisfy $x_{p-1} \geqslant 1$ hence $x_{p-1}+x_n \geqslant 1$). In a similar fashion we see that $TB(n) \cap B(p-1)$ contains the points

$$E(0,n), \ \ldots, \ E(p-2,n-p+2) \quad \text{and} \quad E(p-1,m+1), \qquad 0 \leqslant m \leqslant n-p+1.$$

These $n+1$ points form a nondegenerate simplex. This proves Theorem 4.
If we introduce the local inverses of T, namely the inverse maps of T restricted to $B(i),\ p-1 \leqslant i \leqslant n$, then we can give an alternative description of the vertices of the cylinders which provides some insight how these vertices are generated by the map itself. These inverses are given as follows ($y = Tx$):

(a) $i = p-1$

$$x_1 = \frac{1}{y_{p-1}+y_p}, \quad x_2 = \frac{y_1}{y_{p-1}+y_p}, \quad \ldots, \quad x_{p-1} = \frac{y_{p-2}}{y_{p-1}+y_p},$$
$$x_p = \frac{y_p}{y_{p-1}+y_p}, \quad x_{p+1} = \frac{y_{p+1}}{y_{p-1}+y_p}, \quad \ldots, \quad x_n = \frac{y_n}{y_{p-1}+y_p};$$

(b) $p \leqslant i \leqslant n$

$$x_1 = \frac{1}{y_i+y_{p-1}}, \quad x_2 = \frac{y_1}{y_i+y_{p-1}}, \quad \ldots, \quad x_{p-1} = \frac{y_{p-2}}{y_i+y_{p-1}},$$
$$x_p = \frac{y_{p-1}}{y_i+y_{p-1}}, \quad x_{p+1} = \frac{y_p}{y_i+y_{p-1}}, \quad \ldots, \quad x_i = \frac{y_{i-1}}{y_i+y_{p-1}},$$
$$x_{i+1} = \frac{y_{i+1}}{y_i+y_{p-1}}, \quad x_{i+2} = \frac{y_{i+2}}{y_i+y_{p-1}}, \quad \ldots, \quad x_n = \frac{y_n}{y_i+y_{p-1}}.$$

Then we restate our description as follows:

(a) $B(p-1)$ is a convex polytope with $1+p(n-p+1)$ vertices, namely,

$$E(p-1,0) \quad \text{and} \quad V(p-1)^a E(0,b), \qquad 0 \leqslant a < p, \quad p \leqslant b \leqslant n.$$

(b) $B(i),\ p \leqslant i \leqslant n$, is a convex polytope with $i+1+p(n-1)$ vertices, namely,

$$V(i)^a E(p,0), \qquad 0 \leqslant a \leqslant i,$$

and

$$V(i)^a E(0,b), \qquad 0 \leqslant a < p, \quad i+1 \leqslant b \leqslant n.$$

An alternative description would be

$$V(i)^a E(p,0), \qquad 0 \leqslant a \leqslant i-p,$$

and

$$V(i)^a E(0,b), \qquad 0 \leqslant a < p, \quad i \leqslant b \leqslant n.$$

Remark. Since $E(p-1,0)$ is the only vertex of B which is a fixed point under T there is some hope one can prove that the algorithm is convergent for every $x \in B$. This is known true for the cases $p=1$ and $p=n$ but for $1 < p < n$ new difficulties seem to appear.
It would be nice to know that almost every x contains blocks of digits n of a certain length. At the moment I am only able to prove

THEOREM 5. *For almost every x the orbit $(T^j x)$, $j = 0, 1, 2, \ldots$, enters $B(n)$ infinitely often.*

Proof. We first consider the set

$$A := \{x \in B\colon\ T^j x \in B \setminus B(n) \text{ for all } j \geqslant j(x)\}.$$

If we put

$$E := \{x \in B\colon\ T^j x \in B \setminus B(n) \text{ for all } j \geqslant 0\}$$

then

$$A = \bigcup_{s=0}^{\infty} T^{-s} E.$$

Therefore it is sufficient to show that $\lambda(E) = 0$.
Obviously

$$E = \bigcap_{m=1}^{\infty} \bigcup_{\varepsilon_1, \ldots, \varepsilon_m \neq n} B(\varepsilon_1, \ldots, \varepsilon_m).$$

We shall prove that $E \subseteq \{x \in B\colon\ x_n = 0\}$.
Put $x^{(m)} = (x_1^{(m)}, \ldots, x_n^{(m)}) = V(\varepsilon_1, \ldots, \varepsilon_m)x$, where $\varepsilon_1, \ldots, \varepsilon_m \neq n$.
Then we use the following

LEMMA 2. $\lim_{m \to \infty} x_n^{(m)} = 0.$

Proof. Let $\varphi(q) := q^{-1/p}$ and consider first $m = pq$. We claim that

$$x_n^{(pq)} \leqslant \varphi(q).$$

Since $x_n^{(p)} \leqslant 1/2$ as can be seen by inspection, the case $q = 1$ is settled. Put

$$x^{(pq)} = y = (y_1, \ldots, y_n).$$

Note the following fact.
If $z' = V(\varepsilon)z$ and $\varepsilon \neq n$ then

$$z'_n = \frac{z_n}{z_\gamma + z_{p-1}} \leqslant z_n,$$

where $\gamma = \max(\varepsilon, p)$.

Then we distinguish several cases.

(1.1) If $\quad y_{j_1} + y_{p-1} \geqslant \dfrac{\varphi(q)}{\varphi(q+1)} \quad$ then $\quad y'_n = \dfrac{y_n}{y_{j_1} + y_{p-1}} \leqslant \varphi(q+1).$

Therefore $y_n^{(p)} = x_n^{(pq+p)} \leqslant \varphi(q+1)$.

(1.2) If $\quad y_{j_1} + y_{p-1} < \dfrac{\varphi(q)}{\varphi(q+1)} \quad$ then $\quad y'_1 = \dfrac{1}{y_{j_1} + y_{p-1}} \geqslant \dfrac{\varphi(q+1)}{\varphi(q)}.$

(2.1) If $y'_{j_1} + y'_{p-1} \geqslant \dfrac{\varphi(q)}{\varphi(q+1)}$ then $y''_n = \dfrac{y'_n}{y_{j_1} + y_{p-1}} \leqslant \varphi(q+1)$,

and again $y_n^{(p)} \leqslant \varphi(q+1)$.

(2.2) If $y'_{j_1} + y'_{p-1} \leqslant \dfrac{\varphi(q)}{\varphi(q+1)}$ then

$$y''_2 = \frac{y'_1}{y'_{j_1} + y'_{p-1}} \geqslant \left(\frac{\varphi(q+1)}{\varphi(q)}\right)^2.$$

Iterating we get either $y_n^{(p)} \leqslant \varphi(q+1)$ or, finally,

$(p-1.1)$ If $y_{j_1}^{(p-2)} + y_{p-2}^{(p-2)} \geqslant \dfrac{\varphi(q)}{\varphi(q+1)}$ then $y_n^{(p-1)} \leqslant \varphi(q+1)$,

and hence $y_n^{(p)} \leqslant \varphi(q+1)$.

$(p-1.2)$ If $y_{j_1}^{(p-2)} + y_{p-1}^{(p-2)} < \dfrac{\varphi(q)}{\varphi(q+1)}$ then

$$y_{p-1}^{(p-1)} \geqslant \left(\frac{\varphi(q+1)}{\varphi(q)}\right)^{p-1}.$$

$(p.1)$ If $y_n^{(p-1)} \leqslant \varphi(q+1)$ the work is done.

$(p.2)$ If $y_n^{(p-1)} > \varphi(q+1)$ then we have a fortiori $y_{j_{p-1}}^{(p-1)} > \varphi(q+1)$.
But then

$$y_n^{(p)} = \frac{y_n^{(p-1)}}{y_{j_{p-1}}^{(p-1)} + y_{p-1}^{(p-1)}} \leqslant \frac{\varphi(q)}{\varphi(q+1) + \left(\frac{\varphi(q+1)}{\varphi(q)}\right)^{p-1}} \leqslant \varphi(q+1).$$

Clearly the last inequality needs a proof!
This inequality amounts to

$$\frac{\varphi(q)^p}{\varphi(q+1)\varphi(q)^{p+1} + \varphi(q+1)^{p-1}} \leqslant \varphi(q+1)$$

or, equivalently, to

$$(q+1)^2 q^{p-1} \leqslant q^p (q+1)^p$$

Since

$$1 \leqslant q(q+1)^{p-2} \qquad \text{for} \quad p \geqslant 2,$$

the work is done.
If $m = pq + r$ and $0 \leqslant r < p$ then $q = [m/p]$.
Hence

$$x_n^{(m)} \leqslant x_n^{(pq)} \leqslant \left[\frac{m}{p}\right]^{-\frac{1}{p}}.$$

3. In the last section we present another variant of subtractive algorithms which I call "fully subtractive algorithms".
Again let

$$\Delta^{n+1} = \{b = (b_0, b_1, \dots, b_n) \colon\ 1 \geqslant b_0 \geqslant \dots \geqslant b_n \geqslant 0\}.$$

Then fix $p \in \{1, 2, \dots, n\}$ and define

$$\tau b := (b_0 - b_p, b_1 - b_p, \dots, b_{p-1} - b_p, b_p, \dots, b_n).$$

Then there is a shuffle π such that the permuted $(n+1)$-tuple has its components in descending order of magnitude. As before we obtain a map $T\colon\ \Delta^n \to \Delta^n$ such that the diagram

$$\begin{array}{ccc} \Delta^{n+1} & \xrightarrow{\pi\tau} & \Delta^{n+1} \\ \downarrow p & & \downarrow p \\ \Delta^n & \xrightarrow{T} & \Delta^n \end{array}$$

is commutative. Note that the number of possible shuffles is given as

$$\frac{(n+1)!}{p!(n-p+1)!}.$$

To write down T and its local inverse $V(\pi)$ we distinguish two cases.

(i) $b_0 - b_p$ is still the largest component.

Then

$$p\tau(b_0, b_1, \dots, b_n) = \left(\frac{x_1 - x_p}{1 - x_p}, \dots, \frac{x_{p-1} - x_p}{1 - x_p}, \frac{x_p}{1 - x_p}, \dots, \frac{x_n}{1 - x_p}\right).$$

$Tx = y$ is obtained by a suitable shuffle of the first $p-1$ coordinates into the last $n - p + 1$ coordinates. There is a $g = g(\pi)$, $1 \leqslant g \leqslant p$, such that

$$y_g = \frac{x_p}{1 - x_p}.$$

Furthermore for $1 \leqslant i \leqslant p - 1$ there are indices j such that

$$y_j = \frac{x_i - x_p}{1 - x_p},$$

and for $p + 1 \leqslant i \leqslant n$ there are indices k satisfying $g + 1 \leqslant k$ such that

$$y_k = \frac{x_i}{1 - x_p}.$$

Then the local inverses are given by a selection of numbers

$$1 \leqslant j(1) < \dots < j(p-1) \leqslant n \quad \text{and} \quad g < k(1) < \dots < k(n-p) \leqslant n,$$

namely,

$$V(\pi)y = \left(\frac{y_{j(1)} + y_g}{1 + y_g}, \ldots, \frac{y_{j(p-1)} + y_g}{1 + y_g}, \frac{y_g}{1 + y_g}, \frac{y_{k(1)}}{1 + y_g}, \ldots, \frac{y_{k(n-p)}}{1 + y_g}\right).$$

(ii) b_p is the largest component.

Then

$$p\tau(b_0, b_1, \ldots, b_n) = \left(\frac{1 - x_p}{x_p}, \frac{x_1 - x_p}{x_p}, \ldots, \frac{x_{p-1} - x_p}{x_p}, \frac{x_{p+1}}{x_p}, \ldots, \frac{x_n}{x_p}\right).$$

$Tx = y$ is obtained in a similar manner. There is $g = g(\pi)$, $1 \leqslant g \leqslant n - p + 1$, such that

$$y_g = \frac{1 - x_p}{x_p}.$$

For $1 \leqslant i \leqslant p - 1$ there are indices j with $g + 1 \leqslant j$ and for $p + 1 \leqslant i \leqslant n$ there are indices k such that

$$y_j = \frac{x_i - x_p}{x_p}, \qquad y_k = \frac{x_i}{x_p}.$$

Hence the local inverses are given by a selection of numbers

$$1 \leqslant k(1) < \ldots < k(n - p) \leqslant n \quad \text{and} \quad g < j(1) < \ldots < j(p - 1) \leqslant n,$$

namely,

$$V(\pi)y = \left(\frac{1 + y_{j(1)}}{1 + y_g}, \ldots, \frac{1 + y_{j(p-1)}}{1 + y_g}, \frac{1}{1 + y_g}, \frac{y_{k(1)}}{1 + y_g}, \ldots, \frac{y_{k(n-p)}}{1 + y_g}\right).$$

From this description one sees that any map $V(\pi)$ automatically maps $y \in \Delta^n$ onto a vector $V(\pi)y \in \Delta^n$ and the following theorem is true.

THEOREM 6. *The cylinders of the time-1-partition of any fully subtractive algorithm are proper, i.e. $TB(\pi) = \Delta^n$ for any shuffle π.*

Remark. Since the points $E(s, 0)$, $0 \leqslant s \leqslant p - 1$, are fixed points under the action of T the algorithm is not convergent for $p \geqslant 2$. However I conjecture it is convergent for almost all x.

Acknowledgement. This paper was supported by the hospitality of the University of Maine, USA and Monash University, Australia.

REFERENCES

Brentjes, A. J. (1981). *Multi-dimensional Continued Fraction Algorithms* (*Mathematical Centre Tracts* **145**). Mathematisch Centrum, Amsterdam.

Brun, V. (1957). Algorithmes euclidiens pour trois et quatre nombres. In: *13ième Congr. Math. Scand.*, Helsinki, pp. 45–64.

Greiter, G. (1977). *Mehrdimensionale Kettenbrüche*. Dissertation. Tech. Univ., München.

Schweiger, F. (1982). Ergodische Eigenschaften der Algorithmen von Brun und Selmer. *Sitzungsber. Österr. Akad. Wiss. Math.-naturw. Kl. Abt. II* **191**, 325–329.

Schweiger, F. (1991). Invariant measures for maps of continued fraction type. *J. Number Theory* **39**,162–174.

Selmer, E. S., (1961). Om flerdimensjonal kjedebrøk. *Nord. Mat. Tidskr.* **9**, 37–43.

Toussaint, H.-J. (1986). *Der Algorithmus von Viggo Brun und verwandte Kettenbruchentwicklungen*. Dissertation. Techn. Univ., München.

Part Four

MULTIPLICATIVE NUMBER THEORY

New Trends in Probab. and Statist., pp. 103–110
F. Schweiger and E. Manstavičius (Eds.)
1992 VSP/TEV

SELBERG'S INEQUALITY IN ARITHMETIC PROGRESSIONS

C. CALDERÓN*
Department of Mathematics, University of the Basque Country,
E-48080 Bilbao, Spain

ABSTRACT

We obtain an elementary derivation for the Selberg formula in an arithmetic progression, related to the function $\psi^*_{r,k}(x;q,l)$.

For positive integers r, k let $\psi^*_{r,k}(x;q,l)$ be the summatory function

$$\psi^*_{r,k}(x,q,l) = \sum_{\substack{n \leqslant x \\ n \equiv l \,(\mathrm{mod}\, q)}} \Lambda^*_{r,k}(n) \tag{1}$$

(see (Calderón, 1981, 1989)), where

$$\Lambda^*_{r,k}(n) = \sum_{d\delta = n} \mu^*_r(d) \log^k \delta, \qquad k, r \geqslant 1, \tag{2}$$

and $\mu^*_r(n)$ is the function of Möbius type such that

$$\mu^*(n) = \begin{cases} 1 & \text{if } \; n = 1, \\ 0 & \text{if } \; p^{r+1} \mid n \quad \text{for some prime } p, \\ (-1)^{\Omega} & \text{if } \; n = \prod p_i^{\alpha_1}, \; 0 \leqslant \alpha_i \leqslant r, \; \Omega = \sum \alpha_i. \end{cases}$$

The function $\Lambda^*_{r,k}(n)$ has the property

$$\Lambda^*_{r,k} = \sum_{d\delta = n} \Lambda^*_{1,k}(d) h_r(\delta) \qquad \text{where} \quad h_r(n) = \sum_{d\delta = n} \mu^*_r(d). \tag{3}$$

First we state and prove some asymptotic formulae which are necessary for our Theorem 4. Throughout the paper q denotes a fixed positive integer; l denotes an integer with $(l,q) = 1$. The functions $\varphi(q)$ and $J_\alpha(q)$ denote the totient functions of Euler and Jordan, respectively, i. e.

$$J_\alpha(q) = n^\alpha \prod_{p \mid n} (1 - p^{-\alpha}), \qquad J_1(q) = \varphi(q).$$

---

*Supported in part by the University of the Basque Country

A natural number n is called square r-free ($r \geqslant 2$) if in the canonical decomposition $n = p_1^{\alpha_1} \dots p_s^{\alpha_s}$ we have $\alpha_i \leqslant r-1$ and α_i are even for each $i = 1, \dots, s$. The function $\sigma_s^*(q)$ is the sum of the s-th powers of the square r-free divisors of q, and $\sigma_0^*(q) = \theta(q)$.

THEOREM 1. *For each odd integer $r > 1$,*

$$\sum_{\substack{n \leqslant x \\ (n,q)=1}} \frac{h_r(n)}{n} = A(r,q) - \frac{B(r,q)}{x^{1/2}} + O\left(\frac{J_2(q)\theta(q)\sigma^*_{-1+\varepsilon}(q)}{q} x^{\frac{-r}{r+1}} \delta\left(x^{\frac{1}{r+1}}\right)\right), \quad (4)$$

where

$$A(r,q) = \frac{\zeta(2)J_2(q)}{\zeta(r+1)J_{r+1}(q)} q^{r-1}, \qquad B(r,q) = \frac{\varphi(q) q^{\frac{r-1}{2}}}{\zeta\left(\frac{r+1}{2}\right) J_{\frac{r+1}{2}}(q)}.$$

Moreover, for $r = 1$ the sum (4) is equal to 1.

Proof. Let $h_r(n)$ be the arithmetical function defined by (3). From the definition of $\mu_r^*(n)$ we deduce that

$$h_r(n) = \begin{cases} 1 & \text{if } n \in G_2 \cap Q_r, \\ 0 & \text{otherwise,} \end{cases}$$

where G_2 and Q_r denote the sets of squares and r-free integers, respectively. Then

$$\sum_{\substack{n \leqslant x \\ (n,q)=1}} \frac{h_r(n)}{n} = \sum_{\substack{m^{r+1} \leqslant x \\ (m,q)=1}} \frac{\mu(m)}{m^{r+1}} \sum_{\substack{n^2 \leqslant (x/m^{r+1}) \\ (n,q)=1}} \frac{1}{n^2}. \quad (6)$$

For $\alpha > 0$, $\alpha \neq 1$, $q \geqslant 1$, $x > 1$ we have (Cohen, 1961, Lemma 3.1)

$$\sum_{\substack{n \leqslant x \\ (n,q)=1}} \frac{1}{n^\alpha} = \frac{\zeta(\alpha)J_\alpha(q)}{q^\alpha} - \frac{\varphi(q)}{q(\alpha-1)x^{\alpha-1}} + O\left(\frac{\theta(q)}{x^\alpha}\right), \quad (7)$$

the O-estimate being uniform in x, α, and q. If $\alpha = 1$, then

$$\sum_{\substack{n \leqslant x \\ (n,q)=1}} \frac{1}{n} = \frac{\varphi(q)}{q}(\log x + C_1) + O\left(\frac{1}{x}\right).$$

Using (7) in (6), we deduce

$$\sum_{\substack{n \leqslant x \\ (n,q)=1}} \frac{h_r(n)}{n}$$

$$= \sum_{\substack{m^{r+1} \leqslant x \\ (m,q)=1}} \frac{\mu(m)}{m^{r+1}} \left\{ \frac{\zeta(2)J_2(q)}{q^2} - \frac{\varphi(q)}{q\left(\frac{x}{m^{r+1}}\right)^{1/2}} + O\left(\frac{\theta(q)}{\left(\frac{x}{m^{r+1}}\right)^{1/2}}\right) \right\} \quad (8)$$

For $\alpha > 1$, $x > 3$, it is proved (see (Suryanarayana and Siva Rama Prasad, 1973, Lemmas 3.5, 3.6)) that

$$\sum_{\substack{n\leqslant x\\(n,q)=1}} \mu(n) = O\big(\sigma^{*}_{-1+\epsilon}(q)x\delta(x)\big), \tag{9}$$

where the O-estimate is uniform in x and q. Also

$$\sum_{\substack{n\leqslant x\\(n,q)=1}} \frac{\mu(n)}{n^{\alpha}} = \frac{q^{\alpha}}{\zeta(\alpha)J_{\alpha}(q)} + O\Big(\frac{\sigma_{-1+\epsilon}(q)\delta(x)}{x^{\alpha-1}}\Big) \tag{10}$$

uniformly. The function

$$\delta(x) = \exp\big\{-A\log^{3/5}x(\log\log x)^{-1/5}\big\}$$

is monotonically decreasing. If m is a positive constant, then

$$\log^{m} x\delta(x) = O\big(\exp\{-C\log^{3/5}x(\log\log x)^{-1/5}\}\big), \qquad 0 < C < A.$$

So we replace $\log^{m} x\delta(x)$ appearing in any O-term by $\delta(x)$. Inserting these estimates in (8) we obtain

$$\begin{aligned}\sum_{\substack{n\leqslant x\\(n,q)=1}} \frac{h_r(n)}{n} = {}& \frac{\zeta(2)J_2(q)}{q^2}\left\{\frac{q^{r+1}}{\zeta(r+1)J_{r+1}(q)} + O\left(\frac{\sigma^{*}_{-1+\epsilon}(q)\delta\left(x^{\frac{1}{r+1}}\right)}{\left(x^{\frac{1}{r+1}}\right)^{r}}\right)\right\}\\
&- \frac{\varphi(q)}{q\sqrt{x}}\left\{\frac{q^{\frac{r+1}{2}}}{\zeta\left(\frac{r+1}{2}\right)J_{\frac{r+1}{2}}(q)} + O\left(\frac{\sigma^{*}_{-1+\epsilon}(q)\delta\left(x^{\frac{1}{r+1}}\right)}{\left(x^{\frac{1}{r+1}}\right)^{\frac{r-1}{2}}}\right)\right\}\\
&+ O\left(\theta(q)\sigma^{*}_{-1+\epsilon}(q)x^{\frac{1}{r+1}-1}\delta\left(x^{\frac{1}{r+1}}\right)\right).\end{aligned}$$

This completes the proof of the theorem.
We are now in a position to prove

THEOREM 2. *For each odd integer $r > 1$, $\alpha, \beta \geqslant 0$,*

$$\sum_{\substack{n\leqslant x\\(n,q)=1}} \frac{h_r(n)\log^{\gamma} n}{n}$$

$$= 2^{\gamma}\gamma!\,B(r,q) - B(r,q)\sum_{i=0}^{\gamma}\frac{2^{i}\gamma!}{(\gamma-i)!}\frac{\log^{\gamma-i}x}{\sqrt{x}} + O\left(x^{\frac{-r}{r+1}}\delta\left(x^{\frac{1}{r+1}}\right)\right), \tag{11}$$

$$\sum_{\substack{n\leqslant x\\(n,q)=1}} \frac{h_r(n)\log^{\beta}(x/n)}{n}$$

$$= B(r,q)\sum_{\gamma=0}^{\beta}(-1)^{\gamma}\binom{\beta}{\gamma}2^{\gamma}\gamma!\,\log^{\beta-\gamma}x + O\left(x^{\frac{-r}{r+1}}\delta\left(x^{\frac{1}{r+1}}\right)\right), \tag{12}$$

where $B(r,q)$ is defined as in the preceding theorem. Moreover, for $r = 1$ the sums (11) and (12) are equal to 0 and $\log^\beta x$, respectively.

Proof. By partial summation and Theorem 1, we get (11). The formula (12) follows immediately from (11).

THEOREM 3. *Let $(l,q) = 1$, $r > 1$. Then*

$$\sum_{\substack{n\leqslant x\\(n,q)=1}} \psi^*_{r,k}\left(\frac{x}{n};q,\frac{l}{n}\right) = A_k x\log^k x + A_{k-1}x\log^{k-1}x + \ldots + A_0 x + O(\sqrt{x}), \quad (13)$$

where

$$A_j = (-1)^{k-j}\frac{k!\,\varphi(q)q^{\frac{r-1}{2}}}{qj!\,\zeta\left(\frac{r+1}{2}\right)J_{\frac{r+1}{2}}(q)}\sum_{\delta=j}^{k}2^{\delta-j}, \qquad j = 0,1,\ldots,k.$$

If $r = 1$, then

$$\sum_{\substack{n\leqslant x\\(n,q)=1}} \psi^*_{1,k}\left(\frac{x}{n};q,\frac{l}{n}\right) = \frac{x}{q}\sum_{\delta=0}^{k}(-1)^{k-\delta}\frac{k!}{\delta!}\log^\delta x + O(\log^k x).$$

Proof. For $(l,q) = 1$ and $r = 1$ we have

$$\sum_{\substack{n\leqslant x\\(n,q)=1}} \psi^*_{1,k}\left(\frac{x}{n};q,\frac{l}{n}\right) = \sum_{\substack{n\leqslant x\\n\equiv l\,(\mathrm{mod}\,q)}} \log^k n = \frac{x}{q}\sum_{\delta=0}^{k}(-1)^{k-\delta}\frac{k!}{\delta!}\log^\delta x + O(\log^k x),$$

hence Theorem 3 follows in this case. Now let $r > 1$ and let k be a positive integer,

$$\sum_{\substack{n\leqslant x\\(n,q)=1}} \psi^*_{r,k}\left(\frac{x}{n};q,\frac{l}{n}\right) = \sum_{\substack{n\leqslant x\\(n,q)=1}}\ \sum_{\substack{d\leqslant x/n\\n\equiv l/n\,(\mathrm{mod}\,q)}} \Lambda^*_{r,k}(d).$$

Now we take $nd = m$, and from the properties of $\Lambda^*_{r,k}(n)$ we obtain

$$\begin{aligned}\sum_{\substack{n\leqslant x\\(n,q)=1}} \psi^*_{r,k}\left(\frac{x}{n};q,\frac{l}{n}\right) &= \sum_{\substack{n\leqslant x\\n\equiv l\,(\mathrm{mod}\,q)}}\ \sum_{d\mid n} h_r(d)\log^k(n/d)\\ &= \sum_{\substack{d\leqslant x\\(d,q)=1}} h_r(d) \sum_{\substack{\delta\leqslant x/d\\\delta\equiv l/d\,(\mathrm{mod}\,q)}} \log^k\delta\\ &= \frac{x}{q}\sum_{\delta=0}^{k}(-1)^{k-\delta}\frac{k!}{\delta!}\sum_{\substack{d\leqslant x\\(d,q)=1}}\frac{h_r(d)\log^\delta(x/d)}{d} + O\left(\sum_{\substack{d\leqslant x\\(d,q)=1}} h_r(d)\log^\delta(x/d)\right).\end{aligned}$$

Moreover, if r is an odd integer, $r > 1$, then

$$\sum_{\substack{n\leqslant x\\(n,q)=1}} h_r(n) = \frac{\varphi(q)q^{\frac{r-1}{2}}}{\zeta\left(\frac{r+1}{2}\right)J_{\frac{r+1}{2}}(q)}\sqrt{x} + O\left(x^{\frac{1}{r+1}}\delta\left(x^{\frac{1}{r+1}}\right)\right). \tag{14}$$

Making use of formula (14), we obtain

$$\sum_{\substack{n\leqslant x\\(n,q)=1}} h_r(n)\log^{\beta}(x/n)$$
$$= \frac{\varphi(q)q^{\frac{r-1}{2}}}{\zeta\left(\frac{r+1}{2}\right)J_{\frac{r+1}{2}}(q)}2^{\beta}\beta!\sqrt{x} + O\left(x^{\frac{1}{r+1}}\delta\left(x^{\frac{1}{r+1}}\right)\right), \qquad r > 1. \tag{15}$$

It is clear that $x^{\varepsilon}\delta(x)$ is monotonically increasing for every $\varepsilon > 0$. Hence by Theorem 2 and (15) we deduce the theorem. For $r = 1$ the sums (14) and (15) are equal to 1 and $\log^{\beta} x$, respectively.
By means of the two following lemmas we will obtain an asymptotic formula for $\psi^{*}_{r,k}(x,q,l)$.

LEMMA 1 (Shapiro, 1950). *If $f(x)$ is any complex-valued function, defined for all real $x > 0$, and $g(x)$ is defined by*

$$g(x) = \sum_{\substack{n\leqslant x\\(n,q)=1}} f(x/n),$$

then

$$f(x) = \sum_{\substack{n\leqslant x\\(n,q)=1}} \mu(n)g(x/n).$$

LEMMA 2 (Rieger, 1957). *For all integers $\beta \geqslant 2$*

a)

$$\sum_{\substack{n\leqslant x\\(n,q)=1}} \frac{\log^{\beta}(x/n)}{n} = \frac{\varphi(q)}{q(\beta+1)}\log^{\beta+1}x + \sum_{\rho=0}^{\beta} d_{\rho}^{(\beta)}\log^{\rho}x + O\left(\frac{\log^{\beta}x}{x}\right),$$

where $d_{\rho}^{(\beta)}$ are the constants depending on β, ρ, and q;

b)

$$\sum_{\substack{n\leqslant x\\(n,q)=1}} \frac{\mu(n)}{n}\log^{\beta}(x/n) = \frac{\beta q}{\varphi(q)}\log^{\beta-1}x + \sum_{i=1}^{\beta-2} b_i^{(\beta)}\log^{i}x + O(1),$$

where μ denotes the Möbius function, and $b_i^{(\beta)}$ are constants depending on β, i, and q. If $\beta = 0$ or 1, the sum b) is bounded.

Taking $f(x) = \psi^*_{r,k}(x; q, l)$ in Lemma 1, we see that

$$\psi^*_{r,k}(x; q, l) = \sum_{\substack{n \leqslant x \\ (n,q)=1}} \mu(n) \left\{ \sum_{j=0}^{k} A_j \frac{x}{n} \log^j \left(\frac{x}{n}\right) + O\left(\sqrt{\frac{x}{n}}\right) \right\},$$

hence using Lemma 2 and (9) we get

$$\psi^*_{r,k}(x, q, l) = B_{k-1} x \log^{k-1} x + \ldots + B_1 x \log x + O(x), \qquad (l, q) = 1, \tag{16}$$

where the coefficients are given by

$$B_{k-1} = A_k \frac{kq}{\varphi(q)}, \qquad B_j = A_{j+1} \frac{(j+1)q}{\varphi(q)} + \sum_{i=j+2}^{k} A_i b_j^{(i)}, \qquad i = 1, \ldots, k-2.$$

From Lemma 2, the Abel identity, and the relation

$$\sum_{\substack{n \leqslant x \\ (n,q)=1}} \frac{\log^i n}{n} = \frac{\varphi(q)}{q(i+1)} \log^{i+1} x + O(1), \qquad \forall i \geqslant 0, \tag{17}$$

we deduce that

$$\sum_{\substack{n \leqslant x \\ (n,q)=1}} \frac{\Lambda^*_{1,n}(n)}{n} = \log^i x + \frac{\varphi(q)}{q(i+1)} \sum_{j=1}^{i-1} b_j^{(i)} \log^j x + O(1). \tag{18}$$

Hence

$$\begin{aligned} \sum_{\substack{n \leqslant x \\ (n,q)=1}} \frac{\Lambda^*_{1,i}(n) \log^\delta (x/n)}{n} &= \frac{\Gamma(i+1)\Gamma(\delta+1)}{\Gamma(i+\delta+1)} \log^{i+\delta} x \\ &+ \frac{\varphi(q)}{q(i+1)} \sum_{j=1}^{i-2} b_j^{(i)} \frac{\Gamma(j+1)\Gamma(\delta+1)}{\Gamma(j+\delta+1)} \log^{\delta+j} x + O\left(\log^\delta x\right). \end{aligned} \tag{19}$$

The following lemma is a variant of the formula from (Iseki and Tatuzawa, 1951).

LEMMA 3. *Let F be a real or complex-valued function defined on $(0, \infty)$, and let*

$$G_k(x; q, l) = \log^k x \sum_{\substack{n \leqslant x \\ (n,q)=1}} F\left(\frac{x}{n}; q, \frac{l}{n}\right),$$

where $k \geqslant 1$ is a fixed integer. Then

$$\sum_{\substack{n \leqslant x \\ (n,q)=1}} \mu_r^*(n) G_k\left(\frac{x}{n}; q, \frac{l}{n}\right) = \sum_{\substack{n \leqslant x \\ (n,q)=1}} F\left(\frac{x}{n}; q, \frac{l}{n}\right) \log^k \frac{x}{n} h_r(n)$$
$$+ \sum_{i=1}^{k} \binom{k}{i} \sum_{\substack{n \leqslant x \\ (n,q)=1}} F\left(\frac{x}{n}; q, \frac{l}{n}\right) \log^{k-i}\left(\frac{x}{n}\right) \Lambda_{r,i}^*(n). \tag{20}$$

We are now in a position to prove the following result.

THEOREM 4. *For all $k \geqslant 1$*

$$\psi_{1,k}^*(x; q, l) \log^k x + \sum_{i=1}^{k} \binom{k}{i} \sum_{\substack{n \leqslant x \\ (n,q)=1}} \psi_{1,k}^*\left(\frac{x}{n}; q, \frac{l}{n}\right) \log^{k-i}\left(\frac{x}{n}\right) \Lambda_{1,i}^*(n)$$
$$= \frac{k}{\varphi(q)} \left\{ 1 + \sum_{i=1}^{k} \binom{k}{i} \binom{2k-1}{i}^{-1} \right\} x \log^{2k-1} x + O(x \log^{2k-2} x). \tag{21}$$

Proof. We consider the functions

$$G_1(x; q, l) = \log^k x \sum_{\substack{n \leqslant x \\ (n,q)=1}} \psi_{1,k}^*\left(\frac{x}{n}; q, \frac{l}{n}\right),$$

$$G_2(x; q, l) = \log^k x \sum_{\substack{n \leqslant x \\ (n,q)=1}} F_2\left(\frac{x}{n}; q, \frac{l}{n}\right),$$

where

$$F_2(x; q, l) = H_{k-1} x \log^{k-1} x + \ldots + H_1 x \log x + H_0 x + H_{00}$$

and

$$H_{k-1} = \frac{k}{\psi(q)},$$
$$H_\rho = \frac{(-1)^{k-(\rho+1)} k!}{\rho! \, \varphi(q)} - \frac{(\rho+1) q}{\varphi(q)} \sum_{j=\rho+1}^{k-1} H_j d_{\rho+1}^{(j)}, \qquad 0 \leqslant \rho \leqslant k-2,$$
$$H_{00} = \frac{(-1)^k k!}{\varphi(q)} - \frac{q}{\varphi(q)} \sum_{j=1}^{k-1} H_j d_0^{(j)} - C_1 H_0.$$

From the construction of F_2 and Theorem 3 we deduce that

$$G_1(x; q, l) - G_2(x; q, l) \ll \sqrt{x} \log^k x,$$

and hence

$$\left| \sum_{\substack{n\leqslant x\\(n,q)=1}} \mu_r^*(n)\left(G_1\left(\frac{x}{n};q,\frac{l}{n}\right)-G_2\left(\frac{x}{n};q,\frac{l}{n}\right)\right)\right| \ll x, \qquad \forall r\geqslant 1.$$

Using (20) we deduce that

$$\psi_{1,k}^*(x;q,l)\log^k x\sum_{i=1}^{k}\binom{k}{i}\sum_{\substack{n\leqslant x\\(n,q)=1}}\psi_{1,k}^*\left(\frac{x}{n};q,\frac{l}{n}\right)\log^k\left(\frac{x}{n}\right)\Lambda_{1,i}^*(n)$$

$$= \log^k x\sum_{\alpha=0}^{k-1}H_\alpha x\log^\alpha x \tag{22}$$

$$+\sum_{i=0}^{k}\binom{k}{i}\sum_{i=1}^{k-1}H_\alpha\sum_{\substack{n\leqslant x\\(n,q)=1}}\frac{x}{n}\log^{\alpha+k-i}\left(\frac{x}{n}\right)\Lambda_{1,i}^*(n)+O(x).$$

Applying now the relation (19) we obtain the theorem.
In the particular case $k=1$

$$\psi(x;q,l)\log x+\sum_{\substack{n\leqslant x\\(n,q)=1}}\psi\left(\frac{x}{n};q,\frac{l}{n}\right)\Lambda(n)=\frac{2}{\varphi(q)}x\log x+O(x), \tag{23}$$

so the asymptotic formula from (Rieger, 1965) is obtained.

REFERENCES

Calderón, C. (1981). Relaciones entre funciones aritméticas. In: *Actas VII Jornadas Luso Espanh. Matem.*, III. Univ. Coimbra, pp. 47–54.

Calderón, C. (1989). Sumas en progressiones aritméticas para funciones $\Lambda_{r,k}^*$. In: *Proc. of the Congress on Number Theory, Zarautz, 1984.* (Eds. E. Aparicio *et al.*) U.P.V., pp. 148–162.

Cohen, E. (1961). Unitary products of arithmetical functions. *Acta Arithm.* **7**, 29–38.

Iseki, K. and Tatuzawa, T. (1951). On Selberg's elementary proof of the prime numbers theorem. *Proc. Jap. Acad.* **27**, 340–342.

Rieger, G. J. (1957). Bemerkungen zu einem zahlentheoretischen Satz von Shapiro. *Arch. Math.* **8**, 251–254.

Rieger, G. J (1965). Zur Selbergschen Gleichung für die arithmetische Progression. *Elem. Math.* **20**, 36–39.

Shapiro, S. N. (1950). On primes in arithmetic progression (I). *Ann. Math.* **52**, 1, 217–230.

Suryanarayana, D. and Siva Rama Prasad, V. (1973). The number of k-free and k-ary divisors of m, which are prime to n. *J. reine angew Math.* **264**, 56–75.

New Trends in Probab. and Statist., pp. 111–120
F. Schweiger and E. Manstavičius (Eds.)
1992 VSP/TEV

SMOOTHING AND INVERSION INEQUALITIES

A. S. FAINLEIB
Izmailova 48–20,
700052 Tashkent, Uzbekistan

ABSTRACT

Some inversion inequalities useful for certain problems of analysis and number theory are proved in general setting.

In the present paper we prove two smoothing inequalities connecting the estimation of the function of real variable with the estimation of its mean value. We deduce from the smoothing inequalities the inversion inequalities for Fourier series and integrals, and Laplace integrals, further, we point out some applications to the arithmetical functions. Some special cases of such inequalities have been considered earlier in (Esseen, 1945; Erdös and Turán, 1948; Petrov, 1975; Fainleib, 1968, 1981).

In what follows all functions are understood to be complex-valued. A function of the real variable will be called monotonic if its real and imaginary parts are monotonic in usual sense. If a function φ is defined on certain (finite or infinite) interval I of the real line $\mathbf{R}$, $\sigma \in \mathbf{R}$, $h > 0$, we put

$$M_\varphi(\sigma) = \sup_{x\in I} |\varphi(x)| e^{-\sigma x}, \qquad W_\varphi(\sigma, h) = \sup_{\substack{x,y\in I \\ |x-y|\leqslant h}} |\varphi(x) - \varphi(y)| e^{-\sigma x},$$

$$M_\varphi = M_\varphi(0), \qquad W_\varphi(h) = W_\varphi(0, h).$$

THEOREM 1. *Let the function $\rho(x)$ be defined on $\mathbf{R}$, let it be bounded and integrable on every finite interval. Let $\sigma \in \mathbf{R}$, $h > 0$, and $g(x)$ be such function that the sum $g(x) + \rho(x)e^{\sigma x}$ is monotonic on $\mathbf{R}$. Assume further $q(x)$ is the non-negative function on $\mathbf{R}$ satisfying*

$$\int_{-\infty}^{\infty} q(x)\,dx = 1,$$

and

$$\psi(x) = \int_{-\infty}^{\infty} \rho(u) q(x-u)\,du, \qquad J_q(h) = \int_{|x|\leqslant \frac{h}{2}} q(x)\,dx.$$

Then the following statements hold:
(i) *if* $\sigma \neq 0$, *then*

$$M_\rho \leqslant \frac{\sqrt{2}}{J_q(h)}\left(\frac{e^{|\sigma|h} + J_q(h)}{e^{|\sigma|h} - 1} W_g(\sigma,h) + e^{|\sigma|h} M_\psi\right); \tag{1}$$

(ii) *if* $J_q(h)(1+e^{-|\sigma|h}) > 1$, *then*

$$M_\rho \leqslant \sqrt{2}\,\frac{J_q(h)W_g(\sigma,h) + M_\psi}{J_q(h)(1+e^{-|\sigma|h}) - 1}. \tag{2}$$

Proof. We may suppose that $W_g(\sigma,h) < \infty$. By induction on n we obtain

$$W_g(\sigma,nh) \leqslant \frac{e^{n|\sigma|h} - 1}{e^{|\sigma|h} - 1} W_g(\sigma,h) \qquad (n \in \mathbf{N}) \tag{3}$$

(for $\sigma = 0$ the fraction is considered to be equal to n). If $\sigma \neq 0$, $x, y \in \mathbf{R}$, for

$$n = \left[\frac{|x-y|}{h}\right] + 1$$

it follows from (3) that

$$g(x) - g(y) \ll e^{\sigma x} + e^{\sigma y}$$

(the constant in the sign $\ll$ depends on σ). Therefore, in the case $\sigma > 0$ for $x \to -\infty$, and in the case $\sigma < 0$ for $x \to +\infty$, the function $g(x)$ tends to the finite limit. Furthermore, for $\sigma \neq 0$

$$g(x) = c + O(e^{\sigma x}), \qquad x \in \mathbf{R},$$

where

$$c = \begin{cases} g(-\infty) & \text{for } \sigma > 0, \\ g(+\infty) & \text{for } \sigma < 0. \end{cases}$$

Likewise, in the case $\sigma = 0$ we have that $W_g(h)$ is finite if and only if

$$g(x) - g(y) \ll |x-y| + 1 \qquad \text{for all} \quad x, y \in \mathbf{R}.$$

We put $f(x) = g(x) + \rho(x)e^{\sigma x}$.
(i) At first let be $c = 0$. Then $g(x) \ll e^{\sigma x}$. It follows from this that

$$f(-\infty) = 0 \qquad \text{for} \quad \sigma > 0,$$

$$f(+\infty) = 0 \qquad \text{for} \quad \sigma < 0,$$

and in both cases the monotonic functions $\operatorname{Re} f$ and $\operatorname{Im} f$ have the constant sign. If the function f is non-negative and increasing and the function g is real-valued, then the evident inequality

$$f(x)\int_x^{x+h} e^{-\sigma x} q\left(x + \frac{h}{2} - u\right) du \leqslant \int_x^{x+h} f(u)e^{-\sigma u} q\left(x + \frac{h}{2} - u\right) du \tag{4}$$

implies

$$-g(x)e^{-\sigma x} \leqslant \rho(x) \leqslant -g(x)e^{-\sigma x} + \frac{e^{|\sigma|h}}{J_q(h)}\big(M_g(\sigma) + M_\psi\big),$$

and thus

$$M_\rho \leqslant \frac{1}{J_q(h)}\Big(\big(e^{|\sigma|h} + J_q(h)\big)M_g(\sigma) + e^{|\sigma|h}M_\psi\Big). \tag{5}$$

If f is non-negative and decreasing and g is real-valued, then in a similar manner the relation (5) follows from the inequality

$$f(x)\int\limits_{x-h}^{x} e^{-\sigma u}q\Big(x - \frac{h}{2} - u\Big)\,du \leqslant \int\limits_{x-h}^{x} e^{-\sigma u}f(u)q\Big(x - \frac{h}{2} - u\Big)\,du. \tag{6}$$

If $f \leqslant 0$ and g is real-valued, as earlier, the inequality (5) applies to the functions $(-\rho)$ and $(-g)$ instead of ρ and g. Finally, applying the inequality (5) first to $\operatorname{Re}\rho$, $\operatorname{Re} g$ and then to $\operatorname{Im}\rho$, $\operatorname{Im} g$, and having in mind that $M_g(\sigma) < \infty$ implies

$$W_g(\sigma, h) \geqslant \big(e^{|\sigma|h} - 1\big)M_g(\sigma),$$

we can show that the inequality (1) holds in case $c = 0$.
To obtain (1) in the general case, we apply this result to the functions $\rho(x)$ and $g_0(x) = g(x) - c$.
(ii) We put

$$N_\rho = e^{|\sigma|h}\left(W_g(\sigma, h) + \frac{M_\psi + (1 - J_q(h))M_\rho}{J_q(h)}\right). \tag{7}$$

First, let ρ and g be real valued functions, and f be increasing on $\mathbf{R}$. Following the deduction of the main inequality from the paper (Fainleib, 1968), we obtain from (4) that $\rho(x) \leqslant N_\rho$. By analogy from the inequality opposite to (6) it follows that $\rho(x) \geqslant -N_\rho$. Therefore, $|\rho(x)| \leqslant N_\rho$ for all $x \in \mathbf{R}$, and thus $M_\rho \leqslant N_\rho$. From this in view of (7) we obtain

$$M_\rho \leqslant \frac{J_q(h)W_g(\sigma, h) + W_\psi}{J_q(h)\big(1 + e^{-|\sigma|h}\big) - 1}. \tag{8}$$

If ρ and g are real-valued and f is decreasing, then (8) applies to $(-\rho)$ and $(-g)$ instead of ρ and g. In the general case we apply (8) first to the real and then to the imaginary part of ρ and g and obtain (2).

THEOREM 2. *Let ρ, g, and σ be defined just as in Theorem 1. Furthermore, let ρ be periodic with the period p, N be a natural number. Then*

$$M_\rho \leqslant A\Big(1 + \frac{p|\sigma|}{N}\Big)\Big(W_g\Big(\sigma, \frac{p}{p|\sigma| + N}\Big) + \sum_{n=-(N-1)}^{N-1} |c_{n,\rho}|\Big), \tag{9}$$

where A is an absolute constant,

$$c_{n,\rho} = \frac{1}{p}\int\limits_0^p \rho(x)e\Big(\frac{nx}{p}\Big)\,dx, \qquad e(\alpha) = e^{2\pi i\alpha}.$$

Proof. Let $p = 2\pi$,

$$h = \min\left(\frac{16}{3N}, \frac{2}{3|\sigma|}\right) \leqslant \frac{6}{|\sigma| + N} < 2\pi,$$

$$q(x) = \frac{1 - \cos Nx}{2\pi N(1 - \cos x)} = \frac{1}{2\pi}\left(1 + 2\sum_{m=1}^{N-1}\left(1 - \frac{m}{N}\right)\cos mx\right)$$

for $|x| \leqslant \pi$, $q(x) = 0$ for $|x| > \pi$. Then

$$\psi(x) = \int_{-\pi}^{\pi} \rho(x-u)q(u)\,du = \sum_{n=-(N-1)}^{N-1}\left(1 - \frac{|n|}{N}\right)c_{n,\rho}e^{inx},$$

$$M_\psi \leqslant \sum_{n=-(N-1)}^{N-1} |c_{n,\rho}|,$$

$$J_q(h) = \frac{2}{\pi N^2}\int_0^{Nh/4}\left(\frac{\sin u}{\sin(u/N)}\right)^2 du \geqslant \frac{2}{\pi}\int_0^{Nh/4}\left(1 - \frac{u^2}{6}\right)^2 du \geqslant \frac{3}{23}Nh.$$

It follows from (3) that

$$W_g(\sigma, h) \leqslant \frac{e^{2/3} - 1}{e^{1/9} - 1} W_g\left(\sigma, \frac{1}{|\sigma| + N}\right).$$

Using the inequality (1) for $N \leqslant 8|\sigma|$ and the inequality (2) for $N > 8|\sigma|$, we obtain (9) for $p = 2\pi$ with $A = 270$. The case of an arbitrary period p is covered when we substitute for ρ, g, σ the functions ρ_1, g_1, σ_1, where

$$\rho_1(x) = \rho\left(\frac{px}{2\pi}\right), \qquad g_1(x) = g\left(\frac{px}{2\pi}\right), \qquad \sigma_1 = \frac{p\sigma}{2\pi}.$$

THEOREM 3. *Let $\rho(x)$ be the function defined and properly integrable on a finite interval $[a, b] \subset \mathbf{R}$, $p = b - a$, let the function $g(x)$ on $[a, b]$ be such that the sum $f(x) = g(x) + \rho(x)$ is monotonic. Then for every natural N we have*

$$W_\rho(p) \leqslant B\left(\Delta_\rho + W_g\left(\frac{p}{N}\right) + \sum_{1 \leqslant |n| \leqslant N-1} |c_{n,\rho}|\right), \tag{10}$$

where B is an absolute constant,

$$c_{n,\rho} = \frac{1}{p}\int_a^b \rho(x)e\left(\frac{nx}{p}\right)dx,$$

$\Delta_\rho = 0$, *if*

$$\operatorname{Re}\big(f(b) - f(a)\big)\operatorname{Re}\big(\rho(b) - \rho(a)\big) \leqslant 0$$

and (11)

$$\operatorname{Im}\big(f(b) - f(a)\big)\operatorname{Im}\big(\rho(b) - \rho(a)\big) \leqslant 0,$$

$\Delta_\rho = |\rho(b) - \rho(a)|$ *in other cases.*

Proof. We shall prove that under our conditions

$$M_\rho \leqslant 2A\left(\Delta_\rho + W_g\Big(\frac{p}{N}\Big) + \sum_{|n|\leqslant N-1} |c_{n,\rho}|\right), \tag{12}$$

where A is the constant from Theorem 2. First let $\rho(a) = \rho(b)$. We put

$$\rho_0(x) = \rho\left(a + p\left\{\frac{x-a}{p}\right\}\right),$$

$$g_0(x) = \big(g(b) - g(a)\big)\left[\frac{x-a}{p}\right] + g\left(a + p\left\{\frac{x-a}{p}\right\}\right).$$

Then the function $\rho_0(x)$ is periodic with period p, the function

$$f_0(x) = g_0(x) + \rho_0(x)$$

is monotonic on $\mathbf{R}$. Using Theorem 2 for $\sigma = 0$ and taking into account that $W_{g_0}(h) \leqslant 2W_g(h)$ for $0 < h \leqslant p$, we get (12) in the case $\rho(a) = \rho(b)$. In the general case we define on $[a, b]$ the functions

$$\delta_1(x) = \begin{cases} \rho(b) - \rho(a) & \text{for } x = a, \\ 0 & \text{for } x \neq a, \end{cases}$$

$$\delta_2(x) = \begin{cases} 0 & \text{for } x \neq b, \\ \rho(a) - \rho(b) & \text{for } x = b, \end{cases}$$

$$\rho_i(x) = \rho(x) + \delta_i(x), \qquad g_i(x) = g(x) - \delta_i(x), \qquad i = 1, 2.$$

Since

$$g_i(x) + \rho_i(x) = f(x), \qquad \rho_i(a) = \rho_i(b),$$

$$W_{g_i}(h) \leqslant W_g(h) + \big|\rho(b) - \rho(a)\big|, \qquad i = 1, 2,$$

$$M_\rho = \max\big(M_{\rho_1}, M_{\rho_2}\big),$$

we obtain

$$M_\rho \leqslant 2A\left(\big|\rho(b) - \rho(a)\big| + W_g\Big(\frac{p}{N}\Big) + \sum_{|n|\leqslant N-1} |c_{n,\rho}|\right).$$

Under the conditions (11) the functions $\rho_i(x)+g(x)=f(x)+\delta_i(x)$ are monotonic on $[a,b]$. Consequently, in this case we obtain (12) with $\Delta_\rho=0$, too. If $c_{0,\rho}=0$ then (10) follows immediately from (12), having in mind the evident inequality $W_\rho(p)\leqslant 2M_\rho$. Using this special case to the functions $\rho_3(x)=\rho(x)-c_{0,\rho}$ and $g(x)$, we obtain (10) in the general case.
Theorem 3 includes Erdös–Turán inequality (Erdös and Turán, 1948) and its generalisation in (Fainleib, 1968) as special cases.

THEOREM 4. *Let ρ, g, and σ be defined just as in Theorem 1, $T>0$,*

$$F_\rho(t)=\int_{-\infty}^{\infty} e^{-itx}\rho(x)\,dx$$

and let the integral of $tF_\rho(t)$ converge uniformly on the interval $[-T,T]$. If

$$\liminf_{\substack{x\to\infty\\ H\to+\infty}} \int_{x-H}^{x+H} \left|\rho(u)\right|du=0, \tag{13}$$

then

$$M_\rho\leqslant C\left(1+\frac{|\sigma|}{T}\right)\left(W_g\left(\sigma,\frac{1}{|\sigma|+T}\right)+\int_{-T}^{T}\left|F_\rho(t)\right|dt\right), \tag{14}$$

where C is an absolute constant.

Proof. We put in Theorem 1

$$h=\min\left(\frac{16}{3T},\frac{2}{3|\sigma|}\right)\leqslant\frac{6}{|\sigma|+T},$$

$$g(x)=\frac{1-\cos Tx}{\pi Tx^2}=\frac{1}{\pi}\int_0^T\left(1-\frac{t}{T}\right)\cos tx\,dt.$$

For arbitrary $x,y\in\mathbf{R}$ and $U>0$ we have

$$\int_{|u|\leqslant U}\rho(u)\big(q(x-u)-q(y-u)\big)\,du$$
$$=\frac{1}{2\pi}\int_{|t|\leqslant T}\left(1-\frac{|t|}{T}\right)\left(e^{itx}-e^{ity}\right)dt\int_{|u|\leqslant U}\rho(u)e^{-itu}\,du.$$

In view of the uniform convergence of the integral of $tF_\rho(t)$ we get from this by $U\to+\infty$, that

$$\psi(x)-\psi(y)=\frac{1}{2\pi}\int_{-T}^{T}\left(1-\frac{|t|}{T}\right)F_\rho(t)\left(e^{itx}-e^{ity}\right)dt. \tag{15}$$

We can suppose that

$$\int_{-T}^{T} |F_\rho(t)|\,dt < \infty.$$

In view of (13) there are the sequences of the real numbers y_n, $y_n \to \infty$, and H_n, $H_n \to +\infty$, such that

$$\int_{|u-y_n|\leqslant H_n} |\rho(u)|\,du \longrightarrow 0.$$

Since

$$|\psi(y_n)| \leqslant \frac{T}{2\pi} \int_{|u-y_n|\leqslant H_n} |\rho(u)|\,du + M_\rho \int_{|u|>H_n} q(u)\,du,$$

it follows from (15) for $y = y_n$, $n \to \infty$, that

$$\psi(x) = \frac{1}{2\pi}\int_{-T}^{T}\left(1-\frac{|t|}{T}\right)F_\rho(t)e^{itx}\,dt, \qquad M_\psi \leqslant \frac{1}{2\pi}\int_{-T}^{T}|F_\rho(t)|\,dt.$$

Furthermore, by analogy with the proof of Theorem 2,

$$J_q(h) = \frac{2}{\pi}\int_0^{Th/4}\left(\frac{\sin u}{u}\right)^2 du \geqslant \frac{2}{\pi}\int_0^{Th/4}\left(1-\frac{u^2}{6}\right)^2 du \geqslant \frac{3}{23}Th,$$

$$W_g(\sigma,h) \leqslant \frac{e^{2/3}-1}{e^{1/9}-1}W_g\left(\sigma,\frac{1}{|\sigma|+T}\right).$$

Using (1) for $0 < T \leqslant 8|\sigma|$ and (2) for $T > 8|\sigma|$, we obtain (14) with $C = 270$. In particular, if ρ is bounded and absolutely integrable on $\mathbf{R}$ (or ρ is the function of bounded variation, and $\rho(\pm\infty) = 0$), then the conditions of Theorem 4 are satisfied for any T, $T > 0$.

Theorem 4 includes Esseen's inequality (Esseen, 1945) and its generalisations (Fainleib, 1968; Petrov, 1975) as special cases.

The next inversion inequality for the Laplace transform follows also from Theorem 4.

THEOREM 5. *Let the function $r(x)$ be defined for $x \geqslant 0$ and properly integrable on any finite interval $[0, X]$, let the integral*

$$L_r(s) = \int_0^\infty e^{-sx}r(x)\,dx$$

converge on the half-plane $\operatorname{Re} s > a$, *and let the function $g(x)$ on $[0,\infty]$ be such that the sum $r(x) + g(x)$ is monotonic. Then for $\sigma > a$, $T > 0$*

$$M_r(\sigma) \leqslant D\left(1+\frac{|\sigma|}{T}\right)\left(|r(0)| + W_g\left(\sigma,\frac{1}{|\sigma|+T}\right) + \int_{-T}^{T}|L_r(\sigma+it)|\,dt\right) \qquad (16)$$

converge on the half-plane $\operatorname{Re} s > a$, *and let the function* $g(x)$ *on* $[0, \infty]$ *be such that the sum* $r(x) + g(x)$ *is monotonic. Then for* $\sigma > a$, $T > 0$

$$M_r(\sigma) \leqslant D\Big(1 + \frac{|\sigma|}{T}\Big)\bigg(|r(0)| + W_g\Big(\sigma, \frac{1}{|\sigma| + T}\Big) + \int_{-T}^{T} |L_r(\sigma + it)|\, dt\bigg) \qquad (16)$$

holds, where D *is an absolute constant.*

Proof. Let b be the infimum of those σ, for which $W_g(\sigma, h) < \infty$ for $h > 0$. It is sufficient to consider the cases

(i) $\sigma > \max(a, b)$,

(ii) $\sigma = b > a$, $W_g(b, h) < \infty$.

We put

$$r_0(x) = \begin{cases} r(x) & \text{for} \quad x \geqslant 0, \\ 0 & \text{for} \quad x < 0, \end{cases}$$

$$g_0(x) = \begin{cases} g(x) & \text{for} \quad x \geqslant 0, \\ g(0) + r(0) & \text{for} \quad x < 0. \end{cases}$$

Then the function $f(x) = r_0(x) + g_0(x)$ is monotonic on $\mathbf{R}$ and

$$W_{g_0}(\sigma, h) \leqslant e^{|\sigma| h}\big(|r(0)| + W_g(\sigma, h)\big). \qquad (17)$$

Thus $W_{g_0}(\sigma, h) < \infty$ for $\sigma > b$. Therefore, just as in the proof of Theorem 1 it follows that

$$g(x) = c + O(e^{\sigma x}). \qquad (18)$$

(The condition $\sigma \neq 0$ is superfluous here, because in case $b < 0$ the estimation $g(x) = O(1)$ for $b < \sigma < 0$ follows from (18)). Therefore, the integrals $L_{g-c}(s)$ and $L_{r+g-c}(s)$ converge in the half-plane $\{s\colon \operatorname{Re} s > \max(a, b)\}$. Since the function $r(x) + g(x) - c$ is monotonic, it follows from this that

$$r(x) + g(x) - c \ll e^{\sigma x},$$

and (in view of (18))

$$r(x) \ll e^{\sigma x} \qquad \text{for} \quad \sigma > \max(a, b), \ x \geqslant 0.$$

Applying Theorem 4, in accordance with (17), we obtain (16) for the case (i). Letting here $\sigma \to b + 0$, we get (16) for the case (ii).

Theorem 5 is convenient for proofs of various Tauberian theorems for Laplace and Laplace–Stieltjes transforms (in particular, for Dirichlet and power series) in the complex domain (Fainleib, 1985).

Now we consider some applications of Theorem 5 to the arithmetical functions.

The proof of the following theorem is constructed as a modification of the method of (Halász, 1968).

THEOREM 6. *Let $\alpha(n)$ be the non-negative additive function, let $\beta(n)$ and $\beta^*(n)$ be the multiplicative functions satisfying $|\beta^*(n)| \leqslant \beta(n)$ for all $n \in \mathbf{N}$,*

$$\sigma_p(s) = \sum_{m=1}^{\infty} \beta(p^m)e^{-s\alpha(p^m)}, \qquad \Phi(s) = \prod_p \left(1 + \sigma_p(s)\right)e^{-\sigma_p(s)},$$

where p runs over primes. Assume further

$$\exp\left\{ \sum_{\alpha(p^m)\leqslant x} \beta(p^m)e^{-a\alpha(p^m)} \right\} \sim x^\tau \lambda(x), \qquad x \to \infty, \tag{19}$$

where $a > 0$, $\tau > 0$, and $\lambda(x)$ is a positive slowly oscillating function. If

$$\sum_p \sigma_p(a)|\sigma_p'(a)| < \infty, \qquad \sum_{\alpha(p^m)\leqslant x} \beta(p^m) \ll \frac{e^{ax}}{x},$$

and

$$\sum_p \left(\sigma_p(a) - \operatorname{Re}\sigma_p(a+it)\right) = \infty \tag{20}$$

for any fixed $t > 0$, then

$$\sum_{\alpha(n)\leqslant x} \beta^*(n) = \left(\frac{c\Phi(a)e^{ibx-\tau\gamma}}{(a+bi)\Gamma(\tau)} \mu^i(x) + o(1) \right) e^{ax}x^{\tau-1}\lambda(x), \tag{21}$$

where $0 \leqslant c \leqslant 1$, $b \in \mathbf{R}$, $\mu(x)$ is positive slowly oscillating function, γ is Euler's constant.

In particular, if $\beta^* = \beta$, then $c = 1$, $b = 0$, $\mu(x) \equiv 1$. In this case also (19) and (20) follow from (21), what is just the inversion of Theorem 6.
Theorem 6 includes, for $\alpha(n) = \log n$, the well-known theorems from (Wirsing, 1967; Halász, 1968) and their generalisations in (Levin and Timofeev, 1970; Tulyaganova, 1978; Fainleib, 1981). The other consequence of Theorem 6 concerns the distribution of values of growing multiplicative functions.

THEOREM 7. *Let $\gamma(n)$ be the positive multiplicative function with some τ, $\tau > 0$, satisfying*

$$\exp\left\{ \sum_{\gamma(p^m)\leqslant x} \frac{1}{\gamma(p^m)} \right\} \sim (\log x)^\tau \lambda(\log x), \qquad x \to \infty,$$

where $\lambda(u)$ is slowly oscillating function, and

$$\sum_p \left(\sum_{m=1}^{\infty} \frac{1}{\gamma(p^m)} \sum_{m=1}^{\infty} \frac{|\log \gamma(p^m)|}{\gamma(p^m)} \right) < \infty,$$

$$\sum_{\gamma(p^m)\leqslant x} 1 \ll \frac{x}{\log x}, \qquad \sum_{x<\gamma(p^m)\leqslant(1+\xi(x))x} 1 = o\Big(\frac{x}{\log x}\Big)$$

for some positive decreasing function $\xi(x)$ such that the integral

$$\int_2^\infty \frac{\xi^2(x)}{x\log x}\,dx$$

diverges. Then

$$\sum_{\gamma(n)\leqslant x} 1 \sim Cx(\log x)^{\tau-1}\lambda(\log x), \qquad x\to\infty,$$

where

$$C = \frac{e^{-\tau\gamma}}{\Gamma(\tau)}\prod_p (1+\sigma_p)e^{-\sigma_p}, \qquad \sigma_p = \sum_{m=1}^\infty \frac{1}{\gamma(p^m)}.$$

The special case ($\gamma(p) = p/\tau$, $\gamma(n) > C_1 n^{15/16}$, $C_1 > 0$, $\lambda(u) = \text{const}$) was considered in (Balasubramanian and Ramachandra, 1988).

REFERENCES

Balasubramanian, R. and Ramachandra, K. (1988). On the number of integeres n such that $ng(n) \leqslant x$. *Acta Arithm.* **59**, 313–322.

Erdös, P. and Turán, P. (1948). On a problem in the theory of uniform distribution. *Proc. Acad. Wet. Amsterdam* **51**, 1146–1154.

Esseen, C.-G. (1945). Fourier analysis of distribution functions. *Acta Math.* **77**, 1–125.

Fainleib, A. S. (1968). A generalization of Esseen's inequality and its applications in probabilistic number theory. *Izv. Akad. Nauk SSSR Ser. Math.* **32**, 859–879 (in Russian).

Fainleib, A. S. (1981). Tauberian inequalities and sums of multiplicative functions. In: *Coll. Math. Soc. Janos Bolyai* **34**: *Topics in Classical Number Theory.* North-Holland, Amsterdam pp. 451–489.

Fainleib, A. S. (1985). Tauberian theorems for Laplace–Stiltjes integrals in the complex domain. *Dokl. Akad. Nauk Uz. SSR* 1, 6–8 (in Russian).

Halász, G. (1968). Über die Mittelwerte multiplikativer zahlentheoretischer Funktionen. *Acta Math. Acad. Sci. Hung.* **19**, 365–403.

Levin, B. V. and Timofeev, N. M. (1970). Sums of multiplicative functions. *Dokl. Akad. Nauk SSSR* **103**, 1062–1065 (in Russian).

Petrov, V. V. (1975). *Sums of Independent Random Variables.* Springer-Verlag, Berlin, Heidelberg, New York.

Tulyaganova, M. I. (1978). A generalisation of a theorem of Halász. *Izv. Akad. Nauk Uz. SSR. Ser. Fiz.-Mat.* 4, 35–40 (in Russian).

Wirsing, E. (1967). Das asymptotische Verhalten von Summen über multiplikative Funktionen II. *Acta Math. Acad. Sci. Hung.* **18**, 411–467.

New Trends in Probab. and Statist., pp. 121–133
F. Schweiger and E. Manstavičius (Eds.)
1992 VSP/TEV

THE MEAN VALUES OF MULTIPLICATIVE FUNCTIONS DEFINED ON A SEMIGROUP

A. MAČIULIS
Department of Mathematics, Vilnius University,
Naugarduko 24, 2006 Vilnius, Lithuania

ABSTRACT

We consider the multiplicative and additive functions defined on an arithmetical semigroup. A formula for the mean value of complex-valued multiplicative functions is obtained. This allows us to prove an analogue of the central limit theorem with an estimate of the error term for the real valued additive functions.

1. INTRODUCTION AND RESULTS

We shall investigate the multiplicative and additive functions defined on an arithmetical semigroup. A commutative semigroup $\mathcal{G}$ is called arithmetical if it contains identity element a_0 and a countable subset $\mathfrak{P}$ of primes such that every element $a \neq a_0$ in $\mathcal{G}$ admits a unique decomposition into a finite product of powers of elements of $\mathfrak{P}$. We suppose that the semigroup $\mathcal{G}$ satisfies (see (Knopfmacher, 1979; Indlekofer and Manstavičius, 1990)) the following

AXIOM A*. *A completely additive degree function δ is defined on $\mathcal{G}$ so that $\delta(p) \geqslant 1$ for each $p \in \mathfrak{P}$ and*

$$G(n) := \#\{a \in \mathcal{G}\colon\ \delta(a) = n\} = Aq^n + O(q^{\nu n})$$

for some constants $A > 0$, $q > 1$, $0 \leqslant \nu < 1$.

In what follows we suppose that $n \to \infty$. The constants implied in the symbols O and $\ll$, as well as the constants A, q, ν, c_0, $c_1, \dots$ depend on $\mathcal{G}$ only.
Let $f\colon \mathcal{G} \to \mathbf{C}$ be a multiplicative function, $|f(a)| \leqslant 1$ for each $a \in \mathcal{G}$. We shall investigate the behaviour of

$$M_n(f) := \frac{1}{Aq^n} \sum_{\delta(a)=n} f(a)$$

as $n \to \infty$.
The first results in this direction have been obtained in (Indlekofer and Manstavičius, 1990). They proved some theorems about the asymptotic behaviour of $M_n(f)$ and the distribution of additive functions defined on $\mathcal{G}$. However these theorems give a very poor information about the rate of convergence. The purpose of the

present paper is to obtain a "good" estimate for $M_n(f)$. This allows us to prove an analogue of the central limit theorem with an estimate of the error term for additive functions defined on $\mathcal{G}$.

We shall use the following notations

$$\mu_n := \frac{1}{n}\sum_{k=1}^{m} kq^{-k}\Big|\sum_{\delta(p)=k}(f(p)-1)\Big|,$$

$$\mu := \max\Big(\frac{1}{n}, \mu_n\Big),$$

$$V_n := \exp\Big\{\sum_{k=1}^{n} q^{-k}\sum_{\delta(p)=k}(f(p)-1)\Big\},$$

$$E_n(u) := \exp\Big\{2\sum_{k=1}^{n} q^{-k}\Big|\sum_{\substack{\delta(p)=k\\ |f(p)-1|>u}}(f(p)-1)\Big|\Big\}, \qquad u \geqslant 0.$$

Now we formulate our main result.

THEOREM 1. *Let $f\colon \mathcal{G} \to \mathbf{C}$ be a multiplicative function, $|f(a)| \leqslant 1$. Then*

$$\begin{aligned} M_n(f) = &\prod_{\delta(p)\leqslant n}(1-q^{-\delta(p)})\Big(1+\sum_{k=1}^{\infty} f(p^k)q^{-k\delta(p)}\Big)\\ &+\frac{1}{n}\sum_{k=1}^{m} M_{n-k}(f)kq^{-k}\sum_{\delta(p)=k}(f(p)-1)\\ &+O\Big(\frac{\log n}{n}+|V_n|\min_{0\leqslant u\leqslant 3/8}\big(E_n(u)\mu^{1-8u/\pi}\big)\Big) \end{aligned} \tag{1}$$

and

$$|M_n(f)| \ll |V_n|\Big(1+\sqrt{\mu}E_n\Big(\frac{3}{16}\Big)\Big). \tag{2}$$

The dependence of f on n is possible.

As an example of possible applications of the theorem 1 we shall prove an integral limit theorem with an estimate of the error term. Let $h\colon \mathcal{G} \to \mathbf{R}$ be a strongly additive function, $h_n(a) := h(a)B^{-1}(n)$,

$$F_n(x) := \frac{1}{G(n)}\#\{a \in \mathcal{G}\colon\ \delta(a) = n,\ h_n(a) - \alpha(n) < x\},$$

where

$$B(n) := \Big(\sum_{k=1}^{n} q^{-k}(1-q^{-k})\sum_{\delta(p)=k} h^2(p)\Big)^{1/2},$$

$$\alpha(n) := \sum_{k=1}^{n} q^{-k}\sum_{\delta(p)=k} h_n(p).$$

THEOREM 2. *If*

$$\beta_n := \sum_{k=1}^{n} q^{-k} \sum_{\delta(p)=k} |h_n(p)|^3 \longrightarrow 0 \qquad \text{as} \quad n \to \infty,$$

then

$$F_n(x) = \Phi(x) + O(\beta_n + \rho_n),$$

where $\Phi(x)$ is the distribution function of the standard normal law,

$$\rho_n := \frac{1}{n} \sum_{k=1}^{n} k q^{-k} \sum_{\delta(p)=k} |h_n(p)|.$$

Remark. It is easy to check that $\rho_n \ll \beta_n^{1/3}$.
The proofs of these theorems are based on some ideas and results of (Indlekofer and Manstavičius, 1990; Mačiulis, 1988).

2. PROOF OF THEOREM 1

The values $f(p^k)$ for $\delta(p) > n$ have no influence upon $M_n(f)$. Therefore we shall assume that $f(p^k) = 1$ when $\delta(p) > n$ and $k \geqslant 1$. The method of our proof is based on the properties of the generating function

$$Z_f(y) := A \sum_{n=0}^{\infty} M_n(f) y^n$$

and the connection between $Z_f(y)$ and $M_n(f)$. We define

$$L_f(y) := \sum_{k=1}^{\infty} q^{-k} \sum_{\delta(p)=k} f(p) y^k,$$

$$H_f(y) := Z_f(y) \exp\{-L_f(y)\}.$$

The same functions will be denoted by $Z_1(y)$, $L_1(y)$, and $H_1(y)$ if $f \equiv 1$. The analogue of the prime number theorem (see (Indlekofer *et al.*, 1991)) yields

$$\pi(n) := \{p \in \mathfrak{P},\ \delta(p) = n\} = \frac{q^n}{n}(1 - (-1)^n \varepsilon) + O(q^{c_0 n}) \tag{3}$$

for some $c_0 < 1$. Here $\varepsilon = 1$ if $Z_1(-1) = 0$ and $\varepsilon = 0$ otherwise. Therefore the functions $Z_f(y)$ and $L_f(y)$ are analytic on the disc $|y| < 1$. Moreover, $H_f(y)$ is analytic and $|H_f(y)| + |H_f'(y)| \ll 1$ on the disc $|y| \leqslant 1 + c_1$ with some $c_1 > 0$ (cf. (Indlekofer and Manstavičius, 1990)). We shall also use the functions

$$w_\varepsilon(y) := \sum_{k=1}^{\infty} \frac{1 - (-1)^k \varepsilon}{k} y^k = \log \frac{1 + \varepsilon y}{1 - y} \qquad \text{for} \quad |y| < 1,$$

and

$$\Psi(y) := \exp\{L_1(y) - w_\varepsilon(y)\}.$$

Since

$$\Psi(y) = \exp\left\{\sum_{k=1}^{\infty}\left(\frac{\pi(k)}{q^k} - \frac{1-(-1)^k\varepsilon}{k}\right)y^k\right\},$$

it follows from (3) that the function $\Psi(y)$ is analytic and bounded on the disc $|y| \leqslant 1 + c_1$. Having in mind the representation

$$\Psi(y) = \frac{Z_1(y)(1-y)}{(1+\varepsilon y)H_1(y)}$$

we see that

$$\Psi(1) = \frac{A}{(1+\varepsilon)H_1(1)}.$$

We shall also need the information about the function $Z_f(y)$ near the circle $|y| = 1$. The following auxiliary lemma yields the estimate of $Z_f(y)$ on the circle

$$|y| = r_n := \exp\left\{-\frac{1}{n}\right\}.$$

LEMMA. *For* $y = re^{it}$, $r_n \leqslant r < 1$, $|t| \leqslant \pi$, *we have*

$$\left|\frac{Z_f(y)}{\exp\{w_\varepsilon(y)\}}\right| \leqslant |V_n| \min_{u \geqslant 0}\left(E_n(u)\left|\frac{1-y}{1-r}\right|^{8u/\pi}\right). \tag{4}$$

Proof. By definition,

$$Z_f(y)\exp\{-w_\varepsilon(y)\} = H_f(y)\Psi(y)\exp\{L_f(y) - L_1(y)\}.$$

Having in mind that $H_f(y)$ and $\Psi(y)$ are bounded, we have

$$\left|\frac{Z_f(y)}{\exp\{w_\varepsilon(y)\}}\right| \ll \exp\left\{\operatorname{Re}\sum_{k=1}^{n} q^{-k}\sum_{\delta(p)=k}(f(p)-1)y^k\right\} \leqslant |V_n|e^{\Delta}, \tag{5}$$

where

$$\Delta := \left|\sum_{k=1}^{n} q^{-k}\sum_{\delta(p)=k}(f(p)-1)(y^k-1)\right|.$$

For every $u \geqslant 0$

$$\Delta \leqslant u\sum_{k=1}^{n}\frac{\pi(k)}{q^k}r^k|e^{it}-1| + 2\sum_{k=1}^{n} q^{-k}\left|\sum_{\substack{\delta(p)=k \\ |f(p)-1|>u}}(f(p)-1)\right|r^k$$

$$+ 2\sum_{k=1}^{n}\frac{\pi(k)}{q^k}(1-r^k).$$

Using (3) and definition of $E_n(u)$ we obtain

$$\Delta \leqslant 2u\sigma + \log E_n(u) + c_2, \tag{6}$$

where

$$\sigma := \sum_{k=1}^{\infty} \frac{r^k}{k} \left| e^{itk} - 1 \right|.$$

In order to estimate σ we expand $|e^{itk} - 1|$ in the Fourier series:

$$\left| e^{itk} - 1 \right| = \frac{4}{\pi} \sum_{l=-\infty}^{\infty} \left(1 - 4l^2\right)^{-1} e^{itkl}.$$

Note that

$$\sum_{l \neq 0} \frac{1}{4l^2 - 1} = 1.$$

Hence

$$\frac{\pi}{4}\sigma = \sum_{l=-\infty}^{\infty} \frac{1}{1 - 4l^2} \sum_{k=1}^{\infty} k^{-1} \left(re^{itl}\right)^k = \sum_{l=-\infty}^{\infty} \frac{\log |1 - re^{itl}|}{4l^2 - 1}$$
$$= \log \frac{|1 - y|}{1 - r} + \sum_{l \neq 0} \frac{1}{4l^2 - 1} \log \left| \frac{1 - re^{itl}}{1 - y} \right|.$$

Whence it follows that

$$\sigma \leqslant \frac{4}{\pi} \left(\log \frac{|1 - y|}{1 - r} + \frac{1}{2} \sum_{l \neq 0}{}' \frac{1}{4l^2 - 1} \log \left| \frac{1 - re^{itl}}{1 - y} \right|^2 \right), \tag{7}$$

where $'$ denotes that the summation is restricted to those $l \neq 0$ for which

$$|1 - re^{itl}| > |1 - y|.$$

For those l, after some straightforward estimations, we get

$$\left| \frac{1 - re^{itl}}{1 - y} \right|^2 = 1 + \frac{2r(\cos t - \cos tl)}{(r - 1)^2 + 2r(1 - \cos t)} \leqslant 2l^2.$$

Therefore we deduce from (7) and (6) that

$$\Delta \leqslant \frac{8u}{\pi} \log \left| \frac{1 - y}{1 - r} \right| + \log E_n(u) + c_3.$$

Now the estimate (4) follows from (5). This completes the proof of lemma.
Proof of Theorem 1. Our start point is the Cauchy formula

$$M_n(f) = \frac{1}{2\pi A i} \int_{|y| = r_n} Z_f(y) y^{-n-1} dy.$$

Integrating by parts and having in mind the estimates of $H_f(y)$, we obtain

$$\begin{aligned} M_n(f) &= -\frac{r_n^{-n}}{2\pi Ani}\int_{-\pi}^{\pi} Z_f(r_ne^{it})d(e^{-int}) \\ &= \frac{1}{2\pi Ani}\int_{|y|=r_n} Z'_f(y)y^{-n}dy \\ &= \frac{1}{2\pi Ani}\int_{|y|=r_n} L'_f(y)Z_f(y)y^{-n}dy + O\biggl(\frac{1}{n}\int_{-\pi}^{\pi}\exp\{\operatorname{Re}L_f(r_ne^{it})\}dt\biggr). \end{aligned}$$

A slight modification of Lemma 4 in (Indlekofer and Manstavičius, 1990) yields

$$\int_{-\pi}^{\pi}|Z_f(r_ne^{it})|\,dt \ll \int_{-\pi}^{\pi}\exp\{\operatorname{Re}L_f(r_ne^{it})\}dt \ll \log n.$$

Therefore

$$\begin{aligned} M_n(f) &= \frac{1}{2\pi Ani}\Bigl(J_1 + J_2 + \int_{|y|=r_n}(\log\Psi(y))'Z_f(y)y^{-n}dy\Bigr) + O\Bigl(\frac{\log n}{n}\Bigr) \\ &= \frac{1}{2\pi Ani}(J_1+J_2) + O\Bigl(\frac{\log n}{n}\Bigr), \end{aligned} \tag{8}$$

where

$$J_1 := \int_{|y|=r_n} w'_\varepsilon(y)Z_f(y)y^{-n}dy,$$

$$J_2 := \int_{|y|=r_n}(L'_f(y) - L'_1(y))Z_f(y)y^{-n}dy.$$

Let $0 < K \leqslant n$. We write the integral J_1 in the form

$$\begin{aligned} J_1 &= (1+\varepsilon)\Biggl(\int\limits_{\substack{|y|=r_n \\ |\arg y|\leqslant K/n}} + \int\limits_{\substack{|y|=r_n \\ |\arg y|>K/n}}\Biggr)(1-y)^{-2}\frac{Z_f(y)}{\exp\{w_\varepsilon(y)\}}\,y^{-n}\,dy \\ &=: (1+\varepsilon)(J_{10}+J_{11}). \end{aligned}$$

Now our aim is to extract the main term of the integral J_{10}. Put for $k \geqslant 1,\ r \geqslant 0$

$$b_k := q^{-k}\sum_{\delta(p)=k}(f(p)-1),$$

$$v(r) := \sum_{k\leqslant n/3} k|b_k|r^k,$$

$$\tau(r) := (\max(v(r), 1))^{-1}.$$

Note that

$$|b_k| \ll k^{-1} \quad \text{and} \quad \mu_n = \frac{1}{n}\sum_{k=1}^{n} k|b_k|.$$

Let S be the domain of the complex plane restricted by the lines

$$\begin{aligned}
l_0 &:= \{y\colon\ y = r_n e^{it},\ |t| \leqslant \tau(r_n)\},\\
l_1 &:= \{y\colon\ y = re^{i\tau(r)},\ r_n \leqslant r \leqslant 2\},\\
l_2 &:= \{y\colon\ y = 2e^{it},\ |t| \leqslant \tau(2)\},\\
l_3 &:= \{y\colon\ y = re^{-i\tau(r)},\ r_n \leqslant r \leqslant 2\}.
\end{aligned}$$

We choose $K := n\tau(r_n)$. Then

$$J_{10} = \int_{l_0} (1-y)^{-2} H_f(y)\Psi(y)\exp\{L_f(y) - L_1(y)\} y^{-n}\, dy. \tag{9}$$

It is easy to see that

$$\exp\{L_f(y) - L_1(y)\} = \exp\{T_1(y)\} + \exp\{T_1(y)\}(\exp\{T_2(y)\} - 1), \tag{10}$$

where

$$T_1(y) := \sum_{k\leqslant n/3} b_k y^k, \qquad T_2(y) := \sum_{n/3<k\leqslant n} b_k y^k.$$

We need to estimate $|\exp\{T_1(y)\}|$ on the domain S. Noting that $\operatorname{Re} b_k \leqslant 0$, we have

$$\operatorname{Re} T_1(y) \leqslant \sum_{k\leqslant n/3} r_n^k \operatorname{Re} b_k + \sum_{k\leqslant n/3} |b_k| r^k |e^{itk} - 1|$$

for $y = re^{it} \in S$. Since

$$\sum_{n/3<k\leqslant n} |b_k| \ll 1$$

and

$$\left|\sum_{k\leqslant n/3} (r_n^k - 1)\operatorname{Re} b_k\right| \ll \sum_{k\leqslant n} \frac{1-r_n^k}{k} \ll 1,$$

we get

$$\operatorname{Re} T_1(y) \leqslant \sum_{k\leqslant n} \operatorname{Re} b_k + |t|v(r) + c_4$$

for some $c_4 > 0$. Now, considering that $|t| \leqslant \tau(r)$, we obtain finally

$$|\exp\{T_1(y)\}| \ll |V_n| \qquad \text{for} \quad y \in S. \tag{11}$$

The following simple estimates are valid for $y \in l_0$:

$$H_f(y) = H_f(1) + O(|y-1|),$$

$$\Psi(y) = \Psi(1) + O(|y-1|),$$

and

$$|T_2(y)| \leqslant \sum_{n/3<k\leqslant n} |b_k| \leqslant 3\mu_n.$$

Substituting these estimates, (10), and (11) to the formula (9), we obtain

$$J_{10} = H_f(1)\Psi(1) \int_{l_0} (1-y)^{-2} \exp\{T_1(y)\} y^{-n}\, dy$$
$$+ O\Big(|V_n|\Big(\int_{l_0} |1-y|^{-1}|dy| + \mu_n \int_{l_0} |1-y|^{-2}|dy|\Big)\Big).$$

The first integral in the remainder term is

$$\ll \int_0^{K/n} \Big(\frac{1}{n}+t\Big)^{-1} dt \ll \log(1+K) \ll \log n.$$

The second integral is $\ll n$ analogously. Hence

$$J_{10} = H_f(1)\Psi(1)I + O(|V_n|(\log n + n\mu_n)), \tag{12}$$

where

$$I := \int_{l_0} (1-y)^{-2} \exp\{T_1(y)\} y^{-n}\, dy.$$

According to the residue theorem and (11) we get

$$I = -2\pi i \operatorname*{Res}_{y=1} (1-y)^{-2} \exp\{T_1(y)\} y^{-n} + O(|V_n|(I_1 + I_2)), \tag{13}$$

where

$$I_1 := \int_{l_1 \cup l_3} |1-y|^{-2}|y|^{-n}|dy|, \qquad I_2 := \int_{l_2} |1-y|^{-2}|y|^{-n}|dy|.$$

I_2 may be estimated simply. Considering the definition of l_2 we have that

$$I_2 \ll 2^{-n}.$$

It remains to estimate I_1. Let there exist $r_0, r_0 \in (r_n, 2)$, such that $\tau(r) = 1$ if $r < r_0$ and $\tau(r) = v^{-1}(r)$ otherwise. Then

$$I_1 \ll \int_{r_n}^{r_0} \frac{r^{-n}dr}{|1-re^{i}|^2} + \int_{r_0}^{2} \frac{r^{-n}(1 + rv'(r)v^{-2}(r))}{|1-re^{i/v(r)}|^2}\, dr. \tag{14}$$

By our notation we have for $r \geqslant r_n$

$$r^{-n/2}v(r) = \sum_{k\leqslant n/3} |b_k|kr^{k-n/2} \ll n,$$

$$rv'(r) = \sum_{k\leqslant n/3} |b_k|k^2r^k \leqslant \frac{n}{3}v(r).$$

Therefore

$$I_1 \ll \frac{1}{n} + n\int_{r_n}^{2} r^{-n/2}v(r)\,dr \ll \frac{1}{n} + v(1). \tag{15}$$

If there does not exist r_0 with the conditions mentioned above, then in view of the continuity and monotonicity of $\tau(r)$ we get, that in the whole interval $[r_n, 2]$ the function $\tau(r)$ will be equal either to 1 or to $v^{-1}(r)$. Consequently, in (14) we shall have only one of the two integrals over the whole interval $[r_n, 2]$. Thus the estimate (15) remains true in this case too.
Substituting the estimates I_1 and I_2 into (13) and considering that $v(1) \leqslant n\mu_n$, we get

$$I = -2\pi i \operatorname*{Res}_{y=1}(1-y)^{-2}\exp\{T_1(y)\}y^{-n} + O\Big(|V_n|\Big(\frac{1}{n} + n\mu_n\Big)\Big). \tag{16}$$

The residue is calculated in the standard way:

$$\operatorname*{Res}_{y=1}(1-y)^{-2}\exp\{T_1(y)\}y^{-n} = \big(y^{-n}\exp\{T_1(y)\}\big)'_{y=1}$$
$$= -n\exp\{T_1(1)\} + T_1'(1)\exp\{T_1(1)\} = -nV_n + O(|V_n|n\mu_n).$$

Now the estimates (16) and (12) yield

$$J_{10} = 2\pi i n H_f(1)\Psi(1)V_n + O(|V_n|(\log n + n\mu_n)).$$

As for the integral J_{11}, applying Lemma we have

$$|J_{11}| \ll |V_n|E_n(u)|1-r_n|^{-\lambda}\int_{K/n}^{\pi}\left|1-r_ne^{it}\right|^{\lambda-2}dt$$
$$\ll |V_n|E_n(u)n\int_{K}^{\pi n}(1+x)^{\lambda-2}\,dx \ll n|V_n|E_n(u)\mu^{1-\lambda}$$

for $0 \leqslant u \leqslant 3/8$, $\lambda := 8u/\pi$.
In the final step we have made use of the inequalities $\lambda < 1$ and $K = n\tau(r_n) \geqslant \mu^{-1}$.
The last estimates of integrals J_{10} and J_{11} yield

$$J_1 = 2\pi i n(1+\varepsilon)H_f(1)\Psi(1)V_n + O\big(|V_n|\big(\log n + nE_n(u)\mu^{1-8u/\pi}\big)\big) \tag{17}$$

for each u, $0 \leqslant u \leqslant 3/8$.
It remains to estimate the integral J_2. First of all, applying Lemma, we get

$$\int_0^{\pi} |Z_f(r_n e^{it})|^2\, dt = \int_0^{\pi} \Big|\frac{Z_f(r_n e^{it})}{\exp\{w_\varepsilon(r_n e^{it})\}}\Big|^2 |\exp\{w_\varepsilon(r_n e^{it})\}|^2\, dt$$

$$\leqslant |V_n|^2 E_n^2(u) |1-r_n|^{-2\lambda} \int_0^{\pi} |1-r_n e^{it}|^{2\lambda-2}\, dt \leqslant |V_n|^2 E_n^2(u) n$$

for each u, $0 \leqslant u \leqslant 3/16$.
Moreover by the Parseval equality we have

$$\int_{|y|=r_n} |L_f'(y) - L_1'(y)|^2 |dy| \ll \int_{-\pi}^{\pi} \Big|\sum_{k=1}^{n} k b_k r_n^k e^{itk}\Big|^2 dt$$

$$= 2\pi \sum_{k=1}^{n} k^2 |b_k|^2 r_n^{2k} \ll n\mu_n.$$

Hence

$$'|J_2| \ll \Big(\int_{|y|=r_n} |L_f'(y) - L_1'(y)|^2 |dy|\Big)^{1/2} \Big(\int_0^{\pi} |Z_f(r_n e^{it})|^2\, dt\Big)^{1/2} \qquad (18)$$

$$\ll n\sqrt{\mu_n}\, |V_n| E_n\Big(\frac{3}{16}\Big).$$

On the other hand the Cauchy formula yields

$$J_2 = \int_{|y|=r_n} Z_f(y) y^{-n} \sum_{k=1}^{n} k b_k y^{k-1}\, dy = \sum_{k=1}^{n} k b_k \int_{|y|=r_n} Z_f(y) y^{-(n-k)-1}\, dy$$

$$= 2\pi A i \sum_{k=1}^{n} k b_k M_{n-k}(f).$$

Whence, from (17) and (8), after substituting the value of $\Psi(1)$, we get

$$M_n(f) = R + \frac{1}{n}\sum_{k=1}^{n} k b_k M_{n-k}(f) + O\Big(|V_n| E_n(u) \mu^{1-8u/\pi} + \frac{\log n}{n}\Big)$$

for each u, $0 \leqslant u \leqslant 3/8$, where

$$R := \frac{H_f(1)}{H_1(1)} V_n.$$

To obtain (1) it remains to transform the main term R. Set

$$\chi_p(y;f) := 1 + \sum_{k=1}^{\infty} f(p^k)\left(\frac{y}{q}\right)^{k\delta(p)},$$

$$\chi_p(y;1) := \left(1 - \left(\frac{y}{q}\right)^{\delta(p)}\right)^{-1}.$$

Since $f(p^k) = 1$ for $\delta(p) > n$, we have

$$\frac{H_f(y)}{H_1(y)} = \prod_{\delta(p)\leqslant n} \chi_p(y;f)\chi_p^{-1}(y;1)\exp\left\{-\sum_{k=1}^{n} b_k y^k\right\}$$

on the disc $|y| \leqslant 1 + c_1$. Hence

$$R = \prod_{\delta(p)\leqslant n} \chi_p(1;f)\chi_p^{-1}(1;1) = \prod_{\delta(p)\leqslant n} \left(1 - q^{-\delta(p)}\right)\left(1 + \sum_{k=1}^{\infty} f(p^k)q^{-k\delta(p)}\right).$$

This ends the proof of estimate (1).
We note that

$$|V_n|E_n\left(\frac{3}{16}\right) \geqslant \exp\left\{\sum_{k=1}^{n} \frac{1}{q^k} \sum_{\substack{\delta(p)=k \\ |f(p)-1|\leqslant 3/16}} (\mathrm{Re}\, f(p) - 1)\right\}$$

$$\geqslant \exp\left\{-\frac{3}{16}\sum_{k=1}^{n} q^{-k}\pi(k)\right\} \geqslant c_5 n^{-3/8}.$$

Now the estimate (2) follows from (18), (17), and (8). This completes the proof of Theorem 1.

3. PROOF OF THEOREM 2

We shall use the method of the analogous proof in (Mačiulis, 1988). Let us define the strongly multiplicative function

$$f(a) := \exp\{ith_n(a)\}, \qquad t \in \mathbf{R}, \quad a \in \mathcal{G}.$$

The characteristic function of the distribution $F_n(x)$ will be

$$\varphi_n(t) := \frac{1}{G(n)} e^{-it\alpha(n)} \sum_{\delta(a)=n} f(a) = \frac{Aq^n}{G(n)} e^{-it\alpha(n)} M_n(f).$$

We note that

$$\beta_n \geqslant c_6(\log n)^{-1/2} \qquad \text{if} \quad B(n) > 0.$$

Let

$$|t| \leqslant (8^3\beta_n)^{-1}, \qquad \eta := |t\beta_n|^{1/3}.$$

Then by the first assertion of Theorem 1 and Axiom A^* we get

$$\begin{aligned}\varphi_n(t) = e^{-it\alpha(n)} & \prod_{\delta(p)\leqslant n} \left(1 + \frac{f(p)-1}{q^{\delta(p)}}\right) \\ & + O\left(\frac{1}{n}\sum_{k=1}^{n} k|b_k||M_{n-k}(f)| + |V_n|E_n(\eta)\mu^{1-8\eta/\pi} + \frac{\log n}{n}\right).\end{aligned} \tag{19}$$

First of all we consider the quantities which occur in the error term. By virtue of the definitions of μ and η,

$$\mu^{1-8\eta/\pi} \ll |t|(\beta_n + \rho_n).$$

Further for each $u \geqslant \eta$

$$\begin{aligned}|V_n|E_n(u) &\leqslant \exp\left\{\sum_{k=1}^{n}\frac{1}{q^k}\sum_{\delta(p)=k}(\operatorname{Re} f(p)-1) + 2\sum_{k=1}^{n}\frac{1}{q^k}\sum_{\substack{\delta(p)=k\\ |f(p)-1|>u}}|f(p)-1|\right\} \\ &\leqslant \exp\left\{-\frac{t^2}{2} + \frac{|t|^3\beta_n}{6} + 2|t|\sum_{\substack{\delta(p)\leqslant n\\ |th_n(p)|>\eta}}|h_n(p)|q^{-\delta(p)}\right\} \\ &\leqslant \exp\left\{-t^2\left(\frac{1}{2} - \frac{|t|\beta_n}{6} - 2|t\beta_n|^{1/3}\right)\right\} \leqslant e^{-t^2/5}.\end{aligned}$$

It remains to estimate the first summand in the error term of (19). For

$$k \leqslant n(1-\beta_n^2),$$

to estimate $|M_{n-k}(f)|$ we apply (2), and for other $k \leqslant n$, we shall use the obvious inequality $|M_{n-k}(f)| \ll 1$. Thus,

$$\begin{aligned}\frac{1}{n}\sum_{k=1}^{n}|M_{n-k}(f)b_k|k &\ll \frac{1}{n}\sum_{k\leqslant n(1-\beta_n^2)}|V_{n-k}|E_{n-k}\left(\frac{3}{16}\right)k|b_k| + \frac{1}{n}\sum_{n(1-\beta_n^2)<k\leqslant n}1 \\ &\ll e^{-t^2/5}\frac{1}{n}\sum_{k\leqslant n(1-\beta_n^2)}\left|\frac{V_{n-k}E_{n-k}(3/16)}{V_nE_n(3/16)}\right|k|b_k| + \beta_n^2.\end{aligned}$$

The ratio in the last sum does not exceed

$$\begin{aligned}\exp\left\{\sum_{n-k<m\leqslant n}q^{-m}\sum_{\delta(p)=m}(1-\operatorname{Re} f(p))\right\} &\leqslant \exp\left\{\frac{t^2}{2}\sum_{n\beta_n^2<\delta(p)\leqslant n}h_n^2(p)q^{-\delta(p)}\right\} \\ \leqslant \exp\left\{\frac{t^2}{2}\beta_n^{2/3}\left(\sum_{n\beta_n^2<m\leqslant n}\pi(m)q^{-m}\right)^{1/3}\right\} &\ll \exp\left\{\frac{t^2}{10}\right\}.\end{aligned}$$

The last inequality is valid in view of (3) and the fact that $\beta_n \to 0$ as $n \to \infty$. Substituting all these estimates in the error term of (19) we have that

$$\varphi_n(t) = \prod_{\delta(p) \leqslant n} \psi_p(\tau) + O\Big(\beta_n^2 + |t|(\beta_n + \rho_n)e^{-t^2/10}\Big)$$

for $|t| \leqslant (8^3\beta_n)^{-1}$, where

$$\tau := t/B(n),$$

$$\psi_p(\tau) := \Big(1 + \frac{e^{i\tau h(p)} - 1}{q^{\delta(p)}}\Big) \exp\big\{ - i\tau h(p) q^{-\delta(p)}\big\}$$

are characteristic functions of independent random variables X_p which assume two values with probabilities

$$\mathbb{P}\big(X_p = h(p) - h(p)q^{-\delta(p)}\big) = q^{-\delta(p)},$$

$$\mathbb{P}\big(X_p = -h(p)q^{-\delta(p)}\big) = 1 - q^{-\delta(p)}.$$

After standard calculations (cf. (Petrov, 1987, p. 155)) we get

$$\prod_{\delta(p) \leqslant n} \psi_p(\tau) = e^{-t^2/2} + O\big(\beta_n |t|^3 e^{-t^2/3}\big)$$

for $|t| \leqslant (4\beta_n)^{-1}$. Consequently,

$$\varphi_n(t) = e^{-t^2/2} + O\big(\beta_n^2 + \big(|t|\beta_n + |t|\rho_n + |t|^3\beta_n\big)e^{-t^2/10}\big)$$

for $|t| \leqslant (8^3\beta_n)^{-1}$.

For the values of the characteristic function $\varphi_n(t)$ in a neighborhood of the point $t = 0$, with help of an analogue of the Kubilius inequality one can get the more precise estimate

$$\varphi_n(t) = e^{-t^2/2} + O\big(|t| + t^2\big).$$

Now the proof of Theorem 2 follows from these two estimates of $\varphi_n(t)$ and Esseen's inequality.

REFERENCES

Indlekofer, K.-H. and Manstavičius, E. (1990). Additive and multiplicative functions on arithmetical semigroups. *Preprint.*

Indlekofer, K.-H., Manstavičius, E. and Warlimont, R. (1991). On a certain class of finite products with an application to arithmetical semigroups. *Arch. Math.* **56**, 446–453.

Knopfmacher, J. (1979). *Analytic Arithmetic of Algebraic Function Fields.* Marcel Dekker, New York.

Mačiulis, A. (1988). Mean value of multiplicative functions. *Lithuanian Math. J.* **28**, 221–229.

Petrov, V. V. (1987). *Limit Theorems for Sums of Independent Random Variables.* Nauka, Moscow (in Russian).

New Trends in Probab. and Statist., pp. 135–153
F. Schweiger and E. Manstavičius (Eds.)
1992 VSP/TEV

SEMIGROUP ELEMENTS FREE OF LARGE PRIME FACTORS

E. MANSTAVIČIUS
Department of Mathematics, Vilnius University,
Naugarduko 24, 2006 Vilnius, Lithuania

ABSTRACT

A class of free semigroups including the semigroup of monic polynomials over a finite field and the semigroup of integral divisors in the algebraic function field over a finite constant field is considered. Asymptotic formulae for the number of the semigroup elements without "large" prime factors are obtained.

1. INTRODUCTION

In recent years the asymptotic behavior of the number of natural numbers without large prime factors had been widely investigated (Hensley, 1985, 1986; Hildebrand, 1986a, 1986b; Hildebrand and Tenenbaum, 1986). To our opinion these results can be easily extended to the arithmetical semigroups satisfying Axiom A (Knopfmacher, 1975, 1979). With a hope that in the semigroups with a discrete norming such as the semigroup of monic polynomials over a finite field or the semigroup of integral divisors in the algebraic function field over a finite constant field rather different situation can arise, we will proceed the investigations undertaken in (Indlekofer *et al.*, 1989, 1991).

Let $\mathcal{G}$ be a free commutative multiplicative semigroup with identity element 1 and generated by a countable subset $\mathfrak{P}$ of prime elements. Suppose that a completely additive degree function $\delta\colon \mathcal{G} \to \mathbf{N} \cup \{0\}$ is defined so that $\delta(p) \geqslant 1$ for each $p \in \mathfrak{P}$. We express the main condition accepted in the paper as

AXIOM A*. *There exist constants $A > 0$, $q > 1$ and ν, $0 \leqslant \nu < 1$, such that*

$$G(n) := \#\{a \in \mathcal{G}\colon\ \delta(a) = n\} = Aq^n + O(q^{\nu n}).$$

As it has been shown in (Indlekofer *et al.*, 1991) the prime element theorem for these semigroups $\mathcal{G}$ has the form

$$\pi(k) := \#\{p \in \mathfrak{P}\colon\ \delta(p) = k\} = \big(1 - (-1)^k \varkappa\big)q^k/k + O\big(q^{\mu n}\big) \tag{1.1}$$

where $\varkappa = 1$ or 0 if the function

$$Z(y) = \sum_0^\infty G(n)y^n = \prod_1^\infty \left(1 - y^k\right)^{-\pi(k)}$$

has or not the exceptional zero at the point $y = -q^{-1}$, respectively. Here μ is some constant, $\max\{1/2, \nu\} < \mu < 1$. Different relations between the remainder terms of $G(n)$ and $\pi(k)$ have been considered in (Indlekofer *et al.*, 1989).
The purpose of the present paper is to investigate the quantity

$$\Psi(n,m) = \#\{a \in \mathcal{G}\colon\ \delta(a) = n,\ p \mid n \Longrightarrow \delta(p) \leqslant m\}$$

in different regions for the parameter $u := n/m$.
For $y, \omega \in \mathbf{C}$, $|y| < 1$, $\operatorname{Re}\omega > 0$, we denote

$$Z(y,m) = \prod_{1}^{m} \left(1 - y^k\right)^{-\pi(k)},$$

$$\Phi(\omega) = \log Z\left(q^{-\omega}, m\right),$$

and for $j = 0, 1, \ldots$

$$\Phi_j(\omega) = (-1)^j \Phi^{(j)}(\omega) \log^{-j} q.$$

Let $\alpha = \alpha(u,m)$ be the solution of the equation

$$\Phi_1(\alpha) = n = mu. \tag{1.2}$$

The main result of the paper is the following

THEOREM 1. *If $m \geqslant m_0 \geqslant 3$ and*

$$(\log\log m)^3 \log m \leqslant u \leqslant \exp\left\{\delta q^m / m^2\right\},$$

where $\delta > 0$ is sufficiently small, then

$$\Psi(n,m) = \frac{q^{\alpha n} Z(q^{-\alpha}, m)}{\sqrt{2\pi \Phi_2(\alpha)}} \left(1 + O\left(\frac{m}{n} + \frac{m}{q^m}\right)\right).$$

The constant in the symbol O and m_0 depend only on q, μ, and the constant in the remainder term of the formula (1.1).

Further we present some asymptotic expressions for the quantities involved in Theorem 1.

THEOREM 2. *Suppose that $3 \leqslant m_0 \leqslant m \leqslant n$ and $u \geqslant 2q$. If $\varkappa = 0$, then*

$$\alpha = \alpha(u,m) = \frac{1}{m} \log_q \left(1 + \frac{q^{m+1}}{(q-1)n}\right) \left(1 + O\left(\frac{\log m}{m}\right)\right),$$

$$\Phi_2(\alpha) = mn \left(1 + \frac{(q-1)n}{q^{m+1}}\right) \left(1 + O\left(\frac{1}{\log u} + \frac{1}{m}\right)\right).$$

If $\varkappa = 1$, then for the same m and u we have

$$\alpha = \alpha(u,m) = \frac{1}{m} \log_q \left(1 + \frac{2q^{2\overline{m}+1}}{(q^2-1)n}\right) \left(1 + O\left(\frac{\log m}{m}\right)\right),$$

$$\Phi_2(\alpha) = mn\left(1 + \frac{(q^2-1)n}{2} q^{-2\overline{m}-1}\right)\left(1 + O\left(\frac{1}{\log u} + \frac{1}{m}\right)\right).$$

Here and in what follows $\overline{m} = [(m+1)/2]$.

In a smaller range we can obtain an approximation of $\Psi(n,m)$ expressed in terms of the unique positive solution $\eta(x) = \eta(x, m, \varkappa)$ of the equation

$$\eta(x)^m + \eta(x)^{m-1} + \ldots + \eta(x) = xm$$

or

$$2\left(\eta(x)^{2\overline{m}-1} + \eta(x)^{2\overline{m}-3} + \ldots + \eta(x)\right) = xm$$

with $x > 1$ in the cases $\varkappa = 0$ or 1, respectively. Let further $\eta = \eta(u)$ and η' denotes the derivative at $x = u$.

THEOREM 3. *Suppose that* $3 \leqslant m_0 \leqslant m \leqslant n$ *and*

$$(\log\log m)^3 \log m \leqslant u \leqslant q^{m/2}/m.$$

If $\varkappa = 0$, *then*

$$\Psi(n,m) = \left(\frac{q}{\eta}\right)^n \sqrt{\frac{m\eta'}{2\pi\eta}} \exp\left\{\Gamma + \int_1^\eta \frac{t^m - 1}{t-1}\, dt\right\}\left(1 + O\left(\frac{\log u}{m}\right)\right).$$

If $\varkappa = 1$, *then*

$$\Psi(n,m) = \left(\frac{q}{\eta}\right)^n \sqrt{\frac{m\eta'}{2\pi\eta}} \exp\left\{\Gamma + 2\int_1^\eta \frac{t^{2\overline{m}} - 1}{t^2-1}\, dt\right\}\left(1 + O\left(\frac{\log u}{m}\right)\right).$$

Here $\Gamma = \Gamma(\mathcal{G})$ *is a constant.*

The desire to compare our result with that from (Hildebrand and Tenenbaum, 1986) leads us to the next theorem. Let $\xi(x)$ be the positive solution of the equation

$$e^{\xi(x)} = 1 + x\xi(x), \qquad x \geqslant 1,$$

$\xi = \xi(u)$ and ξ' stands for the derivative at $x = u$. Denote by $\rho(x)$ the continuous solution of the differential difference equation

$$x\rho'(x) + \rho(x-1) = 0, \qquad x > 1,$$

with the initial condition $\rho(x) = 1$ for $0 \leqslant x \leqslant 1$.

THEOREM 4. *Suppose that* $3 \leqslant m_0 \leqslant m \leqslant n$ *and*

$$(\log\log m)^3 \log m \leqslant u \leqslant m^{1-\varepsilon}, \qquad 0 < \varepsilon < 1/2.$$

Then

$$\Psi(n,m) = q^n\sqrt{\frac{\xi'}{2\pi}}\exp\left\{\Gamma + \int_0^{\xi}\frac{e^s-1}{s}\,ds - u\xi\Big(1+O\Big(\frac{1}{m}\Big)\Big)\right\}$$
$$= Aq^n\rho(u)\Big(1+O_\varepsilon\Big(\frac{\log u}{m^\varepsilon}+\frac{1}{u}\Big)\Big).$$

Since $\xi = \xi(u) \sim \log u$ as $u \to \infty$, the remainder term of the first equality does not allow to use it in a wider range. On the contrary to the semigroup **N** (Hildebrand and Tenenbaum, 1986), an extension of the approximation of $\Psi(n,m)$ for larger u seems to be impossible. Most probably, the second expression remains valid even for smaller u. We intend to investigate this problem elsewhere. The upper bound condition for u used in Theorem 1 and Theorem 5 below is superfluous. A bit intricate proof taking into account the lattice structure will be presented in a forthcoming paper.

We shall apply the saddle point method adopted to the analogous problem for the semigroup **N** in (Hildebrand and Tenenbaum, 1986). With a hope that our approach would be useful for those interested in the probability theory, we emphasize the relations with local theorems concerning the large deviations for independent lattice random variables. A new theorem of such kind is obtained in the next chapter.

2. LOCAL THEOREM FOR INTEGER-VALUED RANDOM VARIABLES

Let X_{kj}, $1 \leqslant j \leqslant \pi(k)$, $1 \leqslant k \leqslant m$, be independent random variables having the distribution

$$\mathbb{P}(X_{kj} = lk) = q^{-alk}(1-q^{-ak}), \qquad l = 0, 1, \dots,$$

where, as earlier, $q > 1$, $\pi(k)$ has the asymptotical behaviour (1.1), and a is a positive parameter possibly depending on m or on something else. In this chapter we shall obtain an approximation of the probability

$$P_m(n) := \mathbb{P}(S_m = n) := \mathbb{P}\Big(\sum_1^m \sum_{j=1}^{\pi(k)} x_{kj} = n\Big).$$

It is easy to verify that the random variables X_{kj} do not satisfy the condition (P) used in (Milaševičius, 1980); the first factorial semi-invariant of the sum S_m is considerably greater than the second one, therefore we prefer the version of the saddle point method traditionally used in the normal approximation. At the same time we avoid the Cramér series.

Observe that the characteristic function $f_m(it)$ of the sum S_m equals to

$$Z(q^{-a}e^{it}, m)/Z(q^{-a}, m)$$

and

$$P_m(n) = \frac{e^{-z_0 n}}{2\pi}\int_{-\pi}^{\pi} e^{-itn} f_m(z_0+it)\,dt$$

for each z_0 such that

$$\omega := a - \frac{z_0}{\log q} > 0.$$

Preserving the definition of $\alpha = \alpha(u, m)$ we choose $\omega = \alpha$ to obtain

$$P_m(n) = q^{(\alpha - a)n} \prod_1^m \left(\frac{1 - q^{-ak}}{1 - q^{-\alpha k}} \right)^{\pi(k)} \left\{ \frac{1}{2\pi} \int_{-\pi}^{\pi} e^{-itn} g_m(t)\, dt \right\} \tag{2.1}$$

where

$$g_m(t) = Z(q^{-\alpha} e^{it}, m) / Z(q^{-\alpha}, m)$$

is the characteristic function of the sum of the joint random variables defined as X_{kj} if we take in advance $a = \alpha$.

THEOREM 5. *Let X_{kj}, $1 \leqslant j \leqslant \pi(k)$, $1 \leqslant k \leqslant m$, be the independent random variables defined above. Then for*

$$(\log \log m)^3 \log m \leqslant u \leqslant \exp\left\{\delta q^m / m^2\right\}$$

with sufficiently small $\delta > 0$ and $3 \leqslant m_0 \leqslant m$ we have

$$P_m(n) = q^{(\alpha - a)n} \prod_1^m \left(\frac{1 - q^{-ak}}{1 - q^{-\alpha k}} \right)^{\pi(k)} \frac{1}{\sqrt{2\pi \Phi_2(\alpha)}} \left(1 + O\left(\frac{m}{n} + \frac{m}{q^m}\right)\right).$$

The dependence of the constants is the same as in Theorem 1.

At first we shall prove few lemmas. In what follows all the constants involved in the traditional symbols as well as c, $c_1, \ldots$ and C, $C_1, \ldots$ depend on q, μ, and on the constant in the remainder term of the formula (1.1). For the simplicity we shall consider $m \geqslant m_0$ for sufficiently large m_0.

LEMMA 1. *Let $w \in \mathbf{C}$, $|w| = q^{-\beta}$. Then uniformly in β, $0 \leqslant \beta \leqslant 2$, and w, $|\arg w| \leqslant \pi$, we have*

$$W_m(w) := \sum_1^m k\pi(k) w^k = qw \left(\frac{(qw)^m - 1}{qw - 1} + \varkappa \frac{(-qw)^m - 1}{-qw - 1} \right) + R$$

where the fractions between the parentheses are equal to m, when $qw = 1$ or $qw = -1$ respectively, and

$$R = R(w) \ll m \max\left(1,\ q^{(\mu - \beta)m}\right\} \min\left\{m, |\mu - \beta|^{-1}\right\}.$$

Proof. It is enough to apply the formula (1.1).

In what follows we shall use the abbreviation $\bar{u} = \min\{u, q^m/m\}$.

LEMMA 2. *Let $\alpha = \alpha(u,m)$ be defined by the equality (1.2). Then for $m \geqslant m_0$ the following propositions are true:*

(i_1) *if $2q \leqslant u \leqslant q^m/m$, then*

$$\frac{1}{m} \ll \alpha(u,m) \leqslant 1 - \frac{c_1}{m};$$

(i_2) *if $u > q^m/m$, then*

$$\alpha(u,m) \asymp q^m u^{-1} m^{-2};$$

(i_3) *for all $u \geqslant 2q$ we have*

$$q^{m(1-\alpha(u,m))} \asymp (1-\alpha(u,m))m\bar{u}.$$

Proof. If for some $u \geqslant 2q$ and for a subsequence $m = m' \to \infty$ we had $1 - \alpha = o(1/m)$, then $q^\alpha \geqslant q - \varepsilon$ for each $\varepsilon > 0$ and sufficiently large m. Further, by the definition of α and Lemma 1

$$2qm \leqslant \sum_1^m \frac{k\pi(k)}{q^{\alpha k}(1 - 1/(q-\varepsilon))} = \frac{q-\varepsilon}{q-1-\varepsilon}\,\frac{q^{(1-\alpha)m}-1}{q^{1-\alpha}-1} + O(1) \sim \frac{q-\varepsilon}{q-1-\varepsilon}\,m,$$

as $m = m' \to \infty$. Then the contradiction proves the right-hand side estimate in (i_1). Analogously, for all $u \geqslant 2q$ we have $\alpha \leqslant 1$ and

$$\begin{aligned} um &\geqslant \left(q^{m\alpha}-1\right)^{-1} W_m(1) \\ &= q^m \left(q^{m\alpha}-1\right)^{-1} \frac{q(q+1+\varkappa(q-1)(-1)^m)}{q^2-1} \\ &\quad + O\left(mq^{(\mu-1)m}\right) \gg \frac{q^m}{q^{m\alpha}-1} \end{aligned} \tag{2.2}$$

which for $u \leqslant q^m/m$ yields the lower estimate in (i_1) as well as the inequality

$$\alpha(u,m) \gg q^m/(um^2)$$

when $u > q^m/m$. In order to prove the converse we observe that

$$um \leqslant \sum_1^m \frac{\pi(k)}{\alpha \log q} \ll \frac{q^m}{m\alpha}.$$

When $u > q^m/m$, the last estimate implies $\alpha(u,m) \ll 1/m$, therefore in this case the proposition (i_3) is proved too. If $u \leqslant q^m/m$, one side of the estimate (i_3) follows from the inequality

$$mu \geqslant W_m\left(q^{-\alpha}\right) \gg q^{(1-\alpha)m}/(1-\alpha).$$

At last, having in mind the just proved proposition (i_1) we deduce

$$\begin{aligned} mu &\leqslant \frac{q^\alpha}{q^\alpha - 1} W_{m/2}(q^{-\alpha}) + \frac{q^{m\alpha/2}}{q^{m\alpha/2} - 1} W_m(q^{-\alpha}) \\ &\ll \frac{q^{m(1-\alpha)}}{1-\alpha}\Big(\frac{q^{-m(1-\alpha)/2}}{\alpha} + 1\Big) \ll \frac{q^{m(1-\alpha)}}{1-\alpha}. \end{aligned}$$

Lemma 2 is proved.
We remind that the quantities

$$\sigma_j := \frac{1}{i^j} \frac{d^j}{dt^j} \log g_m(t)\Big|_{t=0} = \Phi_j(\alpha)$$

are called the semiinvariants of the joint random variables. The next lemma gives their estimates.

LEMMA 3. *For $j = 1, 2, 3, 4$ uniformly in $u \geqslant 2q$ we have*

$$0 < \sigma_j \asymp (um)^j \bar{u}^{1-j},$$

and

$$\max\left\{\left|(\log g_m(t))^{(4)}\right|:\ |t| \leqslant \pi\right\} \leqslant \sigma_4.$$

Proof. From the definition of σ_j we derive

$$\sigma_j = \sum_1^m \frac{k^j \pi(k)}{(q^{\alpha k} - 1)^j} Q_{j-1}(q^{k\alpha})$$

where $Q_0(z) = 1$, $Q_1(z) = z$, $Q_2(z) = z^2 + z$, $Q_3(z) = z^3 + 4z^2 + z$. When $u > q^m/m$, then in virtue of Lemma 2 for $1 \leqslant k \leqslant m$ and $1 \leqslant j \leqslant 4$ we have $Q_{j-1}(q^{k\alpha}) \asymp 1$ and $q^{\alpha k} - 1 \asymp \alpha k$. Hence

$$\sigma_j \asymp \alpha^{-j} \sum_1^m \pi(k) \asymp \alpha^{-j} q^m/m \asymp (um)^j \bar{u}^{1-j}.$$

If $2q \leqslant u \leqslant q^m/m$, then from the proposition (i_1) of Lemma 1 we obtain

$$Q_{j-1}(q^{\alpha k}) \asymp q^{(j-1)\alpha k}.$$

Now

$$\begin{aligned} \sigma_j &\asymp \sum_1^m \frac{k\pi(k)}{q^{\alpha k} - 1} \Big(\frac{kq^{\alpha k}}{q^{\alpha k} - 1}\Big)^{j-1} \\ &\leqslant \sum_1^m \frac{k\pi(k)}{q^{\alpha k} - 1} \Big(k + \frac{1}{\alpha \log q}\Big)^{j-1} \ll m^{j-1}(mu) = (mu)^j \bar{u}^{1-j}. \end{aligned}$$

For the lower estimation we observe that uniformly in $k, m/2 \leqslant k \leqslant m$,

$$Q_{j-1}(q^{\alpha k})(q^{\alpha k} - 1)^{1-j} \gg 1.$$

Hence using Lemma 2 we obtain

$$\sigma_j \gg m^{j-1} \sum_{m/2}^{m} \frac{k\pi(k)}{q^{\alpha k}}$$
$$\gg \frac{m^{j-1} q^{(1-\alpha)m}}{1-\alpha} \left(1 - q^{-(1-\alpha)m/2}\right) \gg m^j \bar{u} = (mu)^j \bar{u}^{1-j}.$$

The second proposition of Lemma 3 is easy to verify.
In order to integrate the functions in the formula (2.1) we shall deal with their asymptotic behaviour in different regions of the argument as $m \to \infty$.

LEMMA 4. *Uniformly in* t, $|t| \leqslant T := \bar{u}^{2/3}/(um)$,

$$g_m(t) = \exp\left\{it\sigma_1 - t^2\sigma_2/2\right\}\left(1 - it^3\sigma_3 + O\left(t^6\sigma_3^2 + t^4\sigma_4\right)\right).$$

Proof. By Lemma 3 we have $\sigma_3 T^3 + \sigma_4 T^4 \ll 1$, therefore the desired equality represents a simple asymptotic expansion of the principal branch of the logarithmic function.

LEMMA 5. *Uniformly in* t, $|t| \leqslant 1/m$,

$$|g_m(t)| \ll \exp\left\{ - c_2 q^m m^{-1} \log\left(1 + t^2\sigma_2 m q^{-m}\right)\right\}. \tag{2.3}$$

Proof. From the equality

$$\frac{|1-\rho e^{it}|}{1-\rho} = \left(1 + \frac{2\rho(1-\cos t)}{(1-\rho)^2}\right)^{1/2}, \qquad 0 < \rho < 1, \tag{2.4}$$

we have

$$|g_m(t)| = \exp\left\{ - \frac{1}{2} \sum_{1}^{m} \pi(k) \log\left(1 + \frac{2(1-\cos tk)}{q^{\alpha k}(1-q^{-\alpha k})^2}\right)\right\}. \tag{2.5}$$

For $u > q^m/m$, using the inequality

$$2\theta^2/\pi^2 \leqslant 1 - \cos\theta \leqslant \theta^2/2, \qquad |\theta| \leqslant \pi, \tag{2.6}$$

and the estimate $q^{\alpha m} \ll 1$, we derive

$$|g_m(t)| \leqslant \exp\left\{ - \frac{1}{2} \log\left(1 + c_3 t^2 \alpha^{-2}\right) \sum_{1}^{m} \pi(k)\right\}.$$

But Lemmas 2 and 3 yield

$$\alpha^{-2} \asymp u^2 m^4 q^{-2m} \asymp m\sigma_2 q^{-m},$$

therefore the proposition of Lemma 4 follows from the estimate (1.1) and the last formula.

Now let $2q \leqslant u \leqslant q^m/m$. From the inequalities (2.6) and

$$e^v(1-e^{-v})^2 \geqslant v^2, \qquad v \in \mathbf{R}, \tag{2.7}$$

for $|t| \leqslant 1/m$ we obtain

$$\frac{2(1-\cos tk)}{q^{\alpha k}(1-q^{-\alpha k})^2} \leqslant t^2\alpha^{-2} \leqslant c.$$

Hence in virtue of the inequality $\log(1+x)^{-1} \leqslant -x/(1+c)$ for $0 \leqslant x \leqslant c$, from (2.5) we deduce

$$|g_m(t)| \leqslant \exp\left\{-\frac{1}{1+c}\sum_1^m \pi(k)\frac{1-\cos tk}{q^{\alpha k}(1-q^{\alpha k})^2}\right\}.$$

By (2.6) the right-hand side of the above inequality does not exceed

$$\exp\left\{-\frac{2t^2}{\pi^2(1+c)}\sum_1^m \frac{k^2 q^{\alpha k}\pi(k)}{(q^{\alpha k}-1)^2}\right\} = \exp\{-c_4t^2\sigma_2\}.$$

Since $mt^2\sigma_2 q^{-m} \ll 1$ in the region considered, this estimate implies (2.3).

LEMMA 6. *Uniformly in* t, $1/m \leqslant |t| \leqslant \pi$, *and* $u \geqslant 1$,

$$|g_m(t)| \ll \exp\{-c_5\bar{u}/\log^2 2\bar{u}\}.$$

Proof. Observe that only large u can be considered. At first we use the inequality

$$\left(1+4vx(x-1)^{-2}\right)^{-1} \leqslant \exp\{-4v/x\}, \qquad 0 \leqslant v \leqslant 1 < x,$$

from Lemma 7 in (Hildebrand and Tenenbaum, 1986). Supposing that

$$v = \frac{1-\cos tk}{2} \quad \text{and} \quad x = q^{\alpha k},$$

from the formula (2.5) we obtain

$$\begin{aligned} |g_m(t)| &\leqslant \exp\left\{-\sum_1^m \pi(k)\,\frac{1-\cos tk}{q^{\alpha k}}\right\} \\ &\leqslant \exp\left\{-\frac{1}{m}\left(W_m(q^{-\alpha}) - \operatorname{Re} W_m(q^{-\alpha}e^{it})\right)\right\} \end{aligned} \tag{2.8}$$

where the function $W_m(w)$ has been defined in Lemma 1. At first we estimate the main term given in it. Let $\rho = q^{1-\alpha}$ and $\varkappa = 0$. Then using the estimates (2.4) and (2.6), we have

$$\begin{aligned} A_0(t) &:= \frac{\rho^{m+1}}{\rho-1}\left(1 - \operatorname{Re}\frac{e^{it(m+1)}}{(\rho e^{it}-1)/(\rho-1)}\right) \\ &\geqslant \frac{\rho^{m+1}}{\rho-1}\left(1-\left(1+\frac{2\rho(1-\cos t)}{(\rho-1)^2}\right)^{-1/2}\right) \geqslant \frac{\rho^{m+1}}{\rho-1}\left(1-\frac{\rho-1}{\sqrt{(\rho-1)^2+4\pi^{-2}t^2}}\right) \end{aligned}$$

Further, avoiding the irrationalities we obtain the estimate

$$A_0(t) \gg \frac{\rho^m}{\rho - 1} \frac{t^2}{(\rho-1)^2 + t^2}.$$

Hence using Lemmas 1 and 2 for sufficiently large m and $1/m \leqslant |t| \leqslant \pi$ we have

$$\begin{aligned} m^{-1}\big(W_m(q^{-\alpha}) - \operatorname{Re} W_m(q^{-\alpha}e^{it})\big) &\gg m^{-1}\big(A_0(t)/2 - |R(q^{-\alpha})|\big) \\ &\gg \frac{\rho^m t^2}{m(\rho-1)((\rho-1)^2+t^2)}\big(1 - C_1 q^{(\mu-1)m} m t^{-2}\big) \\ &\gg \frac{\bar{u}t^2}{(1-\alpha)^2 + t^2} \gg \frac{\bar{u}}{(1-\alpha)^2 m^2 + 1}. \end{aligned}$$

But the proposition (i_3) of Lemma 2 implies $(1-\alpha)m \ll \log 2\bar{u}$, and this yields the desired result in the case $\varkappa = 0$.

If $\varkappa = 1$, then by Lemma 1 the difference under the exponent in the formula (2.8) has the main term

$$\begin{aligned} &\rho^{m+1}\Big(\frac{\rho + 1 + (-1)^{m+1}(\rho - 1)}{\rho^2 - 1} \\ &\quad - \operatorname{Re} \frac{e^{it(m+1)}(\rho e^{it} + 1 + (-1)^{m+1}(\rho e^{it} - 1))}{\rho^2 e^{2it} - 1}\Big) \\ &\gg \frac{\rho^m}{\rho - 1}\Big(1 - \frac{\rho^2 - 1}{|\rho^2 e^{2it} - 1|}\Big). \end{aligned} \tag{2.9}$$

Now as in the case $\varkappa = 0$ from the formulae (2.4) and (2.8) we obtain the estimate in Lemma 6, but in the interval $1/m \leqslant |t| \leqslant \pi/2$ only.

Considering $\pi/2 \leqslant |t| \leqslant \pi - 1/m$, we can apply the reduction $t \mapsto \pi - t$ to obtain the estimate

$$|g_m(t)| \ll \exp\Big\{-c_6 \frac{\bar{u}(\pi - t)^2}{(1-\alpha)^2 + (\pi - t)^2}\Big\} \ll \exp\big\{-c_7 \bar{u} \log^{-2} 2\bar{u}\big\}.$$

When $\pi - 1/m \leqslant |t| \leqslant \pi$, directly from (2.8) we derive

$$\begin{aligned} |g_m(t)| &\leqslant \exp\Big\{-\sum_1^m \pi(k) \frac{1 - (-1)^k - (1 - \cos(\pi - t)k)}{q^{\alpha k}}\Big\} \\ &\ll \exp\Big\{-c_8 \frac{q^{(1-\alpha)m}}{m(1-\alpha)}\Big\}. \end{aligned}$$

In virtue of (i_3) of Lemma 2 the last estimate implies the desired result.

Proof of Theorem 5. Using the estimates obtained in Lemmas 4, 5, and 6 we shall integrate separate parts of the integral in the equality (2.1). Lemma 4 implies

$$\begin{aligned} I_1 &:= \frac{1}{2\pi}\int_{-T}^{T} e^{-\sigma_2 t^2/2}\big(1 - it^3\sigma_3 + O(t^6\sigma_3^2 + t^4\sigma_4)\big)\,dt \\ &= \frac{1}{\sqrt{2\pi\sigma_2}}\Big(1 + O\Big(\int_{T\sqrt{\sigma_2}}^{\infty} e^{-v^2/2}\,dv + \frac{\sigma_3^2}{\sigma_2^3} + \frac{\sigma_4}{\sigma_2^2}\Big)\Big) = \frac{1}{\sqrt{2\pi\sigma_2}}\Big(1 + O\Big(\frac{1}{\bar{u}}\Big)\Big). \end{aligned}$$

From Lemma 5 we have

$$I_2 := \int_T^{m^{-1}} |g_m(t)|\, dt \ll \frac{1}{\sqrt{\sigma_2}} \int_{T\sqrt{\sigma_2}}^{\infty} \exp\{-c_2 q^m m^{-1} \log(1 + t^2 m q^{-m})\}\, dt.$$

If $T_1 = \sqrt{2q^m m^{-1}}$, then by the inequality $\log(1+x) \geqslant x/3$ for $0 \leqslant x \leqslant 2$ we obtain

$$I_2 \leqslant \frac{1}{\sqrt{\sigma_2}}\left(\int_{T\sqrt{\sigma_2}}^{\infty} \exp\left\{-\frac{c_2}{3}t^2\right\} dt + \int_{T_1}^{\infty} \left(t^2 m q^{-m}\right)^{-c_2 q^m/m} dt\right) \ll \frac{1}{\sqrt{\sigma_2 \bar{u}}}.$$

When

$$(\log\log m)^3 \log m \leqslant u \leqslant \exp\{c_5 q^m / 2m^2 \log^2 q\}$$

and $m \geqslant m_0$, we shall use Lemmas 6 and 3 to obtain

$$I_3 := \int_{1/m}^{\pi} |g_m(t)|\, dt \ll \exp\{-c_5 \bar{u}/\log^2 2\bar{u}\} \frac{um}{\sqrt{\sigma_2 \bar{u}}} \ll \frac{1}{\bar{u}\sqrt{\sigma_2}}.$$

Returning to the formula (2.1) we have

$$\begin{aligned} P_m(n) &= q^{(\alpha-a)n} \prod_1^m \left(\frac{1-q^{-ak}}{1-q^{-\alpha k}}\right)^{\pi(k)} (I_1 + O(I_2 + I_3)) \\ &= \frac{q^{(\alpha-a)n}}{\sqrt{2\pi\sigma_2}} \prod_1^m \left(\frac{1-q^{-ak}}{1-q^{-\alpha k}}\right)^{\pi(k)} \left(1 + O\left(\frac{1}{\bar{u}}\right)\right). \end{aligned}$$

Proof of Theorem 1. Let $p(a)$ denote an arbitrary prime factor of the element $a \in \mathcal{G}$. The Euler identity

$$\sum_{n \geqslant 0} \Psi(n,m) y^n = \sum_{\substack{a \in \mathcal{G} \\ \delta(p(a)) \leqslant m}} y^{\delta(a)} = Z(y,m)$$

and the Cauchy formula imply

$$\Psi(n,m) = q^{\alpha n} Z(q^{-\alpha}, m) \left\{\frac{1}{2\pi} \int_{-\pi}^{\pi} e^{-itn} \frac{Z(q^{-\alpha}e^{it}, m)}{Z(q^{-\alpha}, m)}\, dt\right\}.$$

Since the quantity in the braces represents the probability $P_m(n)$ for the random variables X_{kj} when we take in advance $a = \alpha$, the proposition of Theorem 1 follows from Theorem 5.

3. ASYMPTOTIC ANALYSIS OF THE PARAMETERS

Preserving all the notations we shall investigate quantities involved in Theorems 1 and 5. The desired results have been formulated as Theorems 2, 3, and 4. At first we prove an auxiliary lemma.
Put for the brevity

$$B_m(\omega, \varkappa) = \sum_1^m q^{k(1-\omega)}\left(1 - (-1)^k \varkappa\right)$$

and

$$D_m(\omega, \varkappa) = \sum_1^m k q^{(1-\omega)k}\left(1 - (-1)^k \varkappa\right).$$

LEMMA 7. *Let $R_1 = R_1(\omega)$ be some remainder term such that $R_1 \ll 1/m$ for $0 \leqslant \omega \leqslant 1$ and $R_1 \ll_\varepsilon q^{-cm}$ for $\varepsilon \leqslant \omega \leqslant 1$. Then for $0 < \omega \leqslant 1$ and $m \geqslant 1$ we have*

$$\Phi_1(\omega) = \sum_1^m \frac{k\pi(k)}{q^{\omega k} - 1} = \frac{1 + R_1}{1 - q^{-m\omega}} B_m(\omega, \varkappa) + O(1) \tag{3.1}$$

and

$$\Phi_2(\omega) = \sum_1^m \frac{k^2\pi(k) q^{\omega k}}{(q^{\omega k} - 1)^2} = \frac{1 + R_1}{(1 - q^{-m\omega})^2} D_m(\omega, \varkappa) + O(1). \tag{3.2}$$

Proof. Observe that

$$R_2 := \sum_1^m \frac{k q^{\mu k}}{q^{\omega k} - 1} \ll m \max\left\{q^{(\mu-\omega)m}, 1\right\} \min\left\{m, |\mu - \omega|^{-1}\right\}$$

when $1/m \ll \omega \leqslant 1$, and

$$R_2 \ll \omega^{-1} q^{\mu m}$$

when $0 < \omega \ll 1/m$.
Then

$$\begin{aligned}\Phi_1(\omega) &= B_m(\omega, \varkappa) + \sum_1^m \frac{q^{k(1-2\omega)}}{1 - q^{-\omega k}}\left(1 - (-1)^k \varkappa\right) + O(R_2) \\ &= B_m(\omega, \varkappa) + O\big(\omega^{-1} \max\left(q^{(1-2\omega)m}, 1\right\} \min\left\{m, |1 - 2\omega|^{-1}\right\} + R_2\big).\end{aligned}$$

While

$$B_m(\omega, \varkappa) \asymp q^{m(1-\omega)} \min\left\{|1 - \omega|^{-1}, m\right\},$$

applying the above estimates separately in the intervals

$$[(2\log m)/(m \log q), \varepsilon], \quad [\varepsilon, 3/4], \quad \text{and} \quad [3/4, 1]$$

we obtain

$$\Phi_1(\omega) = B_m(\omega, \varkappa)(1 + R_1).$$

In the region considered $(1-q^{-\omega}m)^{-1} = 1+R_1$ therefore we have deduced the equivalent to the relation (3.1).
If $0<\omega<(2\log m)/(m\log q)$, we truncate the sums considered. For some $c_9>0$ we have

$$B_m(\omega,\varkappa) = \left(1+O(q^{-c_9 m})\right)\sum_{m/2}^{m} q^{k(1-\omega)}\left(1-(-1)^k\varkappa\right)$$

and

$$S_1 := \sum_{1}^{m}\frac{q^k}{q^{\omega k}-1}\left(1-(-1)^k\varkappa\right) = \left(1+O(q^{-c_9 m})\right)\sum_{m/2}^{m}\frac{q^k}{q^{\omega k}-1}\left(1-(-1)^k\varkappa\right).$$

Further,

$$\begin{aligned}\Delta &:= \left|\sum_{m/2}^{m} q^{k(1-\omega)}\left(1-(-1)^k\varkappa\right)\left(\frac{1}{1-q^{-k\omega}}-\frac{1}{1-q^{-m\omega}}\right)\right| \\ &\leqslant \frac{2\omega\log q}{(1-q^{-m\omega})(1-q^{-m\omega/2})}\sum_{m/2}^{m} q^{(1-2\omega)k}(m-k) \\ &\ll \frac{\omega q^{(1-2\omega)m}}{(1-q^{-m\omega})\min\{1,m\omega\}} \ll \frac{q^{(1-\omega)m}}{1-q^{-m\omega}}\,\frac{1}{m}.\end{aligned}$$

Hence

$$\Phi_1(\omega) = S_1 + O(R_2) = (1+R_1)\left(\frac{B_m(\omega,\varkappa)(1+R_1)}{1-q^{-m\omega}} + O(\Delta)\right) + O(R_2)$$

and the formula (3.1) follows from the above estimates.
In order to prove (3.2) we behave in a similar manner. At first we observe that

$$R_3 := \sum_{1}^{m}\frac{k^2 q^{(\mu+\omega)k}}{(q^{\omega k}-1)^2} \ll R_2\max\{m,\omega^{-1}\}.$$

The goal of the proof is the equality

$$S_2 := \sum_{1}^{m}\frac{kq^{(1+\omega)k}}{(q^{\omega k}-1)^2}\left(1-(-1)^k\varkappa\right) = \frac{1+R_1}{(1-q^{-m\omega})^2}D_m(\omega,\varkappa) + O(1).$$

It is easily to verify this equality on the intervals

$$[(3\log m)/(m\log q),\varepsilon],\quad [\varepsilon,1/4],\quad \text{and}\quad [1/4,1],$$

but for the remaining ω we need an application of the above truncation procedure. As earlier we have

$$D_m(\omega,\varkappa) = (1+R_1)\sum_{m/2}^{m} kq^{(1-\omega)k}\left(1-(-1)^k\varkappa\right) =: (1+R_1)\widetilde{D},$$

$$S_2 = (1+R_1)\sum_{m/2}^{m} \frac{kq^{(1+\omega)k}}{(q^{\omega k}-1)^2}\left(1-(-1)^k\varkappa\right) =: (1+R_1)\widetilde{S},$$

and the estimate

$$\widetilde{S} - \frac{\widetilde{D}}{(1-q^{-\omega m})^2} \ll \frac{mq^{m(1-\omega)}}{(1-q^{-\omega m})^2}\,\frac{1}{m},$$

provided that $0<\omega<(3\log m)/(m\log q)$. The last three estimates lead to the formula (3.2).

LEMMA 8. *Let $\xi(x)$ be the nonzero solution of the equation*

$$e^{\xi(x)} = 1 + x\xi(x).$$

Then for $x \geqslant 2$

$$\xi(x) = \log x + O(\log\log(x+2))$$

and

$$\frac{d}{dx}\,\xi(x) = \frac{1}{x}\Big(1+O\Big(\frac{1}{\log x}\Big)\Big).$$

Proof see in (Hildebrand, 1986b).

Proof of Theorem 2. Let, as previously, $u = n/m \geqslant 2q$ and $\alpha(x) := \alpha(x,m)$ be the solution of the equation $\Phi_1(\alpha(x)) = mx$. Observe that $\alpha(x)$ is strictly decreasing continuous function defined on $(0,\infty)$. Moreover, by Lemma 2 we have

$$\alpha'(x) = -m(\Phi_2(\alpha(x))^{-1}\log q \asymp \frac{\min\{mx, q^m\}}{m^2x^2}. \tag{3.3}$$

In the nearest future we shall use asymptotic formulas for the pre-images of some numbers in the interval $(0,1]$, when the function $\alpha(x)$ is considered. Let

$$q^m \leqslant c_{10}(mu)^2 \tag{3.4}$$

and v be defined by

$$v = \alpha^{-1}\Big(\frac{1}{m}\log_q\Big(1+\frac{q^{m+1}}{(q-1)mu}\Big)\Big), \tag{3.5}$$

provided that c_{10} has been chosen so that the argument of α^{-1} in (3.5) does not exceed $1/2$. Observe that

$$\alpha(v)m \gg \bar{u}/u. \tag{3.6}$$

From Lemma 7 we have the main formula:

$$\begin{aligned} mx &= \Phi_1(\alpha(x)) \\ &= \frac{(1+R_1)q^{1-\alpha(x)}}{1-q^{-m\alpha(x)}}\Big(\frac{q^{m(1-\alpha(x))}-1}{q^{1-\alpha(x)}-1} - \varkappa\,\frac{(-1)^m q^{m(1-\alpha(x))}-1}{q^{1-\alpha(x)}+1}\Big) \\ &\quad + O(1). \end{aligned} \tag{3.7}$$

Hence for $\varkappa = 0$ and $x = v$ we obtain

$$mv = \frac{(q-1)q^{-\alpha(v)}}{q^{1-\alpha(v)}-1} \frac{q^{m+1}}{(q-1)(q^{m\alpha(v)}-1)}(1+R_1) + O(1).$$

By the definition of v we observe that the second fraction in the right-hand side is equal to mu. Hence

$$\begin{aligned} |u-v| &\ll u\Big|1 - \frac{(q-1)q^{-\alpha(v)}}{q^{1-\alpha(v)}-1}(1+R_1)\Big| + O\Big(\frac{1}{m}\Big) \\ &\ll u\alpha(v)(1+R_1) + O\Big(\frac{1}{m}\Big) = u\Big(\alpha(v) + O\Big(\frac{1}{m}\Big)\Big). \end{aligned} \tag{3.8}$$

The last estimate implies that $u \asymp v$. Further, using (3.3) and (3.6), from the mean-value theorem applied in one of the intervals $[u,v]$ or $[v,u]$ we deduce the estimate

$$\alpha(u) - \alpha(v) \ll \Big(\alpha(v) + O\Big(\frac{1}{m}\Big)\Big)\frac{\bar{u}}{mu} \ll \frac{\alpha(v)}{m}. \tag{3.9}$$

This is the desired result when the condition (3.4) is satisfied.
In the case $\varkappa = 1$ from (3.4) we obtain

$$mx = \frac{2q^{2\overline{m}+1}}{(q^{m\alpha(x)}-1)(q^2-1)} \frac{q^{(m-2\overline{m}-1)\alpha(x)}(q^2-1)}{q^{2(1-\alpha(x))}-1}(1+R_1) + O(1) \tag{3.10}$$

where $\overline{m} = [(m+1)/2]$. Now let v be defined by

$$v = \alpha^{-1}\Big(\frac{1}{m}\log_q\Big(1 + \frac{2q^{2\overline{m}+1}}{(q^2-1)um}\Big)\Big),$$

provided that the condition (3.3) with sufficiently small c_{10} is satisfied. As earlier, from (3.10) we deduce the same estimates (3.8) and (3.9).
Now let $q^m > c_{10}(mu)^2$. For $\varkappa = 0$ we apply (3.7) with $x = w$ where w is defined by the equality

$$\alpha(w) = 1 - \frac{\xi(u)}{m\log q}$$

and $\xi(u)$ has been described in Lemma 8. According to it in the range considered we have $1/4 \leqslant \alpha(w) \leqslant 1$ when $m \geqslant m_0$. Now in virtue of the definitions from (3.7) we obtain

$$w = u\frac{\xi(u)/m}{1-e^{-\xi(u)/m}}(1+R_1) + O\Big(\frac{1}{m}\Big) = u\Big(1 + O\Big(\frac{\log u}{m}\Big)\Big) + O\Big(\frac{1}{m}\Big)$$

and

$$w - u \ll \frac{u\log u}{m}.$$

Hence again appealing to (3.3) we have

$$\begin{aligned} \alpha(u) &= \alpha(w) + O\Big(\frac{\log u}{m^2}\Big) = \Big(1 - \frac{\log_q u + O(\log\log(u+2))}{m}\Big)\Big(1 + O\Big(\frac{1}{m}\Big)\Big) \\ &= \frac{1}{m}\Big(\log_q \frac{q^m}{u}\Big)\Big(1 + O\Big(\frac{\log m}{m}\Big)\Big). \end{aligned}$$

For $q^m \geqslant c_{10}(mu)^2$ the last expression is equivalent to that in Theorem 2.
If $\varkappa = 1$, in the last range for u we start from the formula (3.7) to obtain

$$mx = \frac{2q^{\alpha(x)-1}(q^{2\overline{m}(1-\alpha(x))}-1)}{1-q^{2(\alpha(x)-1)}}\,\frac{1+R_1}{1-q^{-m\alpha(x)}} + O(1).$$

We put in it $x = w$, where w is defined by

$$\alpha(w) = 1 - \frac{\xi(u)}{2\overline{m}\log q},$$

and deduce

$$\begin{aligned} w &= \frac{2u\xi(u)}{m}\,\frac{e^{-\xi(u)/2\overline{m}}}{1-e^{-\xi(u)/\overline{m}}}(1+R_1) + O\Big(\frac{1}{m}\Big) \\ &= u\frac{2\overline{m}}{m}\Big(1+O\Big(\frac{\xi(u)}{m}\Big)\Big) + O\Big(\frac{1}{m}\Big) = u + O\Big(\frac{u\log u}{m}\Big). \end{aligned}$$

By (3.3) this leads to the asymptotic formula

$$\begin{aligned} \alpha(u) &= \alpha(w) + O\Big(\frac{\log u}{m^2}\Big) = \Big(1-\frac{\xi(u)}{2\overline{m}\log q}\Big)\Big(1+O\Big(\frac{1}{m}\Big)\Big) \\ &= \frac{1}{m}\Big(\log_q \frac{q^{2\overline{m}}}{u}\Big)\Big(1+O\Big(\frac{\log m}{m}\Big)\Big), \end{aligned} \tag{3.11}$$

equivalent in the region considered to that given in Theorem 2.
The investigation of $\Phi_2(\alpha)$ is based upon the expression (3.2). Summation by parts yields

$$D_m(\omega,\varkappa) = B_m(\omega,\varkappa)\Big(m + O\Big(\frac{1}{1-\omega}\Big)\Big)$$

and

$$\begin{aligned} \Phi_2(\omega) &= \frac{m}{(1-q^{-m\omega})^2}B_m(\omega,\varkappa)\Big(1+O\Big(\frac{1}{m(1-\omega)}\Big)\Big)(1+R_1) + O(1) \\ &= \frac{m}{1-q^{-m\omega}}(\Phi_1(\omega)(1+R_1)+O(1))\Big(1+O\Big(\frac{1}{m(1-\omega)}\Big)\Big)(1+R_1) + O(1). \end{aligned}$$

From the first part of the proof we have

$$\frac{1}{m(1-\alpha)} \ll \frac{1}{m} + \frac{1}{\log u},$$

therefore

$$\Phi_2(\alpha) = \frac{nm}{1-q^{-m\alpha}}(1+O(R_3)) \quad \text{with} \quad R_3 := \frac{1}{m} + \frac{1}{\log u}. \tag{3.12}$$

When $q^m > m^2u$, in both cases $\varkappa = 0$ and $\varkappa = 1$ we have $\alpha \gg m^{-1}\log m$ which implies the asymptotic formula

$$\Phi_2(\alpha) = nm(1+O(R_3)),$$

equivalent to that given in Theorem 2.
If $q^m \leqslant m^2 u$, according to (3.12) it remains to prove that

$$\left(1-q^{-m\alpha}\right)^{-1} = \left(1+\frac{(q-1)mu}{q^{m+1}}\right)\left(1+O\left(\frac{1}{m}\right)\right) \tag{3.13}$$

and

$$\left(1-q^{-m\alpha}\right)^{-1} = \left(1+\frac{(q^2-1)mu}{2}q^{-2\overline{m}-1}\right)\left(1+O\left(\frac{1}{m}\right)\right) \tag{3.14}$$

in the cases $\varkappa = 0$ and $\varkappa = 1$, respectively. In the first of them, from the estimates (3.9) and (3.5) we obtain

$$1-q^{-m\alpha} = \left(1+\frac{(q-1)mu}{q^{m+1}}\right)^{-1} + O\left(m^{-1}\left(\frac{q^{m+1}}{(q-1)mu}\right)^{-1}\log\left(1+\frac{q^{m+1}}{(q-1)mu}\right)\right)$$

which implies (3.13). Similarly, the equalities (3.9) and (3.11) yield the formula (3.14).
Theorem 2 is proved.

Proof of Theorem 3. In the range considered, $1 < \eta \leqslant q^{1/2}$; therefore there exists z such that

$$q^{1-\alpha(z)} = \eta.$$

Now in both cases $\varkappa = 0$ and $\varkappa = 1$, the formula (3.7) with $x = z$ implies the estimate $z = u + O(1/m)$, more exact than that obtained for w in the proof of Theorem 2. In the same way as earlier, applying (3.3) we have

$$\alpha(u) - \alpha(z) \ll \frac{1}{m^2 u}.$$

Hence

$$q^{\eta\alpha} = \left(\frac{q}{\eta}\right)^{n}\left(1+O\left(\frac{1}{m}\right)\right) \tag{3.15}$$

and

$$q^{1-\alpha} = \eta\left(1+O\left(\frac{1}{m^2 u}\right)\right).$$

From Lemma 7 we have the expression

$$Z(q^{-\alpha},m) = Z(q^{-1},m)\exp\left\{(\log q)\int_{\alpha}^{1} B_m(w,\varkappa)\,dw\,(1+R_1) + O(1-\alpha)\right\}$$

which in virtue of $1-\alpha \ll m^{-1}\log u$ and the above estimate implies

$$Z(q^{-\alpha},m) = Z(q^{-1},m)\exp\left\{\int_{1}^{\eta}\frac{t^m-1}{t-1}\,dt + O\left(\frac{\log u}{m}\right)\right\} \tag{3.16}$$

and

$$Z(q^{-\alpha}, m) = Z(q^{-1}, m) \exp\left\{2\int_1^{\eta} \frac{t^{2\overline{m}} - 1}{t^2 - 1}\, dt + O\Big(\frac{\log u}{m}\Big)\right\} \tag{3.17}$$

for $\varkappa = 0$ and 1, respectively. In the routine way from the formula (1.1) we obtain

$$Z(q^{-1}, m) = m \exp\left\{\Gamma + O\Big(\frac{1}{m}\Big)\right\}. \tag{3.18}$$

It remains only to obtain the estimate for $\Phi_2(\alpha)$ sharper than in Theorem 2. From Lemma 7 we have

$$\Phi_2(\alpha) = (1 + R_1)\frac{d}{dx}\left\{\sum_1^m \eta(x)^k (1 - (-1)^k \varkappa)\right\}\Bigg|_{x=u} \frac{\eta}{\eta'} + O(1)$$
$$= m\frac{\eta}{\eta'}(1 + R_1) + O(1) = m\frac{\eta}{\eta'}\Big(1 + O\Big(\frac{1}{m^2 u}\Big)\Big).$$

According to Theorem 1 the estimates (3.15), (3.16) or (3.17), (3.18), and the last one imply the desired result.

Proof of Theorem 4. We shall use the formulas given in Theorem 3 and approximate η and η'/η by suitable values of the function ξ and its derivative. When

$$(\log\log m)^3 \log m \leqslant u \leqslant m^{1-\varepsilon},$$

in virtue of the above estimates, $\eta \sim q^{1-\alpha} \sim \exp\{\xi/m\}$ as $m \to \infty$. So, we can define y such that $m \log \eta = \xi(y)$ when $\varkappa = 0$. Then by the definition of η and Lemma 8

$$u = \frac{e^{\xi(y)} - 1}{\xi(y)} \frac{\xi(y)/m}{1 - \exp\{-\xi(y)/m\}} = y\Big(1 + O\Big(\frac{\log y}{m}\Big)\Big)$$

and

$$\xi(u) - \xi(y) \ll \frac{\log u}{m}. \tag{3.19}$$

Moreover, since for $\varkappa = 0$

$$\frac{\eta'}{\eta}(x)\Big(\frac{m\eta(x)^m}{\eta(x)^m - 1} - \frac{1}{\eta(x) - 1}\Big) = \frac{1}{x},$$

taking $x = u$ and appealing to Lemma 8 we obtain the estimate

$$\frac{\eta'}{\eta} = \frac{y\xi(y)}{mu(y\xi(y) - y + 1)}\Big(1 + O\Big(\frac{1}{m}\Big)\Big)$$

which by (3.19) and $u\xi - u + 1 = \xi/\xi'$ reduces to

$$\frac{\eta'}{\eta} = \frac{\xi'}{m}\Big(1 + O\Big(\frac{\log u}{m}\Big)\Big). \tag{3.20}$$

In the case $\varkappa = 1$, starting from the equalities $2\overline{m}\log\eta = \xi(y)$,

$$mu = 2y\frac{\xi(y)\exp\{\xi(y)/2\overline{m}\}}{\exp\{\xi(y)/\overline{m}\} - 1},$$

and

$$\frac{\eta'}{\eta}\Big(\frac{2\overline{m}\eta^{2\overline{m}}}{\eta^{2\overline{m}} - 1} - \frac{\eta^2 + 1}{\eta^2 - 1}\Big) = \frac{1}{u}.$$

in the same manner we come to (3.19) and (3.20).
In both cases $\varkappa = 0$ and $\varkappa = 1$, from (3.19) and the definition of y we have

$$m\log\eta = \xi\Big(1 + O\Big(\frac{1}{m}\Big)\Big),$$

and hence $\eta^{-n} = \exp\{-u\xi(1 + O(m^{-1}))\}$, as well as the desired expressions with the accuracy $O(u/m)$ for both integrals in Theorem 3. By Lemma 8 the remainder term $O(u\xi m^{-1})$ dominates, and we have the first of the equalities in Theorem 4. Let γ be the classical Euler constant. In the routine way one can verify that $A = \exp\{\Gamma - \gamma\}$, therefore the second assertion of Theorem 4 is the corollary of our result and the formula (2.7) in (Hildebrand and Tenenbaum, 1986).

REFERENCES

Hensley, D. (1985). The number of positive integers $\leqslant x$ and free of prime divisors $> y$. *J. Number Theory* **21**, 286–298.

Hensley, D. (1986). A property of the counting function of integers with no large prime factors. *J. Number Theory* **22**, 46–74.

Hildebrand, A. (1986a). On the number of positive integers $\leqslant x$ and free of prime factors $> y$. *J. Number Theory* **22**. 298–307.

Hildebrand, A. (1986b). On the local behaviour of $\Psi(x,y)$. *Trans. Amer. Math. Soc.* **297**, 729–751.

Hildebrand, A. and Tenenbaum, G. (1986). On integers free of large prime factors. *Trans. Amer. Math. Soc.* **296**, 265–290.

Indlekofer, K.-H., Manstavičius, E. and Warlimont, R. (1989). Remarks on the prime number theorem for algebraic function fields. *Preprint* (to appear in *Analysis*).

Indlekofer, K.-H., Manstavičius, E. and Warlimont, R. (1991). On a certain class of infinite products with an application to arithmetical semigroups. *Arch. Math.* **56**, 446–453.

Knopfmacher, J. (1975). *Abstract Analytic Number Theory.* North-Holand, Amsterdam-Oxford.

Knopfmacher, J. (1979). *Analytic Arithmetic of Algebraic Function Fields.* Marcel Dekker, New York.

Milaševičius, J. (1980). On the convergence of the Grigelionis–Franken asymptotic expansions. In: *Application of Probability Theory and Mathematical Statistics*, **3.** IMC Vilnius, pp. 87–110 (in Russian).

New Trends in Probab. and Statist., pp. 155–163
F. Schweiger and E. Manstavičius (Eds.)
1992 VSP/TEV

ON MULTIPLICATIVE FUNCTIONS WITH REGULARITY PROPERTIES

K. NAGASAKA*, P. JAU-SHYONG SHIUE and X. YU
College of Engineering, Hosei University,
3–7–2 Kajino-Cho, Koganei-Shi, 184 Tokyo, Japan

Department of Mathematical Sciences, University of Nevada,
4505 Maryland Park, Las Vegas, Nevada, 89154–4020 USA

Department of Mathematics, Hangzhou Teacher's College,
96 Wen Yi Road, Hangzhou, Zheijiang, P. R. China

ABSTRACT

The problem to characterize the function n^s in the class of multiplicative arithmetic functions is considered and a new result is obtained.

1. INTRODUCTION

Let $f(n)$ be a complex valued function defined on the set of positive integers $\mathbf{N}$. Then f is called an arithmetical function. If an arithmetical function f satisfies the equation

$$f(mn) = f(m)f(n) \tag{1.1}$$

for each pair of coprime positive integers m, n, then we say that f is multiplicative and denote this by $f \in \mathcal{M}$. If the equality

$$f(mn) = f(m) + f(n) \tag{1.2}$$

holds for every pair of coprime positive integers m, n, then we say that f is an additive function and denote this by $f \in \mathcal{A}$. If the equality holds for any pair of positive integers m, n, then we say that f is completely multiplicative and denote this by $f \in \mathcal{M}^*$. Similarly, f is called completely additive denoting this by $f \in \mathcal{A}^*$, if the equality (1.2) holds for any pair $m, n \in \mathbf{N}$.
It is obvious that $c \log n$ is completely additive and n^s is completely multiplicative, with some constants c and s, respectively, however the converse is not, in general, true. Then the problem to characterize $\log n$ (or n^s) as an additive arithmetical function (multiplicative function) occurs. The first answer was given by P. Erdös in (Erdös, 1946) which was, indeed, a starting point of the characterization problem.

*K. Nagasaka was partially supported by Grant-in-Aid for General Scientific Research, the Ministry of Education, Science and Culture, Japan; No. 3640233.

THEOREM A (P. Erdös). *If f is additive and satisfies one of the following two conditions:*

(i)
$$f(n+1)-f(n)\geqslant 0 \qquad \text{for all} \quad n\in\mathbf{N},$$

(ii)
$$\lim_{n\to\infty}(f(n+1)-f(n))=0,$$

then there exists some constant c such that $f(n)=c\log n$ for $n\in\mathbf{N}$.

P. Erdös posed several conjectures on the characterization of $\log n$. Later one of his conjectures was solved by I. Kátai as follows.

THEOREM B (Kátai, 1970). *If $f\in\mathcal{A}$ and*

$$\sum_{n\leqslant x}|f(n+1)-f(n)|\longrightarrow 0, \qquad x\to\infty,$$

then f is a constant multiple of $\log n$.

There exist many works on the characterization problem of logarithmic function, however, the characterization of power function is seldomly found in the literature. Among them, I. Kátai proved the following theorem in his series of papers in 1984.

THEOREM C (Kátai, 1984). *If $f\in\mathcal{M}$ and*

$$\sum_{n=1}^{\infty}\frac{1}{n}|f(n+1)-f(n)|<\infty, \tag{1.3}$$

then either

$$\sum_{n=1}^{\infty}\frac{|f(n)|}{n}<\infty,$$

or there exists a complex constant $s=\sigma+i\tau$, with $0<\sigma<1$, such that $f(n)=n^s$.

J.-L. Mauclaire and L. Murata got another type of characterization of power functions.

THEOREM D (Mauclaire and Murata, 1984). *The function $f(n)$ is assumed to be multiplicative. Suppose that there exits a positive nondecreasing function $g(x)$ satisfying*

(i)
$$\lim_{x\to\infty}g(x)=\infty \quad \text{and} \quad g(2x)=O(g(x)),$$

(ii)
$$\lim_{n\to\infty}\frac{1}{g(x)}\sum_{n\leqslant x}|f(n+1)-f(n)|=0,$$

(iii)
$$\limsup_{x\to\infty}\frac{1}{g(x)}\left|\sum_{n\leqslant x}f(n)\right|>0.$$

Then $f(n)$ is completely multiplicative, and there exists a constant $\sigma \geqslant -1$ such that $|f(n)| = n^{\sigma}$.

In this note, we discuss the case

$$\sum_{n=1}^{\infty} \frac{|f(n)|}{n} < \infty.$$

Instead of the series (1.3) of Theorem C, we consider the series

$$\sum_{n \geqslant x} \frac{|f(n+1) - f(n)|}{n}.$$

In the next section, we shall show a characterization of n^c as a multiplicative function and remark the relation between our theorem and theorems just stated above.

2. CHARACTERIZATION OF n^c BY REGULARITY PROPERTIES

Our principal aim is to prove the following

THEOREM 1. *Assume that f is multiplicative. We suppose also that there exist a non-negative constant δ and a positive integer $q \in \mathbf{N}$ satisfying*

(i) $$\sum_{n \geqslant x} \frac{|f(n+1) - f(n)|^q}{n^{\delta}} = O(\lambda(x)), \qquad \lambda(x) = O(1),$$

(ii) $$(f(n)) = O\big(n^{\delta}\lambda(n)\big),$$

where $\lambda(x)$ is defined by

(iii) $$\lambda(x) = \sum_{n \geqslant x} \frac{|f(n)|^q}{n^{\delta}},$$

and satisfy $\lambda(x) = O(\lambda(2x))$.
Then $f(n) = n^s t(n)$ where s is a constant and $t(n)$ is completely multiplicative with $|t(n)| = 1$.

As usual in the characterization problem, the implication from multiplicity with some additional properties to complete multiplicity is indispensable. Thus, we prove first Theorem 2 below, then show Theorem 1 as a direct application of Theorem 2.

THEOREM 2. *Suppose that $f(n)$ is multiplicative and $\alpha(n)$ is positive completely multiplicative function. Suppose also that there exist constants C_1 and C_2 such that*

$$C_1\alpha(an + b) \leqslant \alpha(n) \leqslant C_2(an + b) \tag{2.1}$$

holds for fixed $a, b \in \mathbf{N}$ and any $n \in \mathbf{N}$. Then $f(n)$ is completely multiplicative if following conditions are satisfied:

(i)
$$\sum_{n\geqslant N} \alpha(n)|f(n+1)-f(n)|^q = O(\lambda(N)),$$

(ii)
$$\sum_{n\geqslant N} \alpha(n)|f(n)|^q \neq O(\lambda(N)), \qquad \alpha(n)(f(n))^q = O(\lambda(n)),$$

(iii)
$$\lambda(n) = O(1), \qquad |\lambda(n)| \leqslant C_3|\lambda(2n)|,$$

(iv)
$$\sum_{n\geqslant N} |\alpha(n+1)-\alpha(n)||f(n)|^q = O(\lambda(N)),$$

where q is a positive integer and C_3 is a non-negative constant.

Proof. In order to show the complete multiplicativity of $f(n)$, we ought to prove that, for each prime p and $r \in \mathbf{N}$,

$$f(p)f(p^r) = f(p^{r+1}). \tag{2.2}$$

Since $f \in \mathcal{M}$, we have that, for any $m \in \mathbf{N}$ with $p \nmid m$,

$$\begin{aligned} &f(p(p^r m+1)) - f(p^{r+1}m) - f(p)\{f(p^r m+1) - f(p^r m)\} \\ &\quad = \{f(p)f(p^r) - f(p^{r+1})\}f(m). \end{aligned}$$

From the Hölder inequality we get:

$$\begin{aligned} &|f(p(p^r m+1)) - f(p^{r+1}m)|^q + |f(p)|^q|f(p^r m+1) - f(p^r m)|^q \\ &\quad \geqslant C_4|f(p)f(p^r) - f(p^{r+1})|^q \cdot |f(m)|^q. \end{aligned}$$

By multiplying this inequality by $\alpha(m)$ and summing it up for $m \geqslant x$ with $p \nmid m$, we have:

$$\begin{aligned} &\sum_{m\geqslant x, p\nmid m} \alpha(m)|f(p(p^r m+1)) - f(p^{r+1}m)|^q \\ &\quad + |f(p)^q| \sum_{m\geqslant x, p\nmid m} \alpha(m)|f(p^r m+1) - f(p^r m)|^q \\ &\quad \geqslant C_4|f(p)f(p^r) - f(p^{p+1})|^q \sum_{m\geqslant x, p\nmid m} |f(m)|^q. \end{aligned} \tag{2.3}$$

Here C_4, C_5, C_6 (and C_7 which will appear just below) denote constants depending only on C_1, C_2, C_3, p, r, and q. From (2.1) and (i) of Theorem 2 we obtain, again using the Hölder inequality:

$$\sum_{m\geqslant x,p\nmid m} \alpha(m)\left|f(p(p^r m+1)) - f(p^{p+1}m)\right|^q$$

$$= \sum_{m\geqslant x,p\nmid m} \alpha(m)\left|\sum_{i=0}^{p-1} f(p^{r+1}m+i+1) - f(p^{p+1}m+i)\right|^q$$

$$\leqslant C_5 \sum_{m\geqslant x,p\nmid m} \alpha(m)\sum_{i=1}^{p-1}\left|f(p^{r+1}m+i+1) - f(p^{r+1}m+i)\right|^q \tag{2.4}$$

$$\leqslant C_6 \sum_{i=0}^{p-1}\sum_{m\geqslant x,p\nmid m} \alpha(p^{r+1}m+i)\left|f(p^{r+1}m+i+1) - f(p^{r+1}m+i)\right|^q$$

$$= O(\lambda(x)).$$

We obtain also

$$\sum_{m\geqslant x,p\nmid m} \alpha(m)\left|f(p^r m+1) - f(p^r m)\right|^q$$

$$\leqslant C_7 \sum_{m\geqslant x,p\nmid m} \alpha(p^r m)\left|f(p^r m+1) - f(p^r m)\right|^q = O(\lambda(x)). \tag{2.5}$$

Combining (2.3), (2.4), and (2.5), we get

$$\left|f(p)f(p^r) - f(p^{r+1})\right|^q \sum_{m\geqslant x,p\nmid m} \alpha(m)|f(m)|^q = O(\lambda(x)). \tag{2.6}$$

Now, for any $x \in \mathbf{N}$, we have

$$\sum_{m\geqslant px,p\nmid m} \alpha(m)|f(m)|^q = \sum_{m\geqslant px} \alpha(m)|f(m)|^q - p\sum_{m\geqslant px,p|m} \alpha(m)|f(m)|^q$$

$$= \sum_{k\geqslant x}\sum_{i=0}^{p-1}\left\{\alpha(pk+i)|f(pk+i)|^q - \alpha(pk)|f(pk)|^q\right\}$$

$$= \sum_{k\geqslant x}\sum_{i=0}^{p-1}\sum_{j=0}^{i-1}\left\{\alpha(pk+j+1)|f(pk+j+1)|^q - \alpha(pk+j)|f(pk+j)|^q\right\}$$

$$= \sum_{k\geqslant x}\sum_{i=0}^{p-1}\sum_{j=0}^{i-1}\alpha(pk+j+1)\left\{|f(pk+j+1)|^q - |f(pk+j)|^q\right\}$$

$$+ \sum_{k\geqslant x}\sum_{i=0}^{p-1}\sum_{j=0}^{i-1}\left\{\alpha(pk+j+1) - \alpha(pk+j)\right\}|f(pk+j)|^q$$

$$\leqslant C_8 \sum_{k\geqslant x}\sum_{i=0}^{p-1}\sum_{j=0}^{i-1}\alpha(pk+j+1)|f(pk+j+1) - f(pk+j)|^q$$

$$+ \sum_{k\geqslant x}\sum_{i=0}^{p-1}\sum_{j=0}^{i-1}|\alpha(pk+j+1) - \alpha(pk+j)||f(pk+j)|^q.$$

The last two sums are $O(\lambda(x))$ by (i) and (iv), so we get

$$\sum_{m\geqslant px} \alpha(m)|f(m)|^q = p \sum_{m\geqslant px, p|m} \alpha(m)|f(m)|^q + O(\lambda x).$$

Combining this estimate and the assumptions (ii) and (iii) of Theorem 2, we see that

$$\sum_{m\geqslant x} \alpha(m)|f(m)|^q = p \sum_{m\geqslant x, p|m} \alpha(m)|f(m)|^q + O(\lambda(x)) \tag{2.7}$$

holds for any x large enough. Combining (ii), (2.6), and (2.7) we get (2.2). Thus Theorem 2 is proved.

Proof of Theorem 1. From the assumption (iii) of Theorem 1 we have

$$\begin{aligned}\sum_{m\geqslant x} \left(\frac{1}{n^\delta} - \frac{1}{(n+1)^\delta}\right)|f(n)|^q &= O(1) \sum_{n\geqslant x} \frac{|f(n)|^q}{n^{\delta+1}} \\ &= O(1) \lim_{N\to\infty} \sum_{n=x}^{N} \frac{1}{n}(\lambda(n) - \lambda(n+1)) \\ &= O(1) \lim_{N\to\infty} \left\{\frac{\lambda(x)}{x} + O(1) \sum_{n=x+1}^{N} \frac{\lambda(n)}{n^2} - \frac{\lambda(N+1)}{N}\right\} \\ &= O(1)\left\{\frac{\lambda(x)}{x} + O(1)\frac{\lambda(x)}{x}\right\} = O(\lambda(x)).\end{aligned}$$

Setting $\alpha(n) = n^{-\delta}$, the condition (i) of Theorem 1 holds. All other conditions (i), (ii), (iii) of Theorem 1 and (2.1) hold for $\alpha(n) = n^{-\delta}$. Now applying Theorem 1, $f(n)$ is shown to be completely multiplicative. Thus we have

$$\sum_{n\geqslant x, d|n} \frac{|f(n)|^q}{n^\delta} = \frac{1}{d} \sum_{n\geqslant x} \frac{|f(n)|^q}{n^\delta} + O(\lambda(x)), \tag{2.8}$$

for any $n \in \mathbf{N}$ sufficiently large. Since $f \in \mathcal{M}^*$, we get

$$\sum_{n\geqslant x, d|n} \frac{f(n)|^q}{n^\delta} = \sum_{n\geqslant x/d} \frac{|f(dn)|^q}{(dn)^\delta} = \frac{|f(d)|^q}{d^\delta} \sum_{n\geqslant x/d} \frac{|f(n)|^q}{n^\delta} = \frac{|f(d)|^q}{d^\delta} \lambda\left(\frac{x}{d}\right). \tag{2.9}$$

Combining (2.9) with (2.8), we get

$$\frac{|f(d)|^q}{d^{\delta-1}} = \lim_{x\to\infty} \frac{\lambda(x)}{\lambda(x/d)} = \lim_{x\to\infty} \frac{\lambda(dx)}{\lambda(x)}.$$

From the definition of $\lambda(x)$ in (iii) of Theorem 1, it is obvious that $\lambda(d_1x) \leqslant \lambda(d_2x)$ for $d_1 \geqslant d_2$. Then the multiplicative function $f(d)/d^{\delta-1}$ is decreasing. The additive function $\log(|f(d)|/d^{\delta-1})$ is also decreasing. Applying Theorem A, we get $|f(n)| = n^s$ with some constant s. Therefore $f(n) = n^s t(n)$ with $|t(n)| = 1$. Here $t(n) \in \mathcal{M}^*$ because $f(n) \in \mathcal{M}^*$ and $n^s \in \mathcal{M}^*$. This proves Theorem 1.

COROLLARY. *Suppose that $q = 1$ in Theorem 1. Then the condition* (i) *can be deleted.*

Proof. For any $x, N \in \mathbf{N}$ with $N > x$, we have

$$\begin{aligned}\frac{f(x)}{x^\delta} &= \frac{f(N+1)}{(N+1)^\delta} + \sum_{n=x}^{N}\Big(\frac{f(n)}{n^\delta} - \frac{f(n+1)}{(n+1)^\delta}\Big)\\ &= \frac{f(N+1)}{(N+1)^\delta} + \sum_{n=x}^{N}\frac{f(n)-f(n+1)}{n^\delta} + \sum_{n=x}^{N} f(n+1)\Big(\frac{1}{n^\delta} - \frac{1}{(n+1)^\delta}\Big).\end{aligned}$$

From (iii) of Theorem 1 we know that $N^\delta f(N) = o(1)$ as $N \to \infty$, so the above equation gives

$$\begin{aligned}\frac{f(x)}{x^\delta} &= \sum_{n=x}^{\infty}\frac{f(n)-f(n+1)}{n^\delta} + \sum_{n=x}^{\infty} f(n+1)\Big(\frac{1}{n^\delta} - \frac{1}{(n+1)^\delta}\Big)\\ &= O(1)\sum_{n=x}^{\infty}\frac{|f(n)-f(n+1)|}{n^\delta} + \sum_{n=x}^{\infty}\frac{|f(n)|}{n^{\delta+1}}.\end{aligned}$$

Here, the left-hand side is $O(\lambda(x))$ by (ii). In the proof of Theorem 1, we have proved from (iii) that

$$\sum_{n=x}^{\infty}\frac{|f(n)|}{n^{\delta+1}} = O(\lambda(x)).$$

Thus,

$$\sum_{n\geqslant x}^{\infty}\frac{|f(n)-f(n+1)|}{n^\delta} = O(\lambda(x))$$

is deduced, which completes the proof.

THEOREM 3. *Suppose that $f \in \mathcal{M}$, $\delta \geqslant 0$, and the following conditions are satisfied:*

(i) $$\sum_{n\geqslant x}\frac{1}{n^\delta}|f(n+1)-f(n)| = O(\mu(x)\lambda(x)),$$

$$\mu(x) \geqslant 0, \qquad \mu(x) = O(1),$$

(ii) $$f(n) = O(n^\delta\lambda(n)),$$

where $\lambda(x)$ is defined by

(iii) $$\lambda(x) = \sum_{n\geqslant x}\frac{|f(n)|}{n^\delta}$$

and satisfies $\lambda(x) = O(\lambda(2x))$,

(iv)
$$\sum_{n=1}^{\infty} \frac{\mu(n)}{n} < \infty.$$

Then $f(n) = n^{\beta}$ *with a suitable constant* β.

Proof. The conditions (ii) and (iii) are identical with the condition (ii), (iii) of Theorem 1 for $q = 1$. The condition (i) implies the condition (i) of Theorem 1 for $q = 1$. From Theorem 1 we know that

$$f(n) = n^{s} t(n) \quad \text{with} \quad |t(n)| = 1$$

and $t \in \mathcal{M}^{*}$. So, it is enough to prove that $t(n) = n^{i\tau}$ with a constant τ. From (iii) we have

$$\lambda(x) = \sum_{n \geqslant x} \frac{|f(n)|}{n^{\delta}} = \sum_{n \geqslant x} n^{s-\delta} |t(n)| = \sum_{n \geqslant x} n^{s-\delta} = A x^{s+1-\delta}(1 + O(1)),$$

where A is a suitable constant. We notice that $s + 1 - \delta < 0$ because of the convergence of the series (iii). So the equation (i) gives

$$\begin{aligned} O\big(\mu(x) x^{s+1-\delta}\big) &= \sum_{n \geqslant x} \frac{|f(n+1) - f(n)|}{n^{\delta}} \\ &= \sum_{n \geqslant x} \frac{|(n+1)^{s} t(n+1) - n^{s} t(n)|}{n^{\delta}}. \end{aligned} \tag{2.10}$$

Using the equality $|t(n)| = 1$ and simple operations to (2.10), we easily obtain

$$\sum_{n \geqslant x} n^{s-\delta} |t(n+1) - t(n)| = O\Big(x^{s+1-\delta}\Big(\mu(x) + \frac{1}{x}\Big)\Big).$$

Denote

$$\delta - s = 1 + \varepsilon > 1, \qquad U(x) = \mu(x) + \frac{1}{x}.$$

Then the last equality gives

$$\sum_{n \geqslant x} \frac{|t(n+1) - t(n)|}{n^{1+\varepsilon}} = O\Big(\frac{U(x)}{n^{\varepsilon}}\Big). \tag{2.11}$$

From (iv) we have

$$\sum_{n=1}^{\infty} \frac{U(n)}{n} < \infty. \tag{2.12}$$

On the other hand, let us denote

$$S_n = \sum_{k \geqslant n} \frac{|t(k+1) - t(k)|}{k^{1+\varepsilon}}.$$

Then from (2.11) and (2.12) we obtain

$$\sum_{n\geqslant N}\frac{|t(n+1)-t(n)|}{n}=\sum_{n\geqslant N}(S_n-S_{n+1})n^{\epsilon}=O(N^{\epsilon}S_N)+O(1)\sum_{n\geqslant N}S_n n^{\epsilon-1}$$
$$=O(U(N))+O(1)\sum_{n\geqslant N}\frac{U(n)}{n}=O(1),$$

as $N\to\infty$, which means

$$\sum_{n=1}^{\infty}\frac{|t(n+1)-t(n)|}{n}<\infty. \tag{2.13}$$

In the proof of Theorem C (see (Kátai, 1984, p. 108)) we see that the condition (2.13) with $t\in\mathcal{M}^*$ and $|t(n)|=1$ gives $t(n)=n^{i\tau}$ with a suitable constant τ, which completes the proof.

REFERENCES

Erdős, P. (1946). On the distribution of additive functions. *Ann. Math.* **47**, 1–20.

Kátai, I. (1970). On a problem of P. Erdős. *J. Number Theory* **2**, 1–6.

Kátai, I. (1984). Multiplicative functions with regularity properties. II. *Acta Math. Hung.* **43**, 105–130.

Kátai, I. (1989). Characterization of arithmetical functions, problems and results. In: *Theórie des Nombres* (Eds. J.-M. Dekoninck, C. Levesque). Walter de Gruyter, Berlin, New York, pp. 544–555.

Mauclaire, J.-L. and Murata, L. (1984). On a characterization of some arithmetical multiplicative functions. *Acta Math. Hung.* **44**, 191–198.

New Trends in Probab. and Statist., pp. 165–171
F. Schweiger and E. Manstavičius (Eds.)
1992 VSP/TEV

ON AN ANALYTIC PROBLEM CONSIDERED BY SIERPIŃSKI AND RAMANUJAN

A. SCHINZEL
Instytut Matematyczny Polskiej Akademii Nauk,
P. O. 137, 00–950 Warszawa, Poland

ABSTRACT

Let $r(n)$ be the number of representations of n as a sum of two squares. An Ω-estimate for the error term in the summation formula for $r(n)^2$ is obtained.

Let $r(n)$ be the number of representations of n as a sum of two squares.
W. Sierpiński in his doctorate thesis (see Sierpiński, (1908)), written in 1906, has proved the estimate

$$\sum_{n\leqslant x} r(n)^2 = 4x \log x + cx + O(x^{3/4}\log x),$$

where c is a certain constant. The same asymptotic formula with the better error term $O(x^{(3/5)+\varepsilon})$ was stated without proof in (Ramanujan, 1916). B. M. Wilson (1922) indicated without giving the details that $O(x^{(3/5)+\varepsilon})$ can be replaced by $O(x^{(1/2)+\varepsilon})$ for every $\varepsilon > 0$. Recently, W. R. Recknagel (1989) improved the error term to $O(x^{1/2}\log^4 x)$ and W. G. Nowak has improved it further to $O(x^{1/2}(\log x)^{8/3}(\log\log x)^{1/3})$ (unpublished).
As far as I know, there is no Ω-result in the literature, and it is the aim of this paper to prove such a result.

THEOREM. *We have*

$$\sum_{n\leqslant x} r(n)^2 = 4x \log x + cx + \Omega(x^{3/8}).$$

The proof is based on the approach of (Balasubramanian *et al.*, 1988). The possibility of using this approach has been indicated to me by Prof. A. Ivić. In comparison with (Balasubramanian *et al.*, 1988), the present paper contains no new idea, but the work is motivated by historical reasons. I thank Dr. W. G. Nowak for pointing out some obscurities in an early draft of the paper, and the Department of Mathematics in Geneva for their help in preparing the manuscript.

Notation. $\zeta(s)$ is the Riemann zeta function, $\zeta_K(s)$ the Dedekind zeta function of the field $K = \mathbf{Q}(i)$, p denotes a general prime, A is a large constant, not necessarily the same at each occurrence, T is a sufficiently large real number,

$y = T^B$, where B is a sufficiently large constant;

$J = \{T^{2/3} \leqslant t \leqslant 2T$: for any complex number z with $\operatorname{Re} z \geqslant 1/3$ and $|\operatorname{Im} z - t| \leqslant (\log T)^{20}$ we have $\zeta(2z) \neq 0\}$;

$J_1 = \{T^{2/3} - (\log T)^4 \leqslant t \leqslant 2T + (\log T)^4$: for any complex number z with $\operatorname{Re} z \geqslant 1/3$ and $|\operatorname{Im} z - t| \leqslant (\log T)^{15}$ we have $\zeta(2z) \neq 0\}$.

LEMMA 1.

$$F(s) = \sum_{n=1}^{\infty} \frac{r(n)^2}{n^s} = \frac{16\zeta_K(s)^2}{(1+2^{-s})\zeta(2s)} \qquad \textit{for} \quad \operatorname{Re} s > 1.$$

Proof. See (Ramanujan, 1916).

LEMMA 2. *If* $\operatorname{Re} s \geqslant 1/3$ *and* $|\operatorname{Im} s| \geqslant 1$, *then*

$$\zeta_K(s) = O\big((|t|+2)^A\big),$$

where $s = \sigma + it$.

Proof. See Lemma 7 in (Staś, 1959).

LEMMA 3. *If* $\operatorname{Re} s \geqslant 1/3$ *and* $t \in J_1$, *then*

$$\frac{1}{\zeta(2s)} = O(|t|+2).$$

Proof. This is a consequence of Lemma 1 in (Balasubramanian *et al.*, 1988).

LEMMA 4. *If* $\operatorname{Re} s \geqslant 1/3$ *and* $t \in J_1$, *then*

$$F(s) = O\big((|t|+2)^A\big) \qquad \textit{for a suitable} \quad A > 0.$$

Proof. This is a consequence of Lemmas 1–3.

LEMMA 5. *Let*

$$E(x) = \sum_{n \leqslant x} r(n)^2 - 4x \log x - cx.$$

If

$$\int_T^{\infty} \frac{E(u)^2}{u^{7/4}} e^{-u/y}\, du \leqslant \log^2 T,$$

$t = \operatorname{Im} s \in J$, *and* $\operatorname{Re} s = 3/8$, *then we have*

$$F(s) = \sum_{n \leqslant T_0} \frac{r(n)^2}{n^s} e^{-n/y} + s \int_0^1 \sum_{n > T_0} \frac{E(n+u)}{(n+u)^{s+1}} e^{-(n+u)/y} du + O\big((\log T)^{20}\big)$$

for a suitable T_0, $T \leqslant T_0 \leqslant 2T$.

Proof. We start with

$$\sum_{n=1}^{\infty} \frac{r(n)^2 e^{-n/y}}{n^s} = \frac{1}{2\pi i} \int_{\operatorname{Re} w = 2} F(s+w) y^w \Gamma(w)\, dw.$$

Now we break off the portion $|\mathrm{Im}\, w| \geqslant (\log T)^4$ of the integral, with a small error, and move the line of integration to $\mathrm{Re}\, w = -(1/24)$.
Now using the estimate of $F(s)$ given in Lemma 4 (and assuming that B is large enough), and using the fact that, since $t \in J$ and $t + \mathrm{Im}\, w \in J_1$, we see that the value of the integral on the horizontals $t + \mathrm{Im}\, w = c \in J_1$, as well as on the vertical $\mathrm{Re}\, w = -(1/24)$ is small. This proves that

$$\sum_{n=1}^{\infty} \frac{r(n)^2 e^{-n/y}}{n^s}$$

equals nearly the sum of the residues inside the contour. Thus we have

$$F(s) = \sum_{n=1}^{\infty} \frac{r(n)^2}{n^s} e^{-n/y} + O(T^{-10}) = \sum_{n \leqslant T_0} + \sum_{n > T_0} + O(T^{-10}).$$

Now

$$\begin{aligned} \sum_{n > T_0} \frac{r(n)^2 e^{-n/y}}{n^s} &= \int_{T_0}^{\infty} \frac{1}{u^s} e^{-u/y} d\Big(\sum_{n \leqslant u} r(n)^2\Big) \\ &= \int_{T_0}^{\infty} \frac{1}{u^s} e^{-u/y} d(4u \log u + cu + E(u)) \\ &= \int_{T_0}^{\infty} \frac{1}{u^s} e^{-u/y} (4 \log u + 4 + c)\, du + \int_{T_0}^{\infty} \frac{1}{u^s} e^{-u/y} dE(u) \\ &= S_1 + S_2 \quad \text{(say)}. \end{aligned}$$

The first integral does not exceed up to a constant factor

$$\int_{T_0}^{\infty} \frac{1}{u^s} e^{-u/y} u^{1/24}\, du,$$

and hence is small by Lemma 3 of (Balasubramanian *et al.*, 1988). Now S_2 is (after one integration by parts)

$$\left[\frac{E(u) e^{-u/y}}{u^s}\right]_{T_0}^{\infty} + s \int_{T_0}^{\infty} \frac{E(u) e^{-u/y}}{u^{s+1}}\, du + \frac{1}{y} \int_{T_0}^{\infty} \frac{E(u) e^{-u/y}}{u^s}\, du.$$

Hence, using Lemma 4 in (Balasubramanian *et al.*, 1988), and choosing a suitable T_0 in the interval $T \leqslant T_0 \leqslant 2T$, we obtain

$$\begin{aligned} S_2 &= s \int_{T_0}^{\infty} \frac{E(u)}{u^{s+1}} e^{-u/y}\, du + O(\log T) \\ &= s \sum_{n \geqslant T_0} \int_0^1 \frac{E(n+u)}{(n+u)^{s+1}} e^{-(n+u)/y}\, du + O(\log T). \end{aligned}$$

This proves the lemma.

LEMMA 6. *We have*

$$\int_0^T \left|\sum_n b_n n^{it}\right|^2 dt = \sum_n (T+O(n))\left(|b_n|^2\right)$$

for any sequence of complex numbers b_n provided the right hand side is convergent.

Proof. For the proof of this lemma we refer the reader to (Montgomery and Vaughan, 1974).

LEMMA 7. *We have*

$$\int_{\substack{\operatorname{Re} s=3/8 \\ \operatorname{Im} s=t\in J}} \left|\sum_{n\geqslant T_0} \frac{r(n)^2}{n^s} e^{-n/y}\right|^2 \frac{dt}{t^2} \ll 1.$$

Proof. It is sufficient to prove that

$$\int_{T^{2/3}}^{2T} \left|\sum_{n\leqslant T_0} \frac{r(n)^2}{n^{3/8}} e^{-n/y}\right|^2 \frac{dt}{t^2} \ll 1.$$

By Lemma 6, we have

$$\int_{2^m}^{2^{m+1}} \left|\sum_{n\geqslant T_0} \frac{r(n)^2}{n^{3/8}} e^{-n/y}\right|^2 dt = \sum_{n\leqslant T_0} (2^m + O(n))\frac{r(n)^4}{n^{3/4}}$$

$$\ll 2^m T_0^{1/4} \log^7 T_0 + T_0^{5/4} \log^7 T_0.$$

Hence

$$\int_{2^m}^{2^{m+1}} \left|\sum_{n\geqslant T_0} \frac{r(n)^2}{n^s} e^{-n/y}\right|^2 \frac{dt}{t^2} \leqslant \frac{T_0^{1/4}\log^7 T_0}{2^m} + \frac{T_0^{5/4}\log^7 T_0}{2^{2m}}.$$

Now, summing over m satisfying $(1/2)T_0^{2/3} < 2^m \leqslant 2T$, it follows that

$$\int_{T^{2/3}}^{2T} \left|\sum_{n\geqslant T_0} \frac{r(n)^2}{n^s} e^{-n/y}\right|^2 \frac{dt}{t^2} \ll 1.$$

LEMMA 8. *If*

$$\int_T^\infty \frac{E(u)^2}{u^{7/4}} e^{-u/y}\, du \leqslant \log^2 T,$$

then we have

$$\int\limits_{\substack{\operatorname{Re} s=3/8\\ \operatorname{Im} s=t\in J}} \left|\int\limits_0^1 \sum_{n\geqslant T_0} \frac{E(n+u)e^{-(n+u)/y}}{(n+u)^{s+1}}\,du\right|^2 dt \ll \int\limits_T^\infty \frac{|E(u)|}{u^{7/4}}e^{-2u/y}\,du,$$

for a suitable T_0, $T \leqslant T_0 \leqslant 2T$.

Proof. This follows from Lemma 6 in the same way as in (Balasubramanian *et al.*, 1988). Lemma 8 follows from Lemma 6.

LEMMA 9. *If*

$$\int\limits_T^\infty \frac{E(u)^2}{u^{7/4}}e^{-u/y}\,du \leqslant \log^2 T,$$

we have

$$\int\limits_{\substack{\operatorname{Re} s=3/8\\ \operatorname{Im} s=t\in J}} \frac{|F(s)|^2}{|s|^2}\,dt \ll 1 + \int\limits_T^\infty \frac{E(u)^2}{u^{7/4}}e^{-2u/y}du.$$

Proof. This follows from Lemmas 5, 7, and 8.

LEMMA 10. *Let*

$$f(s) = \sum_{n=1}^\infty \frac{b_n}{n^s}$$

be convergent in some half-plane and analytically continuable in $\sigma \geqslant 1/3$, $A \leqslant t \leqslant A+M$. *Then*

$$\int\limits_A^{A+M} |f(3/8+it)|^2 dt \gg M,$$

provided $M \geqslant \log A$, $b_1 = 1$, *and*

$$\max\left\{|f(s)|\colon\ A \leqslant t \leqslant A+M,\ \operatorname{Re} s \geqslant 3/8\right\} \leqslant e^A.$$

Proof. This is a consequence of Lemma 9 in (Balasubramanian *et al.*, 1988) for $\alpha = 3/8$.

DEFINITION 1.

$$f(s) = \frac{\zeta_K(s+1/4)^2}{(1+2^{-s})\zeta(2s)} = \sum_{n=1}^\infty \frac{b_n}{n^s}.$$

LEMMA 11. *If an interval* $[A, A+M]$ *is contained in* J *and* $A \geqslant M \geqslant \log A$, *we have*

$$\int\limits_A^{A+M} |f(3/8+it)|^2 dt \gg M.$$

Proof. It follows from the definition of $f(s)$ that $b_1 = 1$,

$$\sum_{n=1}^{\infty} \frac{b_n}{n^s} \quad \text{is convergent in} \quad \operatorname{Re} s > 3/4$$

and analytically continuable in the whole plane with poles at

$$s = \frac{3}{4}, \frac{(2k+1)\pi i}{\log 2}, \qquad k \in \mathbf{Z},$$

and at the zeros of $\zeta(2s)$. If $\operatorname{Re} s \geqslant 3/8$ and $[A, A+M]$ is contained in J, then by Lemma 2 we have

$$|\zeta_K(s+1/4)| \leqslant (|t|+2)^{A/(2\log(2A+2))} \leqslant e^{A/2},$$

and by Lemma 3,

$$|\zeta(2s)^{-1}| \ll |t| + 2 \leqslant 2A + 2.$$

Hence $|f(s)| \leqslant e^A$ for A large enough. The lemma follows now from Lemma 10.

LEMMA 12. *If $[A, A+M] \subset J$ and $A \geqslant M \geqslant \log A$, we have*

$$\int_A^{A+M} |F(3/8+it)|^2 dt \gg AM.$$

Proof. For $\operatorname{Re} s = 3/8$, we note that $|\zeta_K(1-s)| = |\zeta_K(s+(1/4))|$. Using the functional equation

$$B^s \Gamma(s) \zeta_K(s) = B^{1-s} \Gamma(1-s) \zeta_K(1-s),$$

where B is a constant, we obtain for $\operatorname{Re} s = 3/8$,

$$\begin{aligned} |F(s)| &= 16|f(s)||\zeta_K(s)|^2|\zeta_K(s+1/4)|^{-2} = 16|f(s)||\zeta_K(s)|^2|\zeta_K(1-s)|^{-2} \\ &= 16|f(s)||B^{2-4s}\Gamma(1-s)^2\Gamma(s)^{-2}| \gg |f(s)||t|^{1/2}. \end{aligned}$$

Hence Lemma 12 follows from Lemma 11.

DEFINITION 2. For every x such that $T^{2/3} \leqslant x \leqslant 2T$,

$$J(x) = J \cap [x, 2x].$$

$N(x)$ is the number of zeros of $\zeta(2s)$ with $t \in [x - (\log T)^{20}, 2x + (\log T)^{20}]$ and $\operatorname{Re} s \geqslant 1/3$.

LEMMA 13. *We have*

$$N(x) \ll x^{49/60}.$$

Proof. This follows from Lemma 13 in (Balasubramanian *et al.*, 1988) for $\varepsilon = 1/6$.

DEFINITION 3. $J_2(x)$ is the portion of $J(x)$ obtained by deleting all connected components whose length is not greater than $\log 2x$.

LEMMA 14. *We have*

$$\int_{J_2(x)} |F(3/8+it)|^2\, dt \gg x^2.$$

Proof. The total length of $[x, 2x] \setminus J_2(x)$ is $O(N(x)(\log T)^{20}) = O(x^{5/6})$. Hence the total length of $J_2(x)$ is $\gg x$. Now applying Lemma 12 to each connected component of $J_2(x)$ and adding we obtain Lemma 14.

LEMMA 15. *We have*

$$\int_{\substack{\mathrm{Re}\, s=3/8 \\ \mathrm{Im}\, s=t\in J}} \frac{|F(s)|^2}{|s|^2}\, dt \gg \log T.$$

Proof. From Lemma 14 we have

$$\int_{J(x)} |F(3/8+it)|^2\, dt \gg x^2,$$

and Lemma 15 follows by integration by parts.

Proof of the theorem. By Lemma 9 and Lemma 15 we have

$$\int_T^\infty \frac{E(u)^2}{u^{7/4}} e^{-2u/y}\, du \gg \log T.$$

If for every $\varepsilon > 0$ we had $|E(u)| < \varepsilon u^{3/8}$ for $T \geqslant T_0(\varepsilon)$, it would follow that

$$\int_{T_0(\varepsilon)}^\infty \frac{E(u)^2}{u^{7/4}} e^{-2u/T_0(\varepsilon)^B}\, du \leqslant \varepsilon^2 \left(\int_{T_0(\varepsilon)}^{T_0(\varepsilon)^B} \frac{du}{u} + \int_1^\infty e^{-2v} \frac{dv}{v} \right)$$
$$< \varepsilon^2 (B \log T_0(\varepsilon) + 1),$$

in contrary to the above inequality.

REFERENCES

Balasubramanian, R., Ramachandra, K., and Subarao, M. V. (1988). On the error function in the asymptotic formula for the counting function of k-full numbers. *Acta Arithm.* **50**, 107–118.

Montgomery, H. L. and Vaughan, R. C. (1974). Hilbert's inequality. *J. London Math. Soc.* **8**, 73–82.

Ramanujan, S. (1916). Some formulae in the analytic theory of numbers. *Mess. Math.* **45**, 81–84. Also in: *Collected papers*. Cambridge, 1927, pp. 113–135.

Recknagel, W. (1989). Variantes des Gaußschen Kreisproblems. *Abh. Math. Sem. Univ. Hamburg* **59**, 183–189.

Sierpiński, W. (1908). Sur la sommation de la série $\sum_{n>a}^{n\leqslant b} \tau(n)f(n)$, oú $\tau(n)$ signifie le nombre de décompositions du nombre n en une somme de deux carrés de nombres entiers. *Prace Mat. Fiz.* **18**, 1–59 (in Polish). Also in: *Oeuvres choisies*, I.Varsovie, 1974, pp. 109–154 (in French).

Staś, W. (1959). Über eine Anwendund der Methode von Turán auf die Theorie des Restgliedes im Primidealsatz. *Acta Arithm.* **5**, 179–195.

Wilson, B. M. (1922). Proof of some formulae enunciated by Ramanujan. *Proc. London Math. Soc.* (2) **21**, 235–255.

New Trends in Probab. and Statist., pp. 173–179
F. Schweiger and E. Manstavičius (Eds.)
1992 VSP/TEV

DISTRIBUTION OF PRIME VECTORS IN INTEGRAL LATTICES

M. I. TULYAGANOVA
Visol pass. 6, 5,
700147 Tashkent, Uzbekistan

ABSTRACT

The paper presents a survey of results of the author and her colleagues on the distribution of vectors with prime coordinates in integral lattices of n-dimensional euclidean space. The problem had arisen as a natural generalization of (now classical) Hua Loo Keng's problem. The latter is formulated as follows. Let the system of equations

$$p = ax + by + c,$$
$$p' = a'x + b'y + c'$$

be given where x and y are integers. What conditions should be satisfied by the integers a, b, c, a', b', and c' for the system to have infinitely many solutions p, p' in prime numbers?

The study of prime vectors started after the formulation by Hua Loo Keng of the problem to find the conditions on the integers a, b, c, a', b', and c' necessary and sufficient for the set of pairs of prime numbers p and p' such that

$$p = ax + by + c,$$
$$p' = a'x + b'y + c',$$

where x and y are integers, to be infinite (see (Hua Loo Keng, 1947)).
A. I. Vinogradov (1963) noticed that if the linear forms $ax + by$ and $a'x + b'y$ are independent (the nondegenerate problem) then the Hardy–Littlewood circle method allows to prove the asymptotic formula for special sum over all pairs of primes of such type. In case the principal term of this sum is unbounded, Hua Loo Keng's problem has been solved positively for the numbers a, b, c, a', b', and c'. In the opposite case the problem remained open.
In the beginning of eighties the author managed to connect explicitly the nondegenerate problem with the distribution of prime numbers in an arithmetic progression, and to obtain the complete solution of the problem (see (Tulyaganova, 1981a)).
Hua Loo Keng's problem may be naturally generalized for an arbitrary dimension. In this case we deal with a set of prime vectors in integral lattices of n-dimensional euclidean space. The complete lattice corresponds to the nondegenerate problem, which is solved in the same way as in the initial case $n = 2$. The distribution

of prime vectors in complete lattices is analogous to that of prime numbers in arithmetic progressions (see (Tulyaganova, 1981b, 1983a)).
The problem of the distribution of prime vectors in incomplete lattices is more complicated.
The density of prime vectors in an arbitrary lattice can be estimated from above, and the upper bound coincides with the asymptotics being expected (see (Tulyaganova and Fainleib, 1990)).
The asymptotic formulae are known only for such cases when the problem is reducible to that for a lattice the dimension of which is greater than a half of the space dimension (see (Tulyaganova, 1986; Tulyaganova and Fainleib, 1988; Vinogradov, 1963; Wu Fang, 1957)). Hua Loo Keng's problem, concerning the conditions on the integers a, b, c, a', b', and c' for the set of pairs of prime numbers p, p' such that

$$\begin{aligned} p &= ax + by + c, \\ p' &= a'x + b'y + c' \end{aligned}$$

(x and y are integers) to be infinite, coincides with the problem of prime vectors of two-component lattice, the augmented matrix of which is equal to

$$\begin{pmatrix} a & a' \\ b & b' \\ c & c' \end{pmatrix}.$$

This problem may be naturally generalized for the lattice with an arbitrary number of components. It can be solved for complete lattices (see (Tulyaganova, 1981a)).
Let A^* be the matrix adjoint to A. The complete lattice Λ, with the determinant Δ and the basic matrix A, containing the vector b coincides with the set of solutions of the congruence

$$yA^* \equiv bA^* \pmod{\Delta}. \tag{1}$$

Let us call an integral lattice regular if it contains an infinite set of prime vectors. A vector $l \in \Lambda$ is called regular if some of its components are relatively prime to Δ while the others are congruent mod Δ to the prime divisors of Δ.
The following theorem provides effectively verifiable necessary and sufficient conditions for the regularity of a complete lattice and thus solves Hua Loo Keng's problem for the nondegenerate case.

THEOREM 1. *Let $\Lambda = \mathbf{Z}^n A + b$ be a complete lattice with the determinant Δ. In order that the lattice be regular it is necessary and sufficient that at least one of Δ^{n-1} vectors of n-dimensional cube $[1, \Delta]^n$ satisfying the congruence* (1) *be regular.*

According to Ju. V. Linnik's theorem there exists a constant $C > 1$ (Linnik's constant) such that each one-dimensional primitive progression with the difference $\Delta > 1$ contains the prime number not exceeding Δ^C. The following multidimensional analogue of Linnik's theorem takes place.

THEOREM 2. *Each regular lattice with the determinant $\Delta > 1$ contains a prime vector with the components not exceeding Δ^C, where C is Linnik's constant.*

Let l be a regular vector of the complete lattice Λ. The number of primitive components of the vector l is called the regularity degree of this vector and is

denoted by $r(l)$. For irregular vectors $l \in \Lambda$, we set $r(l) = 0$. So the degree of regularity of each vector of the lattice Λ in $\mathbf{Z}^n$ is a non-negative integer less than or equal to n.
The maximum of the regularity degrees of the vectors of the lattice Λ is called the regularity degree of the lattice Λ, i.e.

$$r(\Lambda) = \max_{l \in \Lambda} r(l).$$

We denote by $\pi(N, \Lambda)$ the number of prime vectors of the lattice Λ all components of which do not exceed N.
The following analogue of the Siegel–Walfish theorem is valid for complete lattices (see (Tulyaganova, 1981b)).

THEOREM 3. *Let Λ be a regular complete lattice in $\mathbf{Z}^n$ with the determinant Δ, and let $S(\Lambda)$ be the number of progressions of the regularity degree $r(\Lambda)$ among Δ^{n-1} progressions with the difference Δ constituting the lattice Λ. If*

$$1 \leqslant \Delta \leqslant (\log N)^{C_1},$$

then

$$\pi(N, \Lambda) = S(\Lambda)\Big(\frac{\operatorname{li} N}{\varphi(\Delta)}\Big)^{r(\Lambda)} + O\Big(N^{r(\Lambda)} e^{-C_2\sqrt{\log n}}\Big),$$

where C_1 is an arbitrary positive constant, $C_2 > 0$, and the constant in the symbol O depends on C_1 and n only.

THEOREM 4 (Tulyaganova and Babanazarov, 1988). *We have that*

$$\pi(N, \Lambda) = S(\Lambda)\Big(\frac{N}{\varphi(\Delta)\log N}\Big)^{r(\Lambda)}\Big(1 + O\Big(\frac{1}{\log N}\Big)\Big), \qquad N \to \infty,$$

uniformly in Δ,

$$\Delta \leqslant \exp\Big(C_3 \frac{\log N}{\log\log N}\Big),$$

with the exception of a multiple of some $\Delta^* = \Delta^*(N) > \log^{C_4} N$, where $C_3 > 0$ is an absolute constant and C_4 is any given positive constant.

If $n = 1$ then this is the Rodosski–Tatuzava theorem on the distribution of prime numbers in short arithmetic progressions.
The case of an incomplete lattice is more complicated A. I. Vinogradov (1963) studied prime vectors of the degenerate lattice $\mathbf{Z}^2 A + b$, $A \in \mathbf{Z}^{2\times 3}$, $b \in \mathbf{Z}^2$. He introduced the special series corresponding to this lattice and obtained an asymptotic formula for it using the circle method. In the case of the positivity of the corresponding special series, the infinity of the set of the prime vectors of the lattice $\mathbf{Z}^2 A + b$ follows from such formulae.
Using the multi-dimensional circle method, Wu Fang (1957) proved asymptotic formulae for the numbers of solutions in prime numbers of some systems of linear equations. These numbers correspond to the function $\pi(N, \Lambda)$ for a certain class of

lattices. Similar results were obtained for wider classes of lattices in (Tulyaganova, 1984, 1986).

Using the methods of (Montgomery and Vaughan, 1975), Liu Ming Chit and Tsang Kai-Man (1989) proved that the system of two linear equations with three unknowns

$$\begin{cases} a_{11}p_1 + a_{12}p_2 + a_{12}p_3 = b_1, \\ a_{21}p_1 + a_{22}p_2 + a_{23}p_3 = b_2, \end{cases}$$

can be solved for almost all pairs (b_1, b_2) under certain conditions for the coefficients a_{ij}.

A. Sultanova (1990) obtained an asymptotic formula for the number of solutions of the system of linear equations in prime numbers in multi-dimensional parallelepiped.

The infinite asymptotic series with respect to the powers of $\log N$ was obtained for $\pi(N, \Lambda)$ in (Tulyaganova and Fainleib, 1988). Moreover, the class of lattices Λ was extended comparing with the previous results of (Tulyaganova, 1984, 1986; Wu Fang, 1957).

The problem on the asymptotics of the function $\pi(N, \Lambda)$ as $N \to \infty$ is reduced to the case when $\Lambda = \Gamma \cap \Psi$, where Γ is the set of integer solutions of the system of linear equations with the matrix $B \in \mathbf{Z}^{r \times n}$, $0 < r < n$, for which all minors of the r-th order are relatively prime, and Ψ is an n-dimensional primitive progression. Let us denote the class of the lattices of such type by $L_{n,r}(\Delta)$ provided that Γ has not immovable components, and $n > 2r$. Suppose, moreover, that after removing of any column from the matrix B two nonintersecting submatrices of the rank r still remain, and the difference of the progression Ψ is Δ.

The following theorem is proved in (Tulyaganova and Fainleib, 1988).

THEOREM 5. *If Λ is a lattice of the class $L_{n,r}(\Delta)$ containing the vector*

$$b = b(N) = Nh + O(N\varepsilon(N)),$$

where h is a fixed vector of $\mathbf{R}^n$, $\varepsilon(N)$ tends to zero more rapidly than each negative power of $\log N$, and $1 \leqslant \Delta \leqslant \log^{C'} N$, where C' is a constant, then the asymptotic expansion

$$\pi(N, \Lambda) = \sigma(\Delta, \Lambda) \frac{N^{n-r}}{\log^n N} \sum_{s=0}^{\infty} \frac{C_s(\Gamma, h)}{\log^s N}, \qquad N \to \infty,$$

is valid, where

$$\sigma(\Delta, \Lambda) = \frac{\Delta^r}{\varphi^n(\Delta)} \prod_{p \nmid \Delta} \frac{p^r \rho(p)}{(p-1)^n},$$

$\rho(p)$ is the number of vectors of the hyperplane Γ_p in the space $\mathbf{Z}_p^n$, all components of which are not equal to zero,

$$C_s(\Gamma, h) = \sum_{\substack{r_1 + \ldots + r_n = s \\ r_j \geqslant 0}} \int_{uA + h \in [0,1]^n} \prod_{j=1}^{n} \log^{r_j} \frac{1}{\lambda_j(uA + h)} \, du,$$

and $\lambda_j(x)$ denotes the j-th component of the vector x.

Special cases of this theorem, giving the first term of the asymptotics, are proved in (Tulyaganova, 1986; Wu Fang, 1957).
This theorem gives the asymptotic expansion with respect to the powers of log N of the number of representations in the Vinogradov–Goldbach theorem and in its generalization to some class of the systems of linear equations with unknown prime variables (Wu Fang's theorem).

COROLLARY 1. *Let $R(N)$ be the number of solutions in prime numbers $p_1, \ldots, p_5$ of the system*

$$\begin{cases} p_1 + p_2 + p_3 = N, \\ p_1 + p_4 + p_5 = N. \end{cases}$$

Then, for odd N,

$$R(N) = \sigma(N) \frac{N^3}{\log^5 N} \sum_{s=0}^{\infty} \frac{C_5}{\log^s N},$$

where the coefficients are defined by the formula

$$C_s = \sum_{\substack{r_1+\ldots+r_5=s \\ r_j \geqslant 0}} \iiint\limits_{\substack{u,v,w>0 \\ u \leqslant v+w=1}} \left(\log \frac{1}{u}\right)^{r_1} \left(\log \frac{1}{v}\right)^{r_2} \left(\log \frac{1}{w}\right)^{r_3}$$

$$\times \left(\log \frac{1}{1-v-w}\right)^{r_4} \left(\log \frac{1}{v+w-u}\right)^{r_5} du\, dv\, dw,$$

and

$$\sigma(N) = \prod_{p|N} \left(1 - \frac{1}{(p-1)^2}\right)^2 \prod_{(p,N)=1} \left(1 + \frac{3p^2 - 5p + 1}{(p-1)^5}\right).$$

It should be noted that the system of linear equations presented here does not belong to the class of systems considered in Wu Fang's theorem.

COROLLARY 2. *Let $R(N)$ be the number of representations of the odd number N by the sum of three prime numbers $p_1 + p_2 + p_3$ such that*

$$p_i \equiv b_i \pmod{\Delta}, \qquad (b_i, \Delta) = 1, \quad i = 1, 2, 3,$$
$$b_1 + b_2 + b_3 \equiv N \pmod{\Delta}, \qquad 1 \leqslant \Delta \leqslant (\log N)^{C''},$$

where C'' is an arbitrary positive constant. Then, for odd number N,

$$R(N) = \sigma(N) \frac{N^2}{\log^3 N} \sum_{s=0}^{\infty} \frac{C_5}{\log^s N},$$

where

$$C_5 = \sum_{\substack{r_1+r_2+r_3=s \\ r_j \geqslant 0}} \iint\limits_{\substack{u+v \leqslant 1 \\ u,v \geqslant 0}} \left(\log \frac{1}{u}\right)^{r_1} \left(\log \frac{1}{v}\right)^{r_2} \left(\log \frac{1}{1-u-v}\right)^{r_3} du\, dv,$$

and

$$\sigma(N) = \frac{\Delta}{\varphi^3(\Delta)} \prod_{\substack{p|N \\ (p,\Delta)=1}} \left(1 - \frac{1}{(p-1)^2}\right) \prod_{(p,N\Delta)=1} \left(1 + \frac{1}{(p-1)^3}\right).$$

In particular, for $\Delta = 1$, we have the asymptotic expansion of the number of representations in the Goldbach–Vinogradov theorem.

All the results of (Vinogradov, 1963; Wu Fang, 1957; Tulyaganova, 1984, 1986; Tulyaganova and Fainleib, 1988) can be applied only to such lattices the dimension of which is greater than a half of the dimension of space, i.e. an integral lattice is not too degenerated.

Let P be the class of lattices $\Lambda = \mathbf{Z}^k A + b$ such that the matrix A has no zero columns, every two columns of the extended matrix are linearly independent over $\mathbf{Q}$, and all the elements of each column of the matrix A' are relatively prime. The problem of the distribution of prime vectors in arbitrary lattices may be reduced to the lattices of the class P.

Theorem given below contains the upper estimate of the function $\pi(N, \Lambda)$ for an arbitrary integer lattice (see (Tulyaganova and Fainleib, 1989, 1990)).

THEOREM 6. *Let $|x|$ be the maximum of the module of components of the vector x,*

$$\Lambda = \mathbf{Z}^k A + b, \qquad A \in \mathbf{Z}^{k\times n},$$

$$m = m(\Lambda) = \inf_{\substack{x \subset \mathbf{R}^k \\ x \neq \bar{0}}} \frac{|xA|}{|x|}.$$

Then, for $N \geqslant 2m$, the inequality

$$\pi(N, \Lambda) \leqslant \frac{c(n)}{\log^n \dfrac{N}{m}} \left(\frac{N}{m}\right)^k \exp\left\{\sum_{p|g} \frac{np^{k-1} - \omega(p)}{p^k}\right\}$$

is valid, where $c(n)$ is a positive constant depending on n only, $g = g(\Lambda)$ is an integer, $\omega(p) = \omega_\Lambda(p)$ is the number of solutions of some polynomial congruence defined by the lattice Λ. There are explicit expressions for the corresponding polynomial and for the number $g(\Lambda)$.

REFERENCES

Hua Loo Keng (1947). Additive theory of prime numbers. *Proc. Steklov Math. Inst.* **22**, 1–180 (in Russian).

Liu Ming Chit and Tsang Kai-Man (1989). On pairs of linear equations in three prime variables and application to Goldbach's problem. *J. reine angew. Math.* **399**, 109–136.

Montgomery, H. L. and Vaughan, R. C. (1975). The exceptional set in Goldbach's problem. *Acta Aritm.* **27**, 353–370.

Sultanova, A. (1990). On a set of solutions in prime vectors of systems of linear equations in multi-dimensional parallelepiped. In: *Abstracts of Commun. Conf. Number Theory and Applications.* Tashkent, pp. 108 (in Russian).

Tulyaganova, M. I. (1981a). On multi-dimensional analogues of Hua's problem. *Izv. Akad. Nauk Uz. SSR, Ser. Fiz-Mat. Nauk* 1, 39–43 (in Russian).

Tulyaganova, M. I. (1981b). The distribution of prime vectors in integral lattices. *Izv. Akad. Nauk Uz. SSR, Ser. Fiz-Mat. Nauk* 3, 37–40 (in Russian).

Tulyaganova, M. I. (1983a). Average law of the distribution of prime vectors in integral lattices. *Izv. Akad. Nauk Uz. SSR, Ser. Fiz-Mat. Nauk* 1, 28–31 (in Russian).

Tulyaganova, M. I. (1983b). Distribution of prime vectors in integral lattice and multi-dimensional analogue of Hua's problem. In: *All-Union Conf. Theory of Transcendental Numbers and its Applications. Abstracts of Commun.* Moscow State University, Moscow, pp. 143–144 (in Russian).

Tulyaganova, M. I. (1984). The distribution of prime vectors in degenerate lattices. *Sov. Math. Dokl.* **29**, 451–453.

Tulyaganova, M. I. (1986). Prime vectors in degenerate lattices. *Math. USSR, Sb.* **54**, 279–295.

Tulyaganova, M. I. and Babanazarov, B. (1988). Distribution of prime vectors in integral lattices in a little multi-dimensional cube. *Dokl. Akad. Nauk Uz. SSR* 7, 3–4 (in Russian).

Tulyaganova, M. I. and Fainleib, A. S. (1988). Asymptotic expansion of functions of density of prime vectors in integral lattices. *Dokl. Akad. Nauk Uz. SSR* 6, 7–9 (in Russian).

Tulyaganova, M. I. and Fainleib, A. S. (1989). *Distribution of Prime Vectors in Integral Lattices.* Fan, Tashkent (in Russian).

Tulyaganova, M. I. and Fainleib, A. S. (1990). The density of prime vectors in integral lattices of common type. In: *Abstracts of Commun. Conf. on Number Theory and Applications.* Tashkent, pp. 119 (in Russian).

Vinogradov, A. I. (1963). On a problem of Hua Loo Keng. *Dokl. Akad. Nauk SSSR* **151**, 2, 255–257 (in Russian).

Wu Fang (1957). On the solutions of the systems of linear equations with prime variables. *Acta Math. Sinica* **7**, 102–121 (in Chinese).

New Trends in Probab. and Statist., pp. 181–188
F. Schweiger and E. Manstavičius (Eds.)
1992 VSP/TEV

MULTIPLICATIVE FUNCTIONS OF SPECIAL TYPE IN SHORT INTERVALS

P. VARBANEC
Department of Mathematics, Odessa University,
Petra Velikovo 2, 270000 Odessa, Ukraine

ABSTRACT

Let

$$f(n) = \sum_{d^2|n} \Phi(d),$$

where $\Phi(d)$ is a multiplicative function, $\Phi(d) = O(d^\varepsilon)$. The asymptotical formula for the summatory function of $f(n)$ in short intervals is obtained.

1. INTRODUCTION

We shall say that a multiplicative function $f(n)$ belongs to the class $\mathfrak{M}_2$ if

$$f(n) = \sum_{d^2|n} \Phi(d) \tag{1}$$

with a suitable multiplicative function $\Phi(d)$. For example, if $\mu(n)$ is the Möbius function, then $\mu^2(n) \in \mathfrak{M}_2$.
We shall estimate the sum

$$\sum_{x \leqslant n \leqslant x+h} f(n), \qquad f(n) \in \mathfrak{M}_2. \tag{2}$$

THEOREM 1. *Let f and Φ be multiplicative functions connected by* (1), *where* $\Phi(n) = O(n^\varepsilon)$. *Then*

$$\sum_{x \leqslant n \leqslant x+h} f(n) = h \sum_{d=1}^{\infty} \frac{\Phi(d)}{d^2} + O(h^{1/2} x^\varepsilon) + O(x^{\theta+\varepsilon}) \tag{3}$$

uniformly in h, $h < x$, *where ε is an arbitrary positive constant and* $\theta = 0.2204$.

Notations: $f(x) \ll g(x)$ means that $f(x) = O(|g(x)|)$, $e(f(x))$ means $e^{2\pi i f(x)}$.

2. PRELIMINARIES

LEMMA 1. *Let $f(x_1, x_2)$ be a real valued function defined on a region D, contained in the rectangle $\{(x_1, x_2)\colon\ a_i \leqslant x_i \leqslant b_i,\ i = 1, 2\}$.*
Then for any q, $1 \leqslant q \leqslant b_1 - a_1$,

$$\sum_{(n_1,n_2)\in D} e(f(n_1, n_2)) \ll q^{-1/2}(b_1 - a_1)(b_2 - a_2)$$

$$+ \left[q^{-1}(b_1 - a_1)(b_2 - a_2) \sum_{1 \leqslant h \leqslant q-1} \left| \sum_{(n_1,n_2)\in D(h)} e(f(n_1 + h, n_2) - f(n_1, n_2))\right|\right]^{1/2}$$

where $D(h) = \{(n_1, n_2) \in D\colon\ (n_1 + h, n_2) \in D\}$.

LEMMA 2. *Let D be a domain restricted by $O(1)$ algebraic curves of bounded degree,*

$$D \subset \{(x_1, x_2)\colon\ X_i \leqslant x_i \leqslant 2X_i,\ i = 1, 2\},$$

and let $f(x_1, x_2)$ be a real valued C^∞-function such that for all $(x_1, x_2) \in D$ we have:

$$\frac{\partial f^{i_1+i_2}}{\partial x_1^{i_1} \partial x_2^{i_2}} \ll F X_1^{-i_1} X_2^{-i_2} \tag{i}$$

for some F and all non-negative integers i_1, i_2;

$$c_1 F_1 \leqslant \left|\frac{\partial^2 f}{\partial x_1^2}\right| \leqslant c_2 F_1, \tag{ii}$$

for some $F_1^{-1} \ll X_1^2 F^{-2/3}$, c_1 and c_2 being constants;
the functions

$$\frac{\partial f}{\partial x_1}, \qquad \frac{\partial f}{\partial x_2}, \qquad \frac{\partial^2 f}{\partial x_1\ \partial x_2}, \tag{iii}$$

do not change sign on D.
Then

$$\sum_{(x_1,x_2)\in D} e(f(x_1, x_2)) \ll F_1^{-1/2} \left| \sum_{(y_1,y_2)\in D_0} e(f_1(y_1, y_2))\right|, + X_1 X_2 F^{-1/2} \log(X_1, X_2),$$

where D_0 is the image of D for the conversion

$$y_1 = \frac{\partial f}{\partial x_1}, \qquad y_2 = x_2,$$

and $f_1(y_1, y_2) = f(\varphi_1(y_1 y_2), y_2) - y_1 \varphi_1(y_1, y_2)$, where $\varphi_1(y_1, y_2)$ is the solution of the equation

$$\left.\frac{\partial f}{\partial x_1}\right|_{\substack{x_1=\varphi_1 \\ x_2=y_2}} = y_1.$$

LEMMA 3. *Let the function $f(x)$ be twice differentiable on $[a,b]$, and let*

$$0 < H_2 \leqslant -f''(x), \qquad f'(a) - f'(b) < 2.$$

Then

$$\sum_{a\leqslant n\leqslant b} e(f(n)) \leqslant 1 + H_2^{-1/2}.$$

Lemmas 1 and 3 are well known, Lemma 2 is the special case of Lemma in (Graham and Kolesnik, 1988).
We denote

$$f_{x_{i_1}\dots x_{i_m}} = \frac{\partial^m f(x_1,\dots,x_k)}{\partial x_{i_1}\dots x_{i_m}}.$$

DEFINITION. A real function $g(x_1,\dots,x_k)$ is said to be the approximation of degree r to a real function $f(x_1,\dots,x_k)$ on D if f and g posses the partial derivatives up to the order r on D, and

$$\left|f_{x_{i_1}\dots x_{i_m}} - g_{x_{i_1}\dots x_{i_m}}\right| < \varepsilon\left|g_{x_{i_1}\dots x_{i_m}}\right|$$

for all $(x_1,\dots,x_k) \in D$, $m \leqslant r$, and for some sufficiently small number $\varepsilon \in [0,1/2]$. Then we write

$$f \overset{r}{\underset{D}{\sim}} g.$$

B. R. Srinivasan proved that for the functions

$$f(x_1+h,x_2) - f(x_1,x_2) \quad \text{and} \quad f_1(x_1,x_2)$$

there exists the approximations of the form $x_1^{\alpha_1}x_2^{\alpha_2}$ of degree ∞, if f can be approximated in a similar manner (see (Srinivasan, 1965, Lemmas 1 and 2)).
The trigonometrical sums

$$\sum_{x\in D} e(f(x))$$

can be estimated by the method of exponent pairs that is a generalized version of the Van der Corput method. For the definitions of a one-dimensional exponent pair and a pair of exponents of dimension 2, see (Richert, 1954; Srinivasan, 1965).

LEMMA 4 (Graham and Kolesnik, 1988). *Let $\psi(u) = u - [u] - 1/2$ and $\Delta > 1$. Then*

$$\sum_{N\leqslant n\leqslant 2N} \psi(yn^{-\gamma}) \ll \frac{N}{\Delta} + \max_{M\leqslant\Delta}\max_{M'\leqslant 2M}\frac{1}{M}\left|\sum_{M\leqslant m\leqslant M'}\sum_{N\leqslant n\leqslant 2N} e(ymn^{-\gamma})\right|.$$

LEMMA 5 (Richert, 1954). *Let α, β, γ be positive numbers, (k,λ) be a one-dimensional exponent pair with $k > 0$. Then*

$$\sum_{n\leqslant x^\alpha}\psi\left(\frac{x^\beta}{n^\gamma}\right) = O\left(x^{\alpha-\frac{\beta-\alpha\gamma}{2}}\right) + \begin{cases} O\left(x^{\frac{\alpha\lambda+(\beta-\alpha\gamma)k}{k+1}}\right) & \text{if} \quad \lambda > \gamma k, \\ O\left(x^{\frac{\beta k}{k+1}}\log x\right) & \text{if} \quad \lambda = \gamma k, \\ O\left(x^{\frac{\beta k}{1+(\gamma+1)k-\lambda}}\right) & \text{if} \quad \lambda > \gamma k. \end{cases}$$

3. ESTIMATES FOR THE SPECIAL TRIGONOMETRIC SUMS

Let

$$S_\gamma = S_\gamma(M,N) = \sum_{M\leqslant m\leqslant M'\leqslant 2M} \sum_{N\leqslant n\leqslant N'\leqslant 2N} e(Xmn^{-\gamma}).$$

THEOREM A. *Let X, M, N, γ be positive numbers, $N\leqslant X^{1/(1+\gamma)}$, and (k,λ) be a one-dimensional exponent pair,*

$$M_0 = \left[X^{-(3+2k)}N^{4+3\gamma+k(4+2\gamma)}\right]^{1/(5+4k-2\lambda)} + 1.$$

Then

$$S_\gamma \ll \begin{cases} \left[X^{-2-2k}M^{\frac{1}{4}(25+20k+\lambda)}N^{(1-\gamma)(1+k)}\right]^{\frac{1}{7+6k}} & \text{if } M\geqslant M_0,\\ X^{1/2}MM_0^{1/2}N^{-\gamma/2} & \text{if } M<M_0. \end{cases}$$

Proof. We apply Lemma 2 (with $x_1=n$, $x_2=m$) and get

$$S_\gamma(M,N) \ll F_1^{-1/2}|S_1| + MNF^{-1/2}\log(MN), \tag{4}$$

where

$$S_1 = \sum_{(m,n)\in D_1} e(f_1(m,n)),$$

$$D_1 \subseteq \{(m,n)\,|\, M\leqslant m\leqslant 2M,\ FN^{-1}\leqslant n\leqslant 2FN^{-1}\},$$

$$f_1(m,n) = c(Xmn^\gamma)^{\frac{1}{1+\gamma}}, \qquad c>0 \quad \text{is a constant},$$

$$F = XMN^{-\gamma}, \qquad F_1 \asymp c_1XMN^{-2-\gamma}, \qquad c_1>0 \quad \text{is a constant}.$$

Using Lemma 1 we obtain

$$S_1 \ll \frac{MFN^{-1}}{q^{1/2}} + \frac{(MFN^{-1})^{3/4}}{q}\left(\sum_{h_1=1}^{q-1}\left(\sum_{h_2=1}^{q^2-1}|S_2|\right)^{1/2}\right)^{1/2}, \tag{5}$$

where

$$S_2 = \sum_{(m,n)\in D(h_1,h_2)} e(f_2(m,n)),$$

$$f_2(m,n) \underset{D_1(h_1,h_2)}{\overset{\infty}{\sim}} h_1h_2(Xmn^{-2-\gamma})^{\frac{1}{1+\gamma}},$$

$$D_1(h_1,h_2) = \{(m,n)\in D_1\colon\ (m,n+h_1+h_2)\in D_1\}.$$

We have

$$\left\{\sum_{h_1=1}^{q-1}\left(\sum_{h_2=1}^{q^2-1}|S_2|\right)^{1/2}\right\}^{1/2} \ll \left\{\sum_{h_1\leqslant q^{1/2}}\left(\sum_{h_2\leqslant q}|S_2|\right)^{1/2}\right\}^{1/2}$$

$$+\left\{\sum_{q^{1/2}\leqslant h_1\leqslant q}\left(\sum_{h_2\leqslant q^2}|S_2|\right)^{1/2}\right\}^{1/2}+\left\{\sum_{h_1\leqslant q}\left(\sum_{q\leqslant h_2\leqslant q^2}|S_2|\right)^{1/2}\right\}^{1/2}$$

$$\ll \left(XMN^{-1-\gamma}(M'-M)\right)^{1/4}q^{1/2}+\left\{\sum_{h_1\leqslant q}\left(\sum_{h_2\leqslant q^2,\, h_1h_2\geqslant q}|S_2|\right)^{1/2}\right\}^{1/2}.$$

Hence

$$S_1 \ll \frac{XMN^{-1-\gamma}(M'-M)}{q^{1/2}} + \frac{(XMN^{-1-\gamma}(M'-M))^{3/4}}{q}\left\{\sum_{h_1\leqslant q}\left(\sum_{h_2\leqslant q^2,\, h_1h_2\geqslant q}|S_2|\right)^{1/2}\right\}^{1/2}. \tag{6}$$

Now we apply Lemma 2 with $x_1=n,\ x_2=m$ to obtain

$$S_2 \ll (h_1h_2)^{-1/2}X^{3/2}M^{3/2}N^{-2-\frac{3\gamma}{2}}|S_3| + M^2XN^{-1-\gamma}(h_1h_2)^{-1/2}X^{1/2}M^{1/2}N^{-1-\frac{\gamma}{2}}\log(MN), \tag{7}$$

$$S_3=\sum_{(m,n)\in D_2} e(f_3(m,n)),$$

$$D_3\subseteq\left\{M\leqslant m\leqslant 2M,\quad c_3h_1h_2X^{-2}M^{-2}N^{3+2\gamma}\leqslant n\leqslant 2c_3h_1h_2X^{-2}M^{-2}N^{3+2\gamma}\right\},$$

$$f_3(m,n)\overset{\infty}{\underset{D_3}{\sim}} c_4(h_1h_2)^{\frac{1+\gamma}{3+2\gamma}}\left(Xmn^{2+\gamma}\right)^{\frac{1}{3+2\gamma}},$$

$c_1,\ c_2>0$ being constants.
Next,

$$\frac{\partial f_3}{\partial m}=(1+O(1))c_4\left[(h_1h_2)^{1+\gamma}Xn^{2+\gamma}m^{-2-2\gamma}\right]^{\frac{1}{3+2\gamma}}.$$

Let

$$Z=\max\left|\frac{\partial f_3}{\partial m}\right|\asymp c_5h_1h_2X^{-1}M^{-2}N^{2+\gamma}.$$

If $Z<1$, we apply Lemma 3 for the inner sum S_3 over m and have

$$S_3\ll h_1h_2X^{-2}M^{-2}N^{3+2\gamma}.\,Z^{-1/2}M^{-1/2}. \tag{8}$$

If $Z\geqslant 1$, then for $f_3(m,n)$ the conditions in the definition of a one-dimensional exponent pair are satisfied, and for every pair (k,λ) we have

$$S_3\ll h_1h_2X^{-2}M^{-2}N^{3+2\gamma}.\,Z^kM^\lambda. \tag{9}$$

Hence from (8) and (9) we have

$$S_3 \ll h_1 h_2 (XM)^{-2} N^{3+2\gamma} (Z^k M^\lambda + Z^{-1/2} M^{1/2}). \tag{10}$$

Now, from (5)–(7) and (10) we obtain

$$\begin{aligned} S_1 &\ll q^{-1/2} X M^2 N^{-1-\gamma} + q^{-1} X^{3/4} M^{3/2} N^{-\frac{3+3\gamma}{4}} \\ &\times \Big[q X^{-1/8} M^{1/4} + q^{\frac{11}{8}+\frac{3k}{4}} X^{-\frac{k}{4}} M^{-\frac{1-2\lambda+4k}{8}} N^{\frac{2+\gamma+4k+2\gamma k}{8}} \\ &+ q X^{\frac{3}{8}} M^{\frac{3}{8}} N^{-\frac{4+3\gamma}{8}} \Big] \log X. \end{aligned} \tag{11}$$

Put

$$q = \left[X^{3+2k} M^{5+4k-2\lambda} N^{-4-3\gamma-k(4+2\gamma)} \right]^{\frac{1}{7+6k}}.$$

If now the exponent pair (k, λ) is such that $M \leqslant M_0$, then we can estimate the sum S_1 trivially

$$S_1 \ll M \cdot XMN^{-1-\lambda} \ll M_0^{1/2} X N^{-1-\gamma} M^{3/2}. \tag{12}$$

Hence,

$$S_\gamma \ll M_0^{1/2} X^{1/2} N^{-\frac{\gamma}{2}} M. \tag{13}$$

If $M \geqslant M_0$, then from (5) and (11) we have

$$S_\gamma \ll \left[X^{-2-2k} M^{\frac{1}{4}(25+20k+\lambda)} N^{-(\gamma-1)(1+k)} \right]^{\frac{1}{7+6k}}. \tag{14}$$

THEOREM B. *Let X, M, N be positive numbers, $N \leqslant X^{\frac{1}{1+\gamma}}$, and (l_0, l_1) be a pair of exponents of dimension 2. Let*

$$q = \left[X^{3+6l_0-l_1} M^{3+9l_0} (M'-M)^{-l_1} N^{4+3\gamma+(9+6\gamma)l_0-(3+2\gamma)l_1} \right]^{\frac{1}{7+17l_0-3l_1}}.$$

If the conditions

$$1 \leqslant q \leqslant (XMN^{-\gamma})^{1/2}, \tag{I}$$

$$1 \leqslant \left(q^{3(l_1-l_0)} X^{-2l_1} N^{(3+2\gamma)l_1-l_0} M^{3l_0} (M'-M)^{-l_1} \right)^{\frac{1}{1+3l_0}} \leqslant M \tag{II}$$

are satisfied, then

$$S_\gamma(M,N) \ll q^{-1/2}(M'-M) XMN^{-\gamma} + q^{\alpha_1} X^{\alpha_2} M^{\alpha_3} (M'-M)^{\alpha_4} N^{\alpha_5},$$

where

$$\alpha_1 = \frac{11+25l_0-3l_1}{8(1+2l_0)}, \qquad \alpha_2 = \frac{-1-2l_0+l_1}{8(1+2l_0)}, \qquad \alpha_3 = \frac{-1-5l_0}{8(1+2l_0)},$$

$$\alpha_4 = \frac{8+4l_0+l_1}{8(1+2l_0)}, \qquad \alpha_5 = \frac{2+\gamma+(5+2\gamma)l_0-(3+2\gamma)l_1}{8(1+2l_0)}.$$

The proof of this theorem is similar to that of Theorem 1 in (Fugelo and Varbanec, 1987).

4. ESTIMATES FOR SUMS OF VALUES OF ψ

LEMMA 6. *For $x \to \infty$*

$$\sum_{n\leqslant x^{1/3}} \psi\left(\sqrt{\frac{x}{n}}\right) \ll x^{0,2}.$$

Proof. We have

$$\sum_{n\leqslant x^{1/3}} \psi\left(\sqrt{\frac{x}{n}}\right) = \sum_{1\leqslant n\leqslant x^{1/4}} \psi\left(\sqrt{\frac{x}{n}}\right) + \sum_{x^{1/4}<n\leqslant x^{1/3}} \psi\left(\sqrt{\frac{x}{n}}\right) = \sum\nolimits_1 + \sum\nolimits_2.$$

We apply Lemma 5 with

$$\alpha = \frac{1}{4}, \qquad \beta = \gamma = \frac{1}{2} \qquad (k, \lambda) = \left(\frac{11}{30}, \frac{16}{30}\right)$$

to the sum $\sum_1$. Then

$$\sum_{n\leqslant x^{1/4}} \psi\left(\sqrt{\frac{x}{n}}\right) \ll x^{0,199}.$$

To give an estimate for $\sum_2$ we use Lemma 4 and Theorem 2 by choosing

$$(k, \lambda) = \left(\frac{1}{14}, \frac{11}{14}\right).$$

Then we obtain

$$\sum_{x^{1/4}\leqslant n\leqslant x^{1/3}} \psi\left(\sqrt{\frac{x}{n}}\right) \ll x^{0,1994} \ll x^{0,2}.$$

LEMMA 7. *For $x \to \infty$*

$$\sum_{n\leqslant x^{1/3}} \psi\left(\frac{x}{n^2}\right) \ll x^{0,2204}.$$

The proof of this estimate is similar to that in (Fugelo and Varbanec, 1987).

5. PROOF OF THE THEOREM 1

For $h \geqslant x/2$ we have

$$\sum_{x\leqslant n\leqslant x+h} f(x) = \sum_{d\leqslant x^{1/2}} \Phi(d)\left(\left[\frac{x+h}{d^2}\right] - \left[\frac{x}{d^2}\right]\right) = h\sum_{d=1}^{\infty} \frac{\Phi(d)}{d^2} + O\left(x^{1/2+\varepsilon}\right).$$

For $x^{3/2} \leqslant h \leqslant x/2$

$$\sum_{x\leqslant n\leqslant x+h} f(n) = \sum_{d\leqslant h^{1/2}} \Phi(d)\left(\left[\frac{x+h}{d^2}\right] - \left[\frac{x}{d^2}\right]\right) + \sum_{\substack{x\leqslant md^2\leqslant x+h\\ d>h^{1/2}}} \Phi(d)$$

$$= h\sum_{d=1}^{\infty} \frac{\Phi(d)}{d^2} + O\left(h^{1/2}x^{\varepsilon}\right).$$

Next, for $h < x^{2/3}$ we get similarly

$$\sum_{x \leqslant n \leqslant x+h} f(n) = h \sum_{d=1}^{\infty} \frac{\Phi(d)}{d^2} + O(h^{1/2} x^{\varepsilon}) + O\left(x^{\varepsilon} \left| \sum_{h \leqslant x^{1/3}} \psi\left(\sqrt{\frac{x}{n}}\right) \right| \right)$$
$$+ O\left(x^{\varepsilon} \left| \sum_{n \leqslant x^{1/3}} \psi\left(\frac{x}{n^2}\right) \right| \right).$$

By Theorems A and B we have

$$\sum_{x \leqslant n \leqslant x+h} f(n) = h \sum_{d=1}^{\infty} \frac{\Phi(d)}{d^2} + O(h^{1/2} x^{\varepsilon}) + O(x^{0,2204+\varepsilon}).$$

Remark. The Theorems A and B can be used for the study of functions $f(n) \in \mathfrak{M}_{a,b,c}$, where $\mathfrak{M}_{a,b,c}$ is the class of multiplicative function such that

$$f(n) = \sum_{n^a = m^b d^c} \Phi(d), \qquad 1 \leqslant a \leqslant b < c.$$

REFERENCES

Fugelo, N. and Varbanec, P. (1987). On a trigonometric sum and its application. *Acta Math. Hung.* **49**, 339–348.

Graham, S. and Kolesnik, G. (1988). On the difference between consecutive squarefree integers. *Acta Arithm.* **49**, 435–447.

Richert, H.-E. (1954). On the difference between consecutive squarefree numbers. *J. London Math. Soc.* **29**, 16–20.

Srinivasan, B. R. (1965). Lattice point problem of many-dimensional hyperboloide. III. *Math. Ann.* **160**, 280–311.

Part Five

DISTRIBUTION OF ADDITIVE FUNCTIONS

New Trends in Probab. and Statist., pp. 191–199
F. Schweiger and E. Manstavičius (Eds.)
1992 VSP/TEV

SMOOTHNESS OF THE DISTRIBUTIONS OF ARITHMETIC FUNCTIONS

G. J. BABU*
Department of Statistics, 219 Pond Laboratory, Pennsylvania State University, University Park, PA 16802, U.S.A.

ABSTRACT

It is well known that the distribution of a real-valued additive arithmetic function f, if it exists, is either purely discrete, purely continuous singular, or purely absolutely continuous. The distribution of f is shown to be absolutely continuous if f is not "too small" on primes. This result is generalized to additive functions on the set of pairs of positive integers. Conditions for existence of singular distributions for additive functions on the set of pairs of positive integers are also given. These are nontrivial extensions, as the convolution of two singular distributions need not be singular. A result on the distributions of multiplicative functions on the set of pairs of positive integers is also given.

1. INTRODUCTION

A real-valued arithmetic function f is called additive if

$$f(mn) = f(m) + f(n),$$

whenever $(m, n) = 1$. A real-valued additive arithmetic function is said to have distribution F, if the density of $(m\colon\ f(m) \leqslant x)$ exists and equals to $F(x)$, whenever x is a continuity point of F. Conditions, under which the distribution of a real-valued additive arithmetic function is absolutely continuous, are given in the next section. This result complements the work of (Babu, 1973), where among other results, it is established that every bounded additive arithmetic function has a singular distribution.

Similar results for the distribution of additive functions on the set of pairs of positive integers are given in Section 3. Though such distributions are related to the distributions of the associated "marginal" additive arithmetic functions, the results of Section 3 are nontrivial extensions. This is because the convolution of two singular distributions need not be singular. A result on multiplicative functions on the set of pairs of positive integers is also stated.

Throughout this paper let p and q denote prime numbers, m and n denote natural numbers, $\sum_p$ denote the sum over prime numbers and let $\prod_p$ denote the product over prime numbers.

---

*Research supported in part by NSA Grant MDA 904-90-M-1001.

2. ABSOLUTE CONTINUITY

To obtain sufficient conditions, under which a real-valued additive arithmetic function has an absolutely continuous distribution, we introduce the following

DEFINITION. A positive real valued function h on the interval $[2,\infty)$ is said to satisfy Condition K, if for some $x_0 \geqslant 3$,

i) h is decreasing in $[x_0,\infty)$, $\lim_{x\to\infty} h(x) = 0$,

and

ii) for some $0 < \alpha < 2$, $h(x)(\log x)^\alpha$ is nondecreasing on $[x_0,\infty)$.

THEOREM 1. *Let Q be a set of primes satisfying*

$$\sum_{p\notin Q} \frac{1}{p} < \infty.$$

Let f be a real-valued additive arithmetic function such that $|f(p)| = h(p)$ for all $p \in Q$, where h satisfies Condition K. Then the distribution of f, if it exists, is absolutely continuous.

Sketch of the Proof. It is well known that f has a distribution F, if and only if the product

$$L(t) = \prod_p \left(1 - \frac{1}{p}\right)\left(1 + \sum_{k=1}^{\infty} \frac{1}{p^k} e^{itf(p^k)}\right)$$

converges. If it converges, then L denotes the characteristic function of F. See Theorem 4.5 in (Kubilius, 1964) or the proof of Theorem 1 in (Babu, 1974). By the Plancherel theorem (see, for instance, (Rudin, 1973, Thm. 7.9)), it is enough to show that L is square integrable. This is established using an estimate of

$$|L(2t)|^2 = \exp\left(-4 \sum_p \frac{1}{p} \sin^2 (th(p)) + O(1)\right), \tag{1}$$

and the well-known estimate that for some $\gamma > 0$ and $d > 0$,

$$\sum_{p\leqslant x} \frac{1}{p} = \log\log x + \gamma + O\Big(\exp\big(- d\sqrt{\log x}\big)\Big). \tag{2}$$

Towards this end, an appropriate lower bound for the sum in the exponent of (1) is obtained by expressing it as several sums over different intervals of primes, and estimating each sum as in (Babu, 1975a). See (Babu, 1991) for details.

Examples

The conditions of Theorem 1 hold if f is a real-valued additive arithmetic function given by one of the following definitions:

a) $|f(p)| = (\log p)^{-\alpha}, \qquad 0 < \alpha < 2,$

b) $|f(p)| = (\log\log p)^{-\alpha}, \qquad \alpha > 1/2,$

c) $|f(p)| = (\log\log p)^{-1/2}(\log\log\log p)^{-\alpha}, \qquad \alpha > 1/2,$

d) $|f(p)| = e^{-d(\log\log p)^{\alpha}}, \qquad 0 < \alpha < 1 \quad \text{and} \quad d > 0.$

The arithmetic function f in the examples above has absolutely continuous distribution, whenever it exists. The special case a) was considered in (Babu, 1975a). Theorem 1 complements the results of (Babu, 1973), where conditions for an additive arithmetic function to have a singular distribution are given.

3. FUNCTIONS ON THE SET OF PAIRS OF POSITIVE INTEGERS

A real-valued arithmetic function g on the set of pairs of positive integers is said to be additive if $g(1,1) = 0$ and

$$g(mi, nj) = g(m,n) + g(i,j),$$

whenever $(mn, ij) = 1$. It is called multiplicative if $g(1,1) = 1$ and

$$g(mi, nj) = g(m,n)g(i,j),$$

whenever $(mn, ij) = 1$. A real-valued arithmetic function g on the set of pairs of positive integers is said to have a distribution if for some probability distribution function F,

$$\frac{1}{MN}\#\{1 \leqslant m \leqslant M,\ 1 \leqslant n \leqslant N\colon\ g(m,n) \leqslant x\} \longrightarrow F(x),$$

as M and N tend to infinity independently, for all continuity points x of F. See (Delange, 1969).

It is well known that the distribution of an additive function g on the set of pairs of positive integers exists if and only if both the marginals $g_1(m) = g(m,1)$ and $g_2(n) = g(1,n)$ have distributions. Further the distribution of g is closely related to the convolution of the distributions of g_1 and g_2. Since convolution of two singular distributions is not necessarily singular the following result is of interest.

THEOREM 2. *Let positive numbers θ and β satisfy the inequality*

$$\theta^{-1} + \beta^{-1} > \frac{1}{2}.$$

Let h_1 and h_2 be two functions satisfying Condition K *with $\alpha = \theta$ and $\alpha = \beta$, respectively. Suppose g is a real-valued additive function on the set of pairs of* *positive integers such that* $|g(p,1)| = h_1(p)$ *and* $|g(1,p)| = h_2(p)$ *for all* $p \in Q$, *where*

$$\sum_{p \notin Q} \frac{1}{p} < \infty.$$

Then the distribution of g, *if it exists, is absolutely continuous.*

Proof. Using the estimates in the proof of Theorem 1, the characteristic function of the distribution of g can be shown to be square integrable under the given conditions on θ and β. Now the result follows from the Plancherel theorem.
If $\theta = 3$ and $\beta = 4$ in Theorem 2, then we cannot conclude absolute continuity of the distributions of g_1 and g_2 from Theorem 1. But on the other hand g in Theorem 2 has an absolutely continuous distribution, as $3^{-1} + 4^{-1} > 2^{-1}$.
Now we consider conditions for singularity of the distributions of additive functions on the set of pairs of positive integers. We require a result from probability theory, which is stated as Lemma 1 below. A proof of the lemma is given for easy reference.

LEMMA 1. *Let* $\{X_n\}$ *and* $\{Y_n\}$ *be two sequences of independent discrete random variables such that* $X = \sum X_n$ *converges almost surely. If*

$$\sum_n \sum_x |\mathbb{P}(X_n = x) - \mathbb{P}(Y_n = x)| < \infty, \tag{3}$$

then $Y = \sum Y_n$ *converges almost surely. Further,* Y *has absolutely continuous distribution if and only if* X *has absolutely continuous distribution. The distribution of* Y *is discrete if and only if the distribution of* X *is discrete.*

Proof. In view of Theorem E of (Babu, 1978), it is enough to prove that $\{X_n\}$ and $\{Y_n\}$ can be defined on a single probability space without changing their distributions and such that

$$\sum_n \mathbb{P}(X_n \neq Y_n) < \infty.$$

To show this, we fix n. Suppose X_n and Y_n take values only in $\{x_1, x_2, \ldots\}$. Define

$$u_i = \mathbb{P}(X_n = x_i) \quad \text{and} \quad v_i = \mathbb{P}(Y_n = x_i).$$

Let λ denote the Lebesgue measure on the unit interval,

$$C = \{i \colon u_i \geqslant v_i\} = \{i_1 < i_2 < \ldots\},$$

$$B = \{i \colon u_i < v_i\} = \{j_1 < j_2 < \ldots\},$$

$$a_0 = 0, \qquad a_j - a_{j-1} = u_{i_j},$$

$$c_0 = \sum_j u_{i_j}, \qquad c_i - c_{i-1} = u_{j_i}$$

and $b_j = a_{j-1} + v_{i_j}$. Note that a_i and c_j make partitions of the unit interval.

If

$$A = \bigcup_j (b_j, a_j],$$

then

$$\lambda(A) = \sum_j \left(u_{i_j} - v_{i_j}\right),$$

and the set A can be partitioned into union of intervals I_i, $i \in B$, such that

$$A = \bigcup_{i \in B} I_i \quad \text{and} \quad \lambda(I_i) = v_i - u_i.$$

Now for $\omega \in (0,1]$ we define

$$X_n'(\omega) = \begin{cases} x_{i_j} & \text{if} \quad \omega \in (a_{j-1}, a_j], \\ x_{j_i} & \text{if} \quad \omega \in (c_{j-1}, c_j], \end{cases}$$

$$Y_n'(\omega) = \begin{cases} x_{i_j} & \text{if} \quad \omega \in (a_{j-1}, b_j], \\ x_{j_i} & \text{if} \quad \omega \in (c_{j-1}, c_j] \bigcup I_{j_i}. \end{cases}$$

Since

$$\sum_{i \in C} (u_i - v_i) = \sum_{i \in B} (v_i - u_i),$$

X_n' and Y_n' have the same distributions as those of X_n and Y_n, respectively. Further,

$$\begin{aligned} \lambda(X_n' \neq Y_n') &= \sum_{i \in C} (u_i - v_i) = \frac{1}{2} \sum_i \left|u_i - v_i\right| \\ &= \frac{1}{2} \sum_x \left|\mathbb{P}(X_n = x) - \mathbb{P}(Y_n = x)\right|. \end{aligned} \tag{4}$$

Let

$$\Omega = (0,1] \times (0,1] \times \dots,$$

and let $\mathbb{P}^0$ denote the product measure on Ω given by the Lebesgue measure on $(0,1]$. For

$$\omega = \left(\omega_1, \omega_2, \dots\right) \in \Omega, \qquad \omega_n \in (0,1],$$

define

$$X_n^0(\omega) = X_n'\left(\omega_n\right) \quad \text{and} \quad Y_n^0(\omega) = Y_n'\left(\omega_n\right).$$

Clearly $\{X_n^0\}$ and $\{Y_n^0\}$ are two sequences of independent random variables such that for each n, X_n^0 and Y_n^0 have the same distributions as those of X_n and Y_n, respectively. From (3) and (4), we have

$$\sum_n \mathbb{P}^0(X_n^0 \neq Y_n^0) < \infty.$$

This completes the proof.

The probabilistic part of the proof of the next theorem is extracted as the following

LEMMA 2. *Let $r \geqslant 2$ be an integer. Let $\{a_{sp}$: p is prime, $1 \leqslant s \leqslant r\}$ be a sequence of real numbers satisfying*

$$\sum_{s=1}^{r} \sum_{p>N} \frac{1}{p} a_{sp}^2 = O(N^{-d}) \tag{5}$$

for some $d > 0$. Let $\{X_{sp}$: p is prime, $1 \leqslant s \leqslant r\}$ be independent random variables satisfying

$$\mathbb{P}(X_{sp} = 1) = 1 - \mathbb{P}(X_{sp} = 0) = p^{-1}.$$

Then

$$\sum_{s=1}^{r} \sum_{p} a_{sp}(X_{sp} - p^{-1})$$

converges almost surely to a random variable X and the distribution of X is singular with respect to the Lebesgue measure on the line.

Proof. The proof is adopted from (Erdös, 1939). Without loss of generality, we assume that $0 < d < 1$. Let $a = d/(4r)$ and $b = d/3$. Define

$$A_N = \sum_{s=1}^{r} \sum_{p \leqslant N} \frac{1}{p} a_{sp},$$

$$B_N = \{m \leqslant N^a\colon \ m \text{ is square-free}\},$$

$$C_N = \left\{ \sum_{s=1}^{r} \sum_{p|m_s} a_{sp} - A_N\colon \ m_s \in B_N \text{ for } s = 1, \ldots, r \right\}.$$

Let $\bigcup'$ and $\sum'$ denote the union and sum over all

$$\{(m_1, \ldots, m_r)\colon \ m_s \in B_N, \ s = 1, \ldots, r\}.$$

Note that

$$\begin{aligned}
&\mathbb{P}\left(\left(\sum_{s=1}^{r} \sum_{p \leqslant N} a_{sp} X_{sp} - A_N\right) \in C_N\right) \\
&\geqslant \mathbb{P}\Bigg(\bigcup{}'\Bigg(\left(\sum_{s=1}^{r} \sum_{p \leqslant N} a_{sp} X_{sp} - A_N\right) \in C_N, \\
&\qquad X_{sp} = 1 \text{ for } p \mid m_s \text{ and } X_{sp} = 0 \text{ for } (p, m_s) = 1; \\
&\qquad \text{for all } s = 1, \ldots, r \text{ and } p \leqslant N\Bigg)\Bigg) \\
&= \mathbb{P}\Bigg(\bigcup{}'(X_{sp} = 1 \text{ for } p \mid m_s \text{ and } X_{sp} = 0 \text{ for } (p, m_s) = 1;
\end{aligned}$$

$$\text{for all } s = 1,\ldots,r \text{ and } p \leqslant N)\Big)$$

$$= \sum{}' \mathbb{P}(X_{sp} = 1 \text{ for } p \mid m_s \text{ and } X_{sp} = 0 \text{ for } (p,m_s) = 1;$$

$$\text{for all } s = 1,\ldots,r \text{ and } p \leqslant N)$$

$$= \prod_{p\leqslant N}\left[\left(1-\frac{1}{p}\right)\left(\sum_{m\in B_N}\frac{1}{m}\prod_{p\mid m}\left(1-\frac{1}{p}\right)^{-1}\right)\right]^r$$

$$\geqslant \prod_{p\leqslant N}\left[\left(1-\frac{1}{p}\right)\sum_{m\in B_N}\frac{1}{m}\right]^r \geqslant \delta$$

for some $\delta > 0$ and for all large N. The second equality above follows from the fact that the events considered are disjoint. The last inequality follows from

$$(a\log N)^{-1}\sum_{m\in B_N}\frac{1}{m} \longrightarrow \prod_p\left(1-\frac{1}{p^2}\right) > 0.$$

Further, observe that by Chebyshev's inequality,

$$\mathbb{P}\left(\left|\sum_{s=1}^{r}\sum_{p>N}a_{sp}\left(X_{sp}-\frac{1}{p}\right)\right| > N^{-b}\right)$$

$$\leqslant N^{2b}\sum_{s=1}^{r}\sum_{p>N}\frac{1}{p}a_{sp}^2 = O(N^{2b-d}) = O(N^{-b})\,.$$

This leads to

$$\mathbb{P}(X\in G_N) \geqslant \delta - O(N^{-b})$$

for all large N, where

$$G_N = \bigcup_{x\in C_N}\left[x-N^{-b}, x+N^{-b}\right].$$

On the other hand the Lebesgue measure of G_N is bounded by

$$2N^{-b}N^{ar} \leqslant 2N^{-d/12} \longrightarrow 0,$$

as $N\to\infty$. It follows that the distribution of X is singular. This completes the proof.

In addition to the conditions of Lemma 2, if

$$\sum_p \frac{1}{p}a_{sp}$$

converges for $1 \leqslant s \leqslant r$, then

$$\sum_{s=1}^{r}\sum_p a_{sp}X_p$$

converges almost surely and it has a singular distribution.

THEOREM 3. *Let g be an additive arithmetic function on the set of pairs of positive integers having a distribution F. If*

$$\sum_{p>N} \frac{1}{p}\left[(g(1,p))^2 + (g(p,1))^2\right] = O(N^{-d})$$

for some $d > 0$ then g has a singular distribution.

Proof. Let X_{1p} and X_{2p} be as in Lemma 2. From Theorem 2 of (Babu, 1976), g has a distribution if and only if

$$\sum_{p} (g(p,1)X_{1p} + g(1,p)X_{2p})$$

converges almost surely. Now the result follows from Lemmas 1 and 2.
Theorem 3 extends the results of (Babu, 1973) to arithmetic functions on the set of pairs of positive integers.

COROLLARY. *If an additive function on the set of pairs of positive integers is bounded, then it has a singular distribution.*

The derivation of the Corollary from Theorem 3 is similar to the proof of Corollary 2 of (Babu, 1973).
We now state a result on the distributions of multiplicative functions on the set of pairs of positive integers. This can be proved following the lines of proof of Theorems 1 and 3 of (Babu, 1980).

THEOREM 4. *Let g be a real-valued multiplicative function on the set of pairs of positive integers. Let g_1 and g_2 be the multiplicative functions given by $g_1(m) = g(m,1)$ and $g_2(n) = g(1,n)$. Then g has a distribution if and only if both g_1 and g_2 have distributions. Further, the distribution of g is concentrated at zero if and only if either the distribution of g_1 or g_2 is concentrated at zero. This occurs if and only if*

$$\sum_{p\in Q} \frac{1}{p} = \infty,$$

where $Q = \{p\colon\ g(p,1)g(1,p) = 0\}$.
The conditional distribution of g given $g \neq 0$ is of pure type. It is absolutely continuous if and only if the conditional distribution of $g_1(m)g_2(n)$ given $g_1(m)g_2(n) \neq 0$ is absolutely continuous.
In addition, if

$$\sum_{g(p,1)<0} \frac{1}{p} + \sum_{g(1,p)<0} \frac{1}{p} = \infty, \tag{6}$$

then g has a symmetric distribution.

Proof of a special case of this result, using characteristic transforms, was given in (Babu, 1975b). We only mention that, in proving that g has a symmetric distribution under (6), we use the following lemma, instead of Lemma 4 of (Babu, 1980).

LEMMA 3. *For $k > 1$, let B_k consist of all pairs (i, j) of positive integers such that $(ij, p) = 1$ for all $p > k$, and A_k consist of all pairs (m, n) of positive integers such that $(mn, p) = 1$ for all $p \leqslant k$. Let*

$$L_k \subseteq A_k, \qquad M_k \subseteq B_k,$$

$$R_k = \{(im, jn)\colon\ (i, j) \in B_k,\ (m, n) \in L_k\}$$

and

$$S_k = \{(im, jn)\colon\ (i, j) \in M_k,\ (m, n) \in A_k\}.$$

Then S_k has a density. If R_k has density, then the density of $R_k \cap S_k$ exists and is given by the product of the densities of R_k and S_k. That is R_k and S_k are "independent."

Proof is similar to that of Lemma 4 of (Babu, 1980).

REFERENCES

Babu, G. J (1973). Some results on the distribution of additive arithmetic functions. III. *Acta Arithm.* **25**, 39–49.

Babu, G. J. (1974). On the characteristic function of the distribution of the values of additive arithmetic functions. *Liet. Matem. Rink.* **14**, 27–31.

Babu, G. J. (1975a). Absolutely continuous distribution functions of additive functions $f(p) = (\log p)^{-\alpha}$, $\alpha > 0$. *Acta Arithm.* **26**, 401–403.

Babu, G. J. (1975b). Some results on the distribution of the values of multiplicative functions. *Sankhyā, Ser. A* **37**, 386–395.

Babu, G. J. (1976). Some results on the distribution of values of additive functions on the set of pairs of positive integers. I. *Acta Arithm.* **29**, 171–179.

Babu, G. J. (1978). *Probabilistic Methods in the Theory of Arithmetic Functions.* Macmillan Comp. of India, New Delhi.

Babu, G. J. (1980). On the distribution of multiplicative functions. *Acta Arithm.* **36**, 331–340.

Babu, G. J. (1991). Absolute continuity of the distributions of arithmetic functions. *Preprint.*

Delange, H. (1969). On some sets of pairs of positive integers. *J. Number Theory* **1**, 261–279.

Erdös, P. (1939). On the smoothness of the asymptotic distribution of additive arithmetical functions. *Amer. J. Math.* **61**, 722–725.

Kubilius, J. (1964). *Probabilistic Methods in the Theory of Numbers.* American Mathem. Soc., Providence, RI.

Rudin, W. (1973). *Functional Analysis.* McGraw-Hill, New York.

New Trends in Probab. and Statist., pp. 201–209
F. Schweiger and E. Manstavičius (Eds.)
1992 VSP/TEV

ON THE ERDÖS–WINTNER THEOREM IN BANACH SPACES

J. DE LA CAL*
Departamento de Matemática Aplicada y Estadística e Investigación Operativa, Facultad de Ciencias, Universidad del País Vasco, Apartado 644, E-48080 Bilbao, Spain

ABSTRACT

The famous Erdös–Wintner theorem can be stated in terms of the a.s. convergence of a random series associated with the real valued additive function. This last statement is extended to additive functions taking values in any real separable Banach space. The classical "three-series" form of the Erdös–Wintner theorem has a perfect infinite-dimensional analogue when the Banach space is isomorphic to a real separable Hilbert space. Some applications to asymptotic density are considered.

1. INTRODUCTION AND THE MAIN RESULT

The Probabilistic Theory of Numbers is concerned with the distribution of values of arithmetic functions, i.e. functions of natural argument. Let f be an arithmetic function with values in a metric space $\mathfrak{X}$, and denote by $\mathbb{P}_n f^{-1}$ the probability distribution of f with respect to the probability measure on the set $\mathbb{N}$ of natural numbers assigning the weight $1/n$ to each $k \leqslant n$, and 0 for $k > n$. If $(\mathbb{P}_n f^{-1})$ converges weakly (as $n \to \infty$) to a probability law μ on $\mathfrak{X}$, then μ is said to be the "limit distribution" of f, and we write $f \Rightarrow \mu$, or $f \Rightarrow Z$ whenever Z is an $\mathfrak{X}$-valued random variable with distribution μ.
One of the most famous results in the Probabilistic Theory of Numbers is the following "three-series theorem" due to Erdös and Wintner, which gives both necessary and sufficient conditions in order that a real additive arithmetic function should have a limit distribution. We recall that f: $\mathbf{N} \to \mathbf{R}$ is additive if $f(mn) = f(m) + f(n)$ whenever m and n are relatively prime, and, in this case, we have $f(1) = 0$ and $f(n) = \sum_p f(p^{\beta_p(n)})$, where $\beta_p(n)$ is the exponent of the prime p in the factorization of n. Obviously, the definition of additive function makes sense when f takes values in a Banach space or, more generally, in an Abelian group.

THEOREM A. *Let f be a real additive arithmetic function.*
(a) *If for some $c > 0$ the three series*

$$\sum_{|f(p)|\geqslant c} \frac{1}{p}, \quad \sum_{|f(p)|<c} \frac{f(p)}{p}, \quad \sum_{|f(p)|<c} \frac{|f(p)|^2}{p} \tag{1}$$

*Supported by the University of the Basque Country.

converge, then

$$f \Longrightarrow \sum_p f(p^{b_p}). \tag{2}$$

(b) *Conversely, if f has a limit distribution, then the three series in* (1) *converge for any $c > 0$.*

Some explanations about the notations are necessary. Following (Billingsley, 1974), the b_p involved in (2) are independent random variables with the geometric distribution

$$\mathbb{P}(b_p = k) = (1 - 1/p)p^{-k}, \qquad k = 0, 1, 2, \dots ,$$

(they are idealized limiting versions of the functions β_p). The indicator function of the event ($b_p \geqslant 1$) will be denoted by d_p. Of course, p ranges over the set of prime numbers.

A polished version of the original proofs of Theorem A can be found in (Billingsley, 1974). Different proofs combining the method of characteristic functions with other analytical tools can be seen in (Elliott, 1979) and (Kubilius, 1964).

In (de la Cal, 1989) we have shown that part (a) of Theorem A extends to additive arithmetic functions taking values in a real separable and r-smoothable Banach space, and part (b) also holds when the Banach space is finite-dimensional. For the concepts and the properties of type r, cotype r, and r-smoothable (i.e. isomorphic to uniformly r-smooth) Banach space see, for instance, (Araujo and Giné, 1980) and (Hoffmann-Jørgensen and Pisier, 1976).

In the present paper we prove that the whole of Theorem A, stated in an equivalent form, actually extends to additive functions with values in any real separable Banach space. More precisely, the following theorem holds.

THEOREM B. *Let $\mathfrak{X}$ be a real separable Banach space. An additive arithmetic function f with values in $\mathfrak{X}$ has a limit distribution if and only if the series $\sum_p f(p^{b_p})$ converges a.s. and, in this case, $f \Rightarrow \sum_p f(p^{b_p})$.*

Note that, by Borel-Cantelli lemma, the sequences $\{f(p^{b_p})\}$ and $\{f(p^{d_p})\}$ are tail-equivalent, and therefore $\sum_p f(p^{b_p})$ converges a.s. if and only if the same thing occurs for $\sum_p f(p^{d_p})$. In the case of real-valued f, by applying to the last series the well known "three-series theorem" of Kolmogorov, we get that the Theorem B is just an equivalent reformulation of Theorem A.

The proof of Theorem B is deferred to the next section. In section 3 we state some immediate corollaries; in particular, the "three-series" Theorem A has a perfect infinite-dimensional analogue when the Banach space is both of type 2 and cotype 2, i.e. when it is isomorphic to a real separable Hilbert space.

Obviously, the class of additive functions with values in a given Banach space (or Abelian group) $\mathfrak{X}$ is very wide. In fact if we choose for each prime p and each positive integer k an element $x_{p,k}$ of $\mathfrak{X}$, then the expression

$$g(n) = \sum_{p^k \| n} x_{p,k}, \qquad n \in \mathbf{N},$$

where $p^k \parallel n$ means that n is divisible by p^k but not by p^{k+1}, defines an $\mathfrak{X}$-valued additive function. Additive functions with values in abstract Abelian groups or

Hilbert spaces have been considered earlier by (Ruzsa, 1977, 1984) and (Nakutis, 1985).
On the other hand a real additive function f generates many interesting additive functions with values in infinite-dimensional spaces in the following way. Let Q_f be the set of prime numbers p such that $f(p^k) \neq 0$ for some k, and let

$$w = \{Q_1, \ldots, Q_j, \ldots\}$$

be a labelled partition of Q_f. Now define

$$f_w(n) = (f_1(n), \ldots, f_j(n), \ldots), \qquad n \in \mathbf{N},$$

where

$$f_j(n) = \sum_{p \in Q_j} f(p^{\beta_p(n)}) \qquad \text{for} \quad j = 1, 2, \ldots.$$

Note that $f = \sum_j f_j$. As the sequence $(f_j(n))_{j \geqslant 1}$ has a finite support we have that f_w is an additive function with values (for instance) in $l_2(\mathbf{R})$. We call it the "w-unfolding" of f. The existence of the limit distribution μ for f provides information about the asymptotic density of the subsets of $\mathbf{N}$ which are definable in terms of f; in particular,

$$\mathbb{D}(m\colon\ f(m) \in B) \stackrel{\text{def}}{=} \lim_n \mathbb{P}_n(m\colon\ f(m) \in B) = \mu(B)$$

for each Borel set B of real numbers such that $\mu(\partial B) = 0$, ∂B being the boundary of B. The corresponding fact for f_w provides information about the asymptotic density of the subsets of $\mathbf{N}$ which are definable in terms of the compounding functions f_j. We shall illustrate this idea in Section 4 (see also (Cal, 1989)).

2. THE PROOF OF THEOREM B

If f is an additive $\mathfrak{X}$-valued function, then, for each y in the dual $\mathfrak{X}^*, y \circ f$ is a real-valued additive function. This observation explains why Theorem A will play an important role in the proof of Theorem B. Some lemmas will be needed in addition. The first one follows from a theorem of de Acosta; see, for instance, (Vakhania *et al.*, 1987, Chap. IV, Thm. 3.3).

LEMMA 1. *Let $\mathfrak{X}$ be a real separable Banach space. In order that a sequence (μ_n) of probability laws on $\mathfrak{X}$ should converge weakly to a probability measure μ it is both necessary and sufficient that the following two conditions be fulfilled:*

(a) $(\mu_n y^{-1})$ *converges weakly to* μy^{-1} *for every y in the dual* $\mathfrak{X}^*$,

(b) *the sequence (μ_n) is flatly concentrated, that is, for any $\varepsilon > 0$ there exists a finite-dimensional subspace F of $\mathfrak{X}$ such that*

$$\inf_n \mu_n(F_\varepsilon) \geqslant 1 - \varepsilon,$$

where F_ε stands for the ε-neighborhood of F.

The following lemma is contained in (Vakhania *et al.*, 1987, Chap. V, Prop. 2.6).

LEMMA 2. *Let* (X_n) *be a sequence of independent random variables taking values in a real separable Banach space* $\mathfrak{X}$, *and let* μ *be a probability measure on* $\mathfrak{X}$. *If* $(\mu_n y^{-1})$ *converges weakly to* μy^{-1}, *for each* $y \in \mathfrak{X}^*$, *where* μ_n *is the probability distribution of* $S_n = X_1 + \ldots + X_n$, *then there exists a sequence* (x_n) *of elements of* $\mathfrak{X}$ *such that the series* $\sum_n (X_n - x_n)$ *converges a.s., and the series* $\sum_n x_n$ *converges in the weak topology of* $\mathfrak{X}$.

Finally, the following results can be found in (Ruzsa, 1984). It is the key for solving the questions concerning flat concentration.

LEMMA 3. *Let* G *be an Abelian group, and let* f *be a* G*-valued additive function. For arbitrary* $A \subset G$ *we have*

$$\mathbb{P}_n(m\colon\ f(m) \notin A + A - A) \leqslant c\mathbb{P}\left(\sum_{p \leqslant n} f(p^{b_p}) \notin A \right),$$

where $A \pm B = \{a \pm b\colon\ a \in A,\ b \in B\}$, *and* c *is an absolute constant, i.e. it does not depend upon* f, n, A, *and* G *(obviously we can suppose* $c > 3$*).*

In what follows μ_n will stand for the probability distribution of $\sum_{p \leqslant n} f(p^{b_p})$. Suppose that the series $\sum_p f(p^{b_p})$ converges a.s., and let μ be the probability distribution of the sum. Then, for each $y \in \mathfrak{X}^*$, the series $\sum_p y \circ f(p^{b_p})$ converges a.s., and the probability distribution of the sum is μy^{-1}. Therefore, by Theorem A, we have that $((\mathbb{P}_n f^{-1}) y^{-1})$ converges weakly to μy^{-1}, for any $y \in \mathfrak{X}^*$. On the other hand, by Lemma 1, the sequence (μ_n) is flatly concentrated. Consequently, given $\varepsilon > 0$ there exists a finite-dimensional subspace F of $\mathfrak{X}$ such that

$$\sup_n \mu_n(F^c_{\varepsilon/3}) \leqslant \varepsilon / c$$

(c is the constant in Lemma 3). As $F_{\varepsilon/3} + F_{\varepsilon/3} - F_{\varepsilon/3} \subset F_\varepsilon$ we deduce from Lemma 3

$$\sup_n \mathbb{P}_n(m\colon\ f(m) \notin F_\varepsilon) \leqslant \varepsilon.$$

This means that $(\mathbb{P}_n f^{-1})$ is flatly concentrated, and, by Lemma 1, we conclude that f has the limit distribution μ.

Now we shall deal with the converse. If f has a limit distribution μ, then for each $y \in \mathfrak{X}^*$ the sequence $((\mathbb{P}_n f^{-1}) y^{-1})$ converges weakly to μy^{-1}, and therefore, by Theorem A, $(\mu_n y^{-1})$ converges weakly to μy^{-1}. Hence, in view of Lemma 2, there exists a sequence (x_p) of elements of $\mathfrak{X}$ such that the series

$$\sum_p (f(p^{b_p}) - x_p)$$

converges a.s., and the series $\sum_p x_p$ converges in the weak topology of $\mathfrak{X}$. Put

$$a_n = \sum_{p \leqslant n} x_p,$$

and let ν_n be the probability distribution of

$$\sum_{p \leqslant n} f(p^{b_p}) - a_n.$$

As (ν_n) is flatly concentrated, given $\varepsilon > 0$ there exists a finite dimensional subspace F of $\mathfrak{X}$ such that

$$\sup_n \nu_n(F^c_{\varepsilon/3}) \leqslant \varepsilon/c,$$

that is

$$\sup_n \mathbb{P}\Big(\sum_{p \leqslant n} f(p^{b_p}) \notin a_n + F_{\varepsilon/3} \Big) \leqslant \varepsilon/c.$$

Therefore, by Lemma 3,

$$\sup_n \mathbb{P}_n(m\colon\ f(m) \notin a_n + F_\varepsilon) \leqslant \varepsilon,$$

or

$$\sup_n \mathbb{P}_n(m\colon\ f(m) - a_n \notin F_\varepsilon) \leqslant \varepsilon.$$

This shows that $(\mathbb{P}_n(f - a_n)^{-1})$ is flatly concentrated. Since $(\mathbb{P}_n f^{-1})$ is also flatly concentrated, it is easy to see that (δ_{a_n}) is flatly concentrated, δ_{a_n} being the unit mass at a_n, and this fact together with the convergence of (a_n) to an element $a \in \mathfrak{X}$ in the weak topology of $\mathfrak{X}$ imply (Lemma 1) that (δ_{a_n}) converges weakly to δ_a. This is equivalent to say that (a_n) converges to a in the norm. Consequently $\sum_p f(p^{b_p})$ converges a.s. and the proof is complete.

3. SOME COROLLARIES

As observed in the introduction, $\sum_p f(p^{b_p})$ converges a.s. if and only if the same thing occurs for $\sum_p f(p^{d_p})$. Thus next corollary follows directly from (Buldygin, 1973, Thm. 3) and the well known results on the a.s. convergence of series of independent, uniformly bounded real valued random variables.

COROLLARY 1. *Let f be an additive arithmetic function with values in a real separable Banach space $\mathfrak{X}$, and let T be a total subset of the dual $\mathfrak{X}^*$. If f has a limit distribution, then, for any $c > 0$, the following conditions are fulfilled:*

(i) $$\sum_{\|f(p)\| \leqslant c} \frac{1}{p} < \infty,$$

(ii) $$\sum_{\|f(p)\| < c} \frac{f(p)}{p} \quad \text{converges in the norm,}$$

(iii) $$\sum_{\|f(p)\| < c} \frac{|y \circ f(p)|^2}{p} < \infty \quad \text{for every} \quad y \in T.$$

If the Banach space satisfies a type or cotype condition, we can get something more. In fact it is easy to see that the following corollary is true.

COROLLARY 2. *Let f be an additive function with values in a real separable Banach space $\mathfrak{X}$.*

(a) *If $\mathfrak{X}$ is of type $r(1 \leqslant r \leqslant 2)$ and for some $c > 0$ the following conditions*

$$\text{(i)} \qquad \sum_{\|f(p)\| \geqslant c} \frac{1}{p} < \infty,$$

$$\text{(ii)} \qquad \sum_{\|f(p)\| < c} \frac{f(p)}{p} \quad \text{converges in the norm,}$$

$$\text{(iii)} \qquad \sum_{\|f(p)\| < c} \frac{\|f(p)\|^r}{p} < \infty$$

are satisfied, then $f \Rightarrow \sum_p f(p^{b_p})$.

(b) *If $\mathfrak{X}$ is of cotype r $(r \geqslant 2)$ and f has a limit distribution then conditions* (i), (ii), *and* (iii) *hold for any $c > 0$.*

In particular, we conclude that Theorem A has a perfect infinite-dimensional analogue when the Banach space is both of type 2 and cotype 2, i.e. if it is isomorphic to a real separable Hilbert space (see (Kwapień, 1972)). Of course, this conclusion also follows from Theorem B and Kolmogorov's three series Theorem in Hilbert spaces (see (Vakhania *et al.*, 1987, p. 289)).

The following corollary is more particular. Let $\mathbf{Z}^{(\infty)}$ be the additive group of all sequences of integer numbers with finite support, and zero element denoted by $\mathbf{0}$, and let $f = (f_1, \ldots, f_j, \ldots)$ be an additive $\mathbf{Z}^{(\infty)}$-valued function. If $\mathbf{Z}^{(\infty)}$ is endowed with any metric which induces the discrete topology, then f has a limit distribution if and only if the asymptotic density

$$\mathbb{D}(m\colon\ f(m) = z) = \lim_n \mathbb{P}_n(m\colon\ f(m) = z)$$

exists for every $z \in \mathbf{Z}^{(\infty)}$. Moreover,

$$\sum_{z \in \mathbf{Z}^{(\infty)}} \mathbb{D}(m\colon\ f(m) = z) = 1.$$

Now by embedding such a metric group in $l_2(\mathbf{R})$ we deduce, from Corollary 2,

COROLLARY 3. *In order that $f = (f_1, \ldots, f_j, \ldots)$ should have a limit distribution it is both necessary and sufficient that*

$$\sum_{f(p) \neq \mathbf{0}} \frac{1}{p} < \infty.$$

We have, in this case, for any subset $S \subset \mathbf{Z}^{(\infty)}$,

$$\mathbb{D}(m\colon\ f(m) \in S) = \mathbb{P}((X_1, \ldots, X_j, \ldots) \in S), \tag{3}$$

where

$$X_j = \sum_p f_j(p^{b_p}).$$

Remark. The first part of this corollary follows from (Ruzsa, 1977, Thm. 3.10) also, but our methods furthermore provide the formula (3) which translates the computation of asymptotic densities into standard probabilities calculations, thus showing that things agree with "intuition". Note that if $Q_{f_i} \cap Q_{f_j} = \varnothing$ whenever $i \neq j$ (i.e. f is essentially unfolding of certain integer-valued additive function), then X_j's are independent, and therefore (3) reduces to

$$\mathbb{D}(m\colon\ f(m) \in S) = \prod_j \mathbb{P}(X_j \in S_j)$$

when the relation

$$((X_1, \ldots, X_j, \ldots) \in S) = \bigcap_j (X_j \in S_j) \quad \text{a.s.}$$

holds for some sets of integer numbers $S_1,\ S_2,\ \ldots$.

4. EXAMPLES

Two examples will serve to illustrate the preceding ideas.
Example 1. Let Q be a set of prime numbers such that

$$\sum_{p \in Q} \frac{1}{p} < \infty, \tag{4}$$

and let $w = \{Q_1, \ldots, Q_j, \ldots\}$ be a labelled partition of Q. Now consider the set A of natural numbers with at most one prime divisor, counting multiplicity, in each Q_j. It is clear that

$$A = \{m\colon\ (f_1(m), \ldots, f_j(m), \ldots) \in S\},$$

where

$$f_j(m) = \sum_{p \in Q_j} \beta_p(m),$$

and S is the set of all the sequences of zeros and units belonging to $\mathbf{Z}^{(\infty)}$. In view of (4) and Corollary 3 the asymptotic density of A is just

$$\mathbb{D}(A) = \mathbb{P}((X_1, \ldots, X_j, \ldots) \in S),$$

where

$$X_j = \sum_{p \in Q_j} b_p.$$

Now as

$$\sum_j \mathbb{P}(X_j = 1) \leqslant \sum_{p \in Q} \frac{1}{p} < \infty,$$

we have from the Borel–Cantelli lemma

$$((X_1, \ldots, X_j, \ldots) \in S) = \bigcap_j (X_j \in \{0,1\}) \quad \text{a.s.},$$

and therefore

$$\mathbb{D}(A) = \prod_j \mathbb{P}(X_j \in \{0,1\}) = \prod_{p \in Q} \left(1 - \frac{1}{p}\right) \prod_j \left(1 + \sum_{p \in Q_j} \frac{1}{p}\right).$$

On the other hand, if B is the set of natural numbers which are divisible by at most one prime from each Q_j, then a similar reasoning leads to

$$\mathbb{D}(B) = \prod_{p \in Q} \left(1 - \frac{1}{p}\right) \prod_j \left(1 + \sum_{p \in Q_j} \frac{1}{p-1}\right).$$

Example 2. Let $w = \{Q_1, \ldots, Q_j, \ldots\}$ be a labelled partition of the set of all prime numbers, and let $\varepsilon_w = (\varepsilon_1, \ldots, \varepsilon_j, \ldots)$ be the w-unfolding of the "excess"

$$\varepsilon(n) = \sum_p (\beta_p(n) - \delta_p(n)),$$

where, for each prime p, δ_p denotes the indicator function of the set

$$\{m\colon\ \beta_p(m) \geqslant 1\}.$$

If C is the set of natural numbers n such that

$$\varepsilon_j(n) \equiv 0 \pmod 2 \qquad \text{for all} \quad j,$$

then the asymptotic density of C is

$$\mathbb{D}(C) = \prod_j \mathbb{P}\left(\sum_{p \in Q_j} (b_p - d_p) \equiv 0 \pmod 2 \right).$$

Now, for each j,

$$\mathbb{P}\left(\sum_{p \in Q_j} (b_p - d_p) \equiv 0 \pmod 2 \right) = \frac{1}{2}(1 + g_j(-1)),$$

where $g_j(z)$ is the probability generating function of $\sum_{p \in Q_j} (b_p - d_p)$, i.e.

$$g_j(z) = \prod_{p \in Q_j} \left(1 - \frac{1}{p}\right)\left(1 + \frac{1}{p-z}\right).$$

Thus we get

$$\mathbb{D}(C) = \prod_j \left\{ \frac{1}{2} + \frac{1}{2} \prod_{p \in Q_j} \left(1 - \frac{2}{p(p+1)}\right) \right\}.$$

In particular, the set of natural numbers whose prime factorization is composed only by odd powers has the asymptotic density

$$\prod_p \left(1 - \frac{1}{p(p+1)}\right).$$

REFERENCES

Araujo, A and Giné, E. (1980). *The Central Limit Theorems for Real and Banach Valued Random Variables*. John Wiley and Sons, New York.

Billingsley, P. (1974). The probability theory of additive arithmetic functions. *Ann. Probab.* **2**, 749–791.

Buldygin, V. V. (1973). On random series in a Banach space. *Theory Probab. Appl.* **18**, 464–475.

de la Cal, J. (1989). On the three series theorem in number theory. *Ann. Probab.* **17**, 357–361.

Elliott, P. D. T. A. (1979). *Probabilistic Number Theory*, I. Springer-Verlag, Berlin, Heidelberg, New York.

Hoffmann-Jørgensen, J. and Pisier, G. (1976). The law of large numbers and the central limit theorem in Banach spaces. *Ann. Probab.* **4**, 587-599.

Kubilius, J. (1964). *Probabilistic Methods in the Theory of Numbers*. American Math. Soc., Providence, RI.

Kwapień, S. (1972). Isomorphic characterization of inner product spaces by orthogonal series with vector valued coefficients. *Studia Math.* **44**, 583–595.

Nakutis, V. (1985). On the distribution of a certain class of additive arithmetic functions with values in a Hilbert space. *Liet. Matem. Rink.* **25**, 101–111 (in Russian).

Ruzsa, I. (1977). General multiplicative functions. *Acta Arithm.* **32**, 313-347.

Ruzsa, I. (1984). Generalized moments of additive functions. *J. Number Theory* **18**, 27–33.

Vakhania, N. N., Tarieladze, V. I. and Chobanyan, S. A. (1987). *Probability Distributions on Banach Spaces*. D. Reidel Publishing Company, Dordrecht.

New Trends in Probab. and Statist., pp. 211–221
F. Schweiger and E. Manstavičius (Eds.)
1992 VSP/TEV

ADDITIVE ARITHMETIC FUNCTIONS AND PROCESSES WITH INDEPENDENT INCREMENTS

B. GRIGELIONIS
Institute of Mathematics and Informatics,
Akademijos 4, 2600 Vilnius, Lithuania

ABSTRACT

Stochastic processes constructed from a Kubilius-type sequence of Hilbert space valued additive arithmetic functions are discussed. Necessary and sufficient conditions for such processes to converge to a given limiting stochastically continuous process with independent increments are obtained.

INTRODUCTION

Let (G, ρ) be an Abelian complete separable metric group, with the group operation being denoted by $g_1 + g_2$, $g_1, g_2 \in G$, and the metrics by ρ, $\Omega := \mathcal{D}_{[0,1]}(G)$ be the space of all G-valued right continuous and having left limits functions ω: $[0,1] \to G$, endowed with the Skorokhod topology, generated by the metrics

$$d(\omega_1, \omega_2) = \inf_{\lambda \in \Lambda} \left[\sup_{0 \leqslant t \leqslant 1} |\lambda(t) - t| + \sup_{0 \leqslant t \leqslant 1} \rho(\omega_1(\lambda(t)), \omega_2(t)) \right], \qquad \omega_1, \omega_2 \in \Omega,$$

where Λ is the class of all homomorphisms

$$\lambda\colon [0,1] \longrightarrow [0,1], \qquad \lambda(0) = 0, \qquad \lambda(1) = 1.$$

Let $\mathfrak{F} := \mathfrak{B}(\Omega)$ be the σ-algebra of all Borel subsets of Ω.
Consider

$$\Omega_n = \{1, \dots, n\}, \qquad \mathfrak{F}_n = 2^{\Omega_n},$$

$$\nu_n(A) = \frac{1}{n}\text{card}\,(A), \qquad A \in \mathfrak{F}_n,$$

$$f_n(\,\cdot\,)\colon \Omega_n \longrightarrow G,$$

satisfying $f_n(1) = 0$ and

$$f_n(m_1 m_2) = f_n(m_1) + f_n(m_2) \tag{1}$$

for each pair of relatively prime numbers $m_1, m_2 \in \Omega_n$. Consider a sequence $\{y_n(t)\colon n \geqslant 1\}$ $(0 \leqslant t \leqslant 1)$ of right continuous increasing functions,

$$y_n(0) = 1, \qquad y_n(1) = n, \qquad n \geqslant 1.$$

For each prime p and integer m, let $\alpha_p(m)$ denote the largest power of p dividing m. Define the sequence of stochastic processes

$$X_n(t,m) = \sum_{p \leqslant y_n(t)} \left(f_n\left(p^{\alpha_p(m)}\right) - g_n(p)\right),$$

$$m \in \Omega_n, \qquad 0 \leqslant t \leqslant 1, \qquad n \geqslant 1,$$

where $g_n(p) \in G,\ p \leqslant n,\ n \geqslant 1$, are some centralizing elements.
In this paper we shall investigate conditions, under which the sequence of probability measures

$$\mathbb{P}_n(B) := \nu_n(\{m\colon X_n(\cdot, m) \in B\}), \qquad B \in \mathfrak{F}, \quad n \geqslant 1,$$

converges weakly to the probability measure, corresponding to the definite G-valued stochastically continuous process with independent increments. This problem, which can be interpreted as a problem of arithmetical modelling of stochastic processes with independent increments, starting with the paper (Kubilius, 1955), was almost completely solved when $G = \mathbf{R}^1$ (see (Babu, 1973; Billingsley, 1974; Grigelionis and Mikulevičius, 1984; Kryžius, 1981; Kubilius, 1986; Manstavičius, 1984, 1985, 1986, 1988; Phillipp, 1973; Timofeev and Usmanov, 1978, 1984, 1986)). Some general results were obtained in (Usmanov, 1979) for $G = \mathbf{R}^d,\ d > 1$. Here we shall discuss the case, when G is a normed linear space.
We say that the sequence $\{f_n\colon n \geqslant 1\}$ satisfying (1) is of Kubilius type if for each prime number p and $\alpha = 0, 1, \ldots$

$$\lim_{n\to\infty} \|f_n(p^\alpha)\| = 0, \tag{2}$$

$$\lim_{n\to\infty} \max_{p\leqslant n} \frac{\|f_n(p)\|}{p} = 0, \tag{3}$$

and there exist $r_n',\ r_n' \geqslant 1,\ \log r_n' = o(\log n)$, such that

$$\lim_{n\to\infty} \sum_{r_n' < p \leqslant n} \frac{\|f_n(p)\|^2}{p} = 0. \tag{4}$$

Choosing $g_n(p) = p^{-1} f_n(p)$ and using the fundamental Kubilius lemma (see (Kubilius, 1964)) we shall prove in § 1 that for Kubilius-type sequence $\{f_n\colon n \geqslant 1\}$ the weak convergence $\mathbb{P}_n \Rightarrow \mathbb{P}$, as $n \to \infty$, is equivalent to the weak convergence $\widetilde{\mathbb{P}}_n \Rightarrow \mathbb{P}$, as $n \to \infty$, where the probability measures $\widetilde{\mathbb{P}}_n$ correspond to the certain random walks on G. Assuming that G is a Hilbert space H with the inner product $(\cdot, \cdot)$, in § 2 we shall find necessary and sufficient conditions for the weak convergence of $\{\mathbb{P}_n\colon n \geqslant 1\}$ to the measure $\mathbb{P} = \mathbb{P}_{m,S,M}$, corresponding to the stochastically continuous H-valued process with independent increments characterized by the following properties (see, e.g. (Gikhman and Skorokhod, 1975)):

i) $\mathbb{P}(\{\omega\colon \omega(0) = 0\}) = 1$;

ii) for each t_i and x_i satisfying $0 = t_0 < t_1 < \ldots < t_k \leqslant 1,\ x_1, \ldots, x_k \in H_k$, $k \geqslant 1$,

$$\int_\Omega \exp\left\{i \sum_{j=1}^{k} (\omega(t_j) - \omega(t_{j-1}), x_j)\right\} \mathbb{P}(d\omega)$$
$$= \prod_{j=1}^{k} \int_\Omega \exp\{i(\omega(t_j) - \omega(t_{j-1}), x_j)\} \mathbb{P}(d\omega);$$

iii) for each $x \in H$ and $t,\ 0 < t \leqslant 1$,

$$\varphi_t(x) := \int_\Omega e^{i(\omega(t),x)} \mathbb{P}(d\omega)$$
$$= \exp\left\{i(m(t), x) - \frac{1}{2}(S(t)x, x) + \int_{H\setminus\{0\}} K(x,y) M(t; dy)\right\},$$

where

$$m(\,\cdot\,)\colon\ [0,1] \longrightarrow H,$$

$$S(\,\cdot\,)\colon\ [0,1] \longrightarrow \mathcal{L}(H) \quad \text{(the space of linear operators on } H),$$

$$M(\cdot\,;\cdot)\colon\ [0,1] \times \mathcal{B}(H \setminus \{0\}) \longrightarrow \mathbf{R}_+$$

have the following properties: for each $x,\ x \in H$, the function $\varphi_t(x)$ is continuous in t; for each $u,\ t,\ 0 \leqslant u < t \leqslant 1$, the difference $S(t) - S(u)$ is symmetric positive semidefinite finite trace operator (S-operator) and $M(t;\,\cdot\,) - M(u;\,\cdot\,)$ is σ-finite measure on $\mathcal{B}(H \setminus \{0\})$, satisfying

$$\int_{H\setminus\{0\}} (\|x\|^2 \wedge 1) M(t; dx) < \infty,$$

$$K(x,y) = e^{i(x,y)} - 1 - \frac{i(x,y)}{1 + \|y\|^2}, \qquad x, y \in H.$$

1. REDUCTION TO RANDOM WALK

Denote

$$r_n = r_n' \vee n^{\delta_n}, \qquad \gamma_p = \left[\frac{\log r_n}{\log p}\right], \qquad \beta_p(m) = \alpha_p(m) \wedge \gamma_p,$$

where

$$\delta_n = \sum_{r_n' < p \leqslant n} \frac{\|f_n(p)\|^2}{p}.$$

Following (Kubilius, 1964), consider sub-σ-algebra $\mathcal{F}_n^{(r_n)}$ of $\mathcal{F}_n$, generated by random variables β_p, $2 \leqslant p \leqslant r_n$, and define the probability measure μ_n on $\mathcal{F}_n^{(r_n)}$ by means of the equality

$$\mu_n\left(\bigcap_{j=1}^{k}\{\beta_{p_j} = \alpha_j\}\right) = \prod_{j=1}^{k} \pi(p_j^{\alpha_j}), \tag{5}$$

where

$$2 \leqslant p_1 < \ldots < p_k \leqslant r_n,$$

$$0 \leqslant \alpha_j \leqslant \gamma_{p_j}, \qquad j = 1, \ldots, k, \quad k \geqslant 1,$$

$$\pi(p^\alpha) = \begin{cases} \dfrac{1}{p^\alpha}\Big(1 - \dfrac{1}{p}\Big) & \text{for } \ 0 \leqslant \alpha < \gamma_p, \\ \dfrac{1}{p^\alpha} & \text{for } \ \alpha = \gamma_p. \end{cases}$$

Let

$$X_n^{(r_n)}(t,m) = \sum_{p \leqslant y_n(t) \wedge r_n} \left[f_n(p^{\beta_p(m)\wedge 1}) - \frac{f_n(p)}{p}\right], \qquad 0 \leqslant t \leqslant 1, \quad n \geqslant 1, \tag{6}$$

and

$$\widehat{\mathbb{P}}_n(B) := \mu_n\big(\{m\colon\ X_n^{(r_n)}(\,\cdot\,, m) \in B\}\big), \qquad B \in \mathcal{F}, \quad n \geqslant 1.$$

Remark that, according to (5), the summands in (6) are independent random variables with respect to the measure μ_n.

THEOREM 1 (Grigelionis and Mikulevičius, 1984). *For a Kubilius-type sequence $\{f_n\colon\ n \geqslant 1\}$ the sets of limiting points of $\{\mathbb{P}_n\colon\ n \geqslant 1\}$ and $\{\widetilde{\mathbb{P}}_n\colon\ n \geqslant 1\}$ coincide.*

Proof. Denote

$$\mathbb{P}_n^{(r_n)}(B) := \nu_n\big(\{m\colon\ X_n^{(r_n)}(\,\cdot\,, m) \in B\}\big), \qquad B \in \mathcal{F}, \quad n \geqslant 1.$$

According to the fundamental Kubilius lemma there exists an absolute constant $c > 0$ such that

$$\max_{A \in \mathcal{F}_n^{(r_n)}} |(\nu_n(A) - \mu_n(A)| = O\Big(\exp\Big\{-c\frac{\log n}{\log r_n}\Big\}\Big). \tag{7}$$

From (6) and (7) we find that

$$\lim_{n\to\infty} \sup_{B \in \mathcal{F}} \big|\mathbb{P}_n^{(r_n)}(B) - \widehat{\mathbb{P}}_n(B)\big| = 0.$$

Thus it is enough to check that the sets of limiting points of $\{\mathbb{P}_n\colon\ n \geqslant 1\}$ and $\{\mathbb{P}_n^{(r_n)}\colon\ n \geqslant 1\}$ coincide. For this purpose we derive that for each $0 < \varepsilon < 1$

$$\lim_{n\to\infty} \nu_n\Big(\max_{0 \leqslant t \leqslant 1} \big\|X_n(t,\cdot) - X_n^{(r_n)}(t,\cdot)\big\| > \varepsilon\Big) = 0. \tag{8}$$

Indeed,

$$\max_{0\leqslant t\leqslant 1}\|X_n(t,m)-X_n^{(r_n)}(t,m)\|\leqslant\sum_{r_n<p\leqslant n}\|f_n(p^{\alpha_p(m)})\|$$
$$+\sum_{p\leqslant r_n}\|f_n(p^{\alpha_p(m)})-f_n(p^{\alpha_p(m)\wedge 1})\|+\sum_{r_n<p\leqslant n}\frac{\|f_n(p)\|}{p}. \tag{9}$$

Noting that

$$\nu_n(\alpha_p=\alpha)=\frac{1}{n}\left(\left[\frac{n}{p^\alpha}\right]-\left[\frac{n}{p^{\alpha+1}}\right]\right)\leqslant\frac{1}{p^\alpha},$$

we have the following inequalities:

$$\nu_n\left(\sum_{r_n<p\leqslant n}\|f_n(p^{\alpha_p})\|>\frac{\varepsilon}{3}\right)=\nu_n\left(\left(\sum_{r_n<p\leqslant n}\|f_n(p^{\alpha_p})\|\right)\wedge 1>\frac{\varepsilon}{3}\right)$$
$$\leqslant\nu_n\left(\sum_{r_n<p\leqslant n}(\|f_n(p^{\alpha_p})\|\wedge 1)>\frac{\varepsilon}{3}\right)\leqslant\frac{3}{\varepsilon}\sum_{r_n<p\leqslant n}\sum_{\alpha=1}^{\infty}\frac{\|f_n(p^\alpha)\|\wedge 1}{p^\alpha} \tag{10}$$
$$\leqslant\frac{3}{\varepsilon}\left[\sum_{r_n<p\leqslant n}\frac{\|f_n(p)\|}{p}+\sum_{p,\alpha\geqslant 2}\frac{\|f_n(p^\alpha)\|\wedge 1}{p^\alpha}\right],$$

$$\nu_n\left(\sum_{p\leqslant r_n}\|f_n(p^{\alpha_p})-f_n(p^{\alpha_p\wedge 1})\|>\frac{\varepsilon}{3}\right)$$
$$\leqslant\nu_n\left(\left(\sum_{p\leqslant r_n}\mathbf{1}_{\{\alpha_p>1\}}(\|f_n(p^{\alpha_p})\|+\|f_n(p)\|)\right)\wedge 1>\frac{\varepsilon}{3}\right)$$
$$\leqslant\nu_n\left(\sum_{p\leqslant r_n}\mathbf{1}_{\{\alpha_p>1\}}(\|f_n(p^{\alpha_p})\|\wedge 1+\|f_n(p)\|\wedge 1)>\frac{\varepsilon}{3}\right) \tag{11}$$
$$\leqslant\frac{3}{\varepsilon}\sum_{p,\alpha\geqslant 2}\left(\frac{\|f_n(p^\alpha)\|\wedge 1}{p^\alpha}+\frac{\|f_n(p)\|\wedge 1}{p^\alpha}\right).$$

Having in mind that

$$\sum_{p,\alpha\geqslant 2}\frac{1}{p^\alpha}=\sum_{p\geqslant 2}\frac{1}{p(p-1)}<\infty$$

and using (2) and (4), we have (as $n\to\infty$)

$$\sum_{r_n<p\leqslant n}\frac{\|f_n(p)\|}{p}\leqslant\left(\sum_{r_n<p\leqslant n}\frac{\|f_n(p)\|^2}{p}\sum_{r_n<p\leqslant n}\frac{1}{p}\right)^{1/2}$$
$$\leqslant\left(\delta_n\sum_{n^{\delta_n}<p\leqslant n}\frac{1}{p}\right)^{1/2}\longrightarrow 0 \tag{12}$$

and

$$\sum_{p,\alpha\geqslant 2}\left(\frac{\|f_n(p^\alpha)\|\wedge 1}{p^\alpha}+\frac{\|f_n(p)\|\wedge 1}{p^\alpha}\right)\longrightarrow 0. \tag{13}$$

Now from (9)–(13) we derive (8) easily. Theorem 1 is proved.

For identification of the limiting points of $\{\widehat{\mathbb{P}}_n\colon\ n\geqslant 1\}$ we may apply known results on the weak convergence of probability measures in Banach spaces (see, e.g. (Araujo and Giné, 1980)), the structure of Banach space valued processes with independent increments (see, e.g. (Gikhman and Skorokhod, 1975)), and the suitable criteria of the relative weak compactness of sets of probability measures on the Skorokhod space $\mathcal{D}_{[0,1]}(G)$ (see, e.g. (Aldous, 1978; Ethier and Kurtz, 1986; Gikhman and Skorokhod, 1974; Grigelionis and Mikulevičius, 1981)).

2. FUNCTIONAL LIMIT THEOREMS FOR HILBERT SPACE VALUED ADDITIVE ARITHMETIC FUNCTIONS

Assume now that G is a separable Hilbert space H. Denote for $0\leqslant t\leqslant 1$ and $n\geqslant 1$:

$$m_n(t)=-\sum_{p\leqslant y_n(t)}\frac{\|f_n(p)\|^2}{p(1+\|f_n(p)\|^2)}f_n(p), \tag{14}$$

$$(T_n(t)x,y)=\sum_{\substack{p\leqslant y_n(t)\\ \|f_n(p)\|\leqslant 1}}\frac{(f_n(p),x)(f_n(p),y)}{p},\qquad x,y\in H, \tag{15}$$

$$M_n(t;\Gamma)=\sum_{\substack{p\leqslant y_n(t)\\ f_n(p)\in\Gamma}}\frac{1}{p},\qquad \Gamma\in\mathcal{B}(H\setminus\{0\}). \tag{16}$$

Remark that the operators $T_n(t)$, $0\leqslant t\leqslant 1$, $n\geqslant 1$, are S-operators, and

$$\operatorname{Tr}T_n(t)=\sum_{\substack{p\leqslant y_n(t)\\ \|f_n(p)\|\leqslant 1}}\frac{\|f_n(p)\|^2}{p}. \tag{17}$$

We shall prove the following statements, denoting by $\Rightarrow$ the weak convergence of finite measures.

THEOREM 2. *For a Kubilius-type sequence* $\{f_n\colon\ n\geqslant 1\}$,

$$\mathbb{P}_n\Longrightarrow\mathbb{P}_{m,S,M}\qquad\text{as}\quad n\to\infty \tag{18}$$

iff the sequence $\{T_n(1)\colon\ n\geqslant 1\}$ *is compact,*

$$\sup_{0\leqslant t\leqslant 1}\|m_n(t)-m(t)\|\longrightarrow 0,\qquad n\to\infty, \tag{19}$$

for each $\varepsilon>0$ *and* $0<t\leqslant 1$,

$$M_n(t;\cdot\cap\{x\colon\ \|x\|>\varepsilon\})\Longrightarrow M(t;\cdot\cap\{x\colon\ \|x\|>\varepsilon\}),\qquad n\to\infty, \tag{20}$$

and for each $0 < t \leqslant 1$ *and* $x \in H$

$$\lim_{\varepsilon\to 0} \overline{\lim_{n\to\infty}} \sum_{\substack{p \leqslant y_n(t) \\ \|f_n(p)\| \leqslant \varepsilon}} \frac{1}{p}(f_n(p), x)^2 = (S(t)x, x). \tag{21}$$

Remark that in the case, when $M \equiv 0$, the limiting measure $\mathbb{P}_{m,S,0}$ is Gaussian and the condition (20) is equivalent to the Lindeberg-type condition: for each $\varepsilon > 0$

$$\lim_{n\to\infty} \sum_{\substack{p \leqslant n \\ \|f_n(p)\| > \varepsilon}} \frac{1}{p} = 0. \tag{22}$$

Denote

$$A_n = \sum_{p \leqslant n} \frac{f_n(p)}{p},$$

$$\mathbb{F}_n(\Gamma) = \frac{1}{n}\,\mathrm{card}\,(\{m\colon\ f_n(m) - A_n \in \Gamma\}), \qquad \Gamma \in \mathcal{B}(H),$$

$$\mathbb{F}_{m(1),S(1),M(1)}(\Gamma) = \mathbb{P}_{m,S,M}(\{\omega\colon\ \omega(1) \in \Gamma\}), \qquad \Gamma \in \mathcal{B}(H).$$

THEOREM 3. *For a Kubilius-type sequence* $\{f_n\colon\ n \geqslant 1\}$,

$$\mathbb{F}_n \Longrightarrow \mathbb{F}_{m(1),S(1),M(1)} \qquad \text{as} \quad n \to \infty \tag{23}$$

iff the sequence $\{T_n(1),\ n \geqslant 1\}$ *is compact,*

$$\|m_n(1) - m(1)\| \longrightarrow 0, \qquad n \to \infty \tag{24}$$

for each $\varepsilon > 0$

$$M_n(1;\ \cdot \cap \{x\colon\ \|x\| > \varepsilon\}) \Longrightarrow M(1;\ \cdot \cap \{x\colon\ \|x\| > \varepsilon\}), \qquad n \to \infty, \tag{25}$$

and for each $x \in H$

$$\lim_{\varepsilon\to 0} \overline{\lim_{n\to\infty}} \sum_{\substack{p \leqslant n \\ \|f_n(p)\| \leqslant \varepsilon}} \frac{1}{p}(f_n(p), x)^2 = (S(1)x, x). \tag{26}$$

Recall that a sequence $\{S_n\colon\ n \geqslant 1\}$ of S-operators is compact, iff

$$\sup_{n \geqslant 1} \mathrm{Tr}\, S_n < \infty$$

and for some orthonormal basis $\{e_i\colon\ i \geqslant 1\}$ in H

$$\lim_{N\to\infty} \sup_{n \geqslant 1} \sum_{i=N}^{\infty} (S_n e_i, e_i) = 0.$$

Thus the sequence $\{T_n(1)\colon\ n \geqslant 1\}$ is compact, iff

$$\sup_{n\geqslant 1} \sum_{\substack{p\leqslant n \\ \|f_n(p)\|\leqslant 1}} \frac{\|f_n(p)\|^2}{p} < \infty \tag{27}$$

and for some orthonormal basis $\{e_i\colon\ i \geqslant 1\}$ in H

$$\lim_{N\to\infty} \sup_{n\geqslant 1} \sum_{i=N}^{\infty} \sum_{p\leqslant n} \frac{(f_n(p), e_i)^2}{p} = 0. \tag{28}$$

When $H = \mathbf{R}^d$, the sequence $\{T_n(1)\colon\ n \geqslant 1\}$ is compact, iff (27) is satisfied. In the case, when

$$f_n(m) = \left(\frac{f^{(1)}(m)}{B^{(1)}(n)}, \ldots, \frac{f^{(d)}(m)}{B^{(d)}(n)}\right),$$

where

$$B^{(j)}(n) = \left(\sum_{p\leqslant n} \frac{(f^{(j)}(p))^2}{p}\right)^{1/2}, \qquad j = 1, \ldots, d,$$

the condition (27) is obviously satisfied, because

$$\sum_{\substack{p\leqslant n \\ \|f_n(p)\|\leqslant 1}} \frac{\|f_n(p)\|^2}{p} \leqslant d.$$

In this situation Theorem 3 reduces to the classical Kubilius theorem (see (Kubilius, 1964)).

COROLLARY 1. *For a Kubilius-type sequence* $\{f_n\colon\ n \geqslant 1\}$,

$$\mathbb{F}_n \Longrightarrow \mathbb{F}_{m(1),S(1),0} \qquad \text{as} \quad n \to \infty$$

iff the conditions (22), (24), and (26)–(28) hold.
In the finite-dimensional case (28) is trivial, and (27) follows from (22) and (26).

Proof. Consider a probability space $(\widetilde{\Omega}_n, \widetilde{\mathfrak{F}}_n, \widetilde{\mathbb{Q}}_n)$ with a triangular array of independent random H-valued elements $\widetilde{X}_{np}$, $2 \leqslant p \leqslant n$, such that

$$\widetilde{\mathbb{Q}}_n\left(\widetilde{X}_{np} = -\frac{f_n(p)}{p}\right) = 1 - \frac{1}{p}$$

and

$$\widetilde{\mathbb{Q}}_n\left(\widetilde{X}_{np} = \left(1 - \frac{1}{p}\right) f_n(p)\right) = \frac{1}{p}.$$

Remark that the triangular array of distributions, corresponding to

$$\{\widetilde{X}_{np}\colon\ 2 \leqslant p \leqslant n\}$$

is uniformly infinitesimal, because, according to (3), for each $\varepsilon > 0$

$$\lim_{n\to\infty} \max_{p\leqslant n} \widetilde{\mathbb{Q}}_n\left(\|\widetilde{X}_{np}\| > \varepsilon\right) \leqslant \frac{1}{\varepsilon} \lim_{n\to\infty} \max_{p\leqslant n} \frac{\|f_n(p)\|}{p} = 0.$$

Define

$$\widetilde{X}_n(t) = \sum_{p\leqslant y_n(t)} \widetilde{X}_{np}, \qquad 0\leqslant t\leqslant 1, \quad n\geqslant 1,$$

and

$$\widetilde{\mathbb{P}}_n(B) := \widetilde{\mathbb{Q}}_n(\widetilde{X}_n(\cdot)\in B), \qquad B\in\mathcal{F}, \quad n\geqslant 1.$$

Taking into consideration Theorem 1, and that for each $\varepsilon > 0$,

$$\widetilde{\mathbb{Q}}_n\left(\left\|\sum_{r_n<p\leqslant n} \widetilde{X}_{np}\right\| > \varepsilon\right) \leqslant \frac{1}{\varepsilon} \sum_{r_n<p\leqslant n} \frac{\|f_n(p)\|}{p} \longrightarrow 0 \qquad \text{as} \quad n\to\infty$$

we have that the sets of limiting points of sequences $\{\mathbb{P}_n\colon n\geqslant 1\}$ and $\{\widetilde{\mathbb{P}}_n\colon n\geqslant 1\}$ coincide.

In order to prove the weak convergence of finite-dimensional distributions of $\widetilde{X}_n$ to the corresponding limiting ones we shall apply the classical limit theorems (see, e.g. (Parthasarathy, 1967, Chap. VI). Accompanying laws for distributions of $\widetilde{X}_n(t)$ are infinitely divisible ones, defined by the triplets $(\widetilde{m}_n(t),\ 0,\ \widetilde{M}_n(t;\Gamma))$, where

$$\widetilde{m}_n(t) = -\sum_{p\leqslant y_n(t)} \frac{f_n(p)}{p}\left(1-\frac{1}{p}\right)\left(\frac{1}{1+\|f_n(p)\|^2p^{-2}} - \frac{1}{1+\|f_n(p)\|^2(1-1/p)^{-2}}\right)$$

and

$$\widetilde{M}_n(t;\Gamma) = \sum_{p\leqslant y_n(t)} \left[\left(1-\frac{1}{p}\right)\varepsilon_{\left\{-\frac{f_n(p)}{p}\right\}}(\Gamma) + \frac{1}{p}\varepsilon_{\left\{f_n(p)\left(1-\frac{1}{p}\right)\right\}}(\Gamma)\right],$$

$$\Gamma\in\mathcal{B}(H\setminus\{0\}),$$

$$\varepsilon_{\{a\}}(\Gamma) = \begin{cases} 1 & \text{for} \quad a\in\Gamma, \\ 0 & \text{for} \quad a\notin\Gamma. \end{cases}$$

Easy calculations show that for each $0 < t\leqslant 1$

$$\lim_{n\to\infty} \|\widetilde{m}_n(t) - m_n(t)\| = 0, \tag{29}$$

for each $\varepsilon > 0$ and $0 < t \leqslant 1$

$$\lim_{n\to\infty} \sup_{\Gamma\in\mathcal{B}(H\setminus\{0\})} \left|\widetilde{M}_n(t;\Gamma\cap\{x\colon \|x\|>\varepsilon\}) - M_n(t;\Gamma\cap\{x\colon \|x\|>\varepsilon\})\right| = 0, \tag{30}$$

and for each $x\in H$ and $0 < t\leqslant 1$

$$\lim_{n\to\infty}\left|\int_{\|y\|\leqslant 1} (x,y)^2 \widetilde{M}_n(t;dy) - \int_{\|y\|\leqslant 1} (x,y)^2 M_n(t;dy)\right| = 0. \tag{31}$$

From the above mentioned classical limit theorems and (29)–(31) it follows that the conditions (20), (21), (27), (28), and the assumption, that for each $0 < t \leqslant 1$

$$\lim_{n\to\infty} \|m_n(t) - m(t)\| = 0, \tag{19'}$$

are necessary and sufficient for the weak convergence of finite-dimensional distributions of $\widetilde{X}_n(t)$, $0 \leqslant t \leqslant 1$, to the corresponding projections of the measure $\mathbb{P}_{m,S,M}$.

The case, when $t = 1$, corresponds to the statement of Theorem 3.

To complete the proof of Theorem 2 it is enough to remark that the conditions (19)–(21), (27), and (28) imply the relative weak compactness of the sequence $\{\mathbb{P}_n\colon n \geqslant 1\}$ (see, e.g. (Grigelionis and Mikulevičius, 1983; Lavrent'ev, 1983)).

Necessity of the assumption (19) can be derived analogously to the functional limit theorems for finite-dimensional stochastic process with independent increments (see (Jacod, 1983)). We shall omit the technical details here.

As a simple illustration, let us assume that $H = \mathbf{l}^2$, $a = (a_1, a_2, \ldots) \in \mathbf{l}^2$,

$$f_n(p) = \frac{a}{\sqrt{\log\log n}},$$

$$\|f_n(p^\alpha)\| \longrightarrow 0 \qquad \text{as} \quad n \to \infty \qquad \text{for each} \quad p, \alpha \geqslant 2,$$

$$y_n(t) = \left[e^{(\log n)^t}\right], \qquad 0 \leqslant t \leqslant 1,$$

$$r'_n = n^{(\log\log n)^{-1}}.$$

The limiting measure $\mathbb{P}_{0,S,0}$ will be Gaussian, where for each $x \in \mathbf{l}^2$ and $0 \leqslant t \leqslant 1$

$$(S(t)x, x) = t(a, x)^2,$$

i.e.

$$S(t) = t(S_{jk})_{j,k\geqslant 1}, \qquad S_{jk} = a_j a_k, \qquad j, k = 1, 2, \ldots.$$

Concluding let us formulate the conjecture that under the mild regularly properties of the sequence $\{y_n(\cdot)\colon n \geqslant 1\}$ from the assumptions (2), (3), and

$$\mathbb{P}_n \Longrightarrow \mathbb{P}_{m,S,M} \qquad \text{as} \quad n \to \infty$$

it follows that the sequence $\{f_n\colon n \geqslant 1\}$ is of Kubilius type. In the case, when $G = \mathbf{R}^1$, this assertion was proved in (Manstavičius, 1985, 1986, 1988; Timofeev and Usmanov, 1984, 1986).

REFERENCES

Aldous, D. (1978). Stopping times and tightness. *Ann. Probab.* **6**, 2, 335–340.

Araujo, A. and Giné, E. (1980). *The Central Limit Theorem for Real and Banach Valued Random Variables.* John Wiley and Sons, New York.

Babu, G. J. (1973). *Probabilistic Methods in the Theory of Arithmetical Functions.* Indian Statistical Institute, Calcutta.

Billingsley, P. (1974). The probability theory of additive arithmetic functions. *Ann. Probab.* **2**, 5, 749–791.

Ethier, S. N. and Kurtz, T. G. (1986). *Markov Processes: Characterization and Convergence.* John Wiley and Sons, New York.

Gikhman, I. I. and Skorokhod, A. V. (1974). *The Theory of Stochastic Processes,* I. Springer-Verlag, Berlin, Heidelberg, New York.

Gikhman, I. I. and Skorokhod, A. V. (1975) *The Theory of Stochastic Processes,* II. Springer-Verlag, Berlin, Heidelberg, New York.

Grigelionis, B. and Mikulevičius, R. (1981). Weak convergence of semimartingales. *Lithuanian Math. J.* **21**, 213–224.

Grigelionis, B. and Mikulevičius, R. (1983). On stably weak convergence of semimartingales and point processes. *Theory Probab. Appl.* **28**, 2, 337–350.

Grigelionis, B. and Mikulevičius, R. (1984). On the functional limit theorems of the probabilistic number theory. *Liet. Matem. Rink.* **24**, 2, 72–81 (in Russian).

Jacod, J. (1983). Processus à accroissements indépendants: one condition necéssaire et suffisante de convergence en loi. *Z. Warsch. verw. Geb.* **63**, 109–136.

Kryžius, Z.(1981). Application of limit theorems for random processes in probabilistic number theory. In: *Abstracts of Commun. 3rd Vilnius Conf. Probab. Theory and Math. Statist.,* III. IMC, Vilnius, pp. 165–166.

Kubilius, J. (1955). An analog of A. N. Kolmogorov's theorem on Markov processes in the theory of prime numbers. *Dokl. Akad. Nauk SSSR* **103**, 361–363 (in Russian).

Kubilius, J. (1964). *Probabilistic Methods in the Theory of Numbers.* American Mathem. Soc. Providence, RI.

Kubilius, J. (1976). Additive arithmetic functions and Brownian motion. In: *Proc. III Japan–USSR Symp. Probab. Theory (Lecture Notes in Math.* **505**). Springer-Verlag, Berlin, Heidelberg, New York pp. 335–350.

Lavrent'ev, V. V. (1983). On weak convergence of Hilbert space valued semimartingales to stochastically continuous processes with conditionally independent increments. *Uspekhi Mat. Nauk* **38**, 3, 183–184 (in Russian).

Manstavičius, E. (1984). Arithmetic simulation of stochastic processes. *Lithuanian Math. J.* **24**, 276–285.

Manstavičius, E. (1985). Additive functions and stochastic processes. I. *Lithuanian Math. J.* **25**, 52–61.

Manstavičius, E. (1986). Additive functions and stochastic processes. II. *Liet. Matem. Rink.* **26**, 3, 478–481 (in Russian).

Manstavičius, E. (1988). An invariance principle for additive arithmetic functions. *Soviet Math. Dokl.* **37**, 1, 259–263.

Parthasarathy, K. R. (1967). *Probability Measures on Metric Spaces.* Academic Press, New York, London.

Philipp, W. (1973). Arithmetic functions and Brownian motion. *Proc. Symp. Pure Math.* **24**, 233–246.

Timofeev, N. M. and Usmanov, Ch. Ch. (1978). Additive functions and limit theorems in functional spaces. *Izv. Akad. Nauk Tadzhik. SSR* 4, 25–33 (in Russian).

Timofeev, N. M. and Usmanov, Ch. Ch. (1986). On a class of arithmetical models of stochastic processes. *Dokl. Akad. Nauk. Tadzhik. SSR* **26**, 6, 330–334 (in Russian).

Usmanov, Ch. Ch. (1979). On arithmetical modelling of multi-dimensional stochastic processes. *Dokl. Akad. Nauk Tadzhik. SSR* **22**, 5, 292–295 (in Russian).

New Trends in Probab. and Statist., pp. 223–234
F. Schweiger and E. Manstavičius (Eds.)
1992 VSP/TEV

ON THE MODULUS OF CONTINUITY OF THE DISTRIBUTION OF SOME ARITHMETICAL FUNCTIONS

K.-H. INDLEKOFER and I. KÁTAI*
Universität–GH Paderborn, Fachbereich Mathematik-Informatik
Warburger Str. 100, D-4790 Paderborn, Germany

Eötvös Loránd University, Computer Center
Bogdánfy ut 10/b, Budapest H-1117, Hungary

ABSTRACT

The lower and upper bounds for the continuity modulus of the distribution function connected with the arithmetical functions are given as the main result of the paper. Let a sequence of real numbers l_n has the limit distribution function F_l. The estimates of the continuity moduli of F_l are proved in the cases $l_n = \sigma(n)/n + \sigma(n+1)/(n+1)$, $\phi(n)/n + \phi(n+1)/(n+1)$, $\sigma(n)/n + \phi(n)/n$.

1. INTRODUCTION

We say that an infinite sequence L_n has the limit distribution $F(x)$, where $F(x)$ is a distribution function, if

$$\lim_{N\to\infty} N^{-1} \# \{n \leqslant N:\ L_n < x\} = F(x)$$

at every continuity point x of F.
In this paper our purpose is to estimate the modulus of continuity of the limit distribution function of the sequences

$$L_n = f_1(a_n^{(1)}) + f_2(a_n^{(2)}) + \ldots + f_r(a_n^{(r)})$$

under some conditions stated for the real-valued additive functions f_j and for the subsequences $a_n^{(j)}$ of positive integers (see Theorem 1). We apply the method for sequences

$$\frac{\phi(n)}{n} + \frac{\sigma(n)}{n}, \qquad \frac{\sigma(n)}{n} + \frac{\sigma(n+1)}{n+1}, \qquad \frac{\phi(n)}{n} + \frac{\phi(n+1)}{n+1}$$

(see Theorem 2).

*The paper was written while the second named author held a visiting professorship at the University of Paderborn. Financial support had been provided by DFG.

2. ABBREVIATIONS AND NOTATIONS

Abbreviations: DF means a distribution function, LDF means a limit DF, CHF means a characteristic function, CM means a continuity modulus, RV means a random variable, IRV means independent RVs.
Notations: $\mathcal{A}$ ($\mathcal{A}_s$) is the set of real-valued (strongly) additive functions. $\mathbf{P}$ is the set of primes, $\mathbf{P}^{(l)} := \{p + l : p \in \mathbf{P}\}$. The letters p, p_1, p_2, ... stand for primes. If ζ is a RV with DF $F_\zeta = F$, then φ_ζ or φ_F will denote its CHF, furthermore

$$Q_\zeta(h) = Q_F(h) := \sup_x \left(F(x+h) - F(x)\right)$$

is called the CM of F.
$\sigma(n)$ is the sum of divisors of n and $\phi(n)$ is Euler's totient function.
The letters c, c_1, c_2, ... denote suitable positive constants, not the same at every occurrence.
$P(n)$ and $p(n)$ mean the largest and the smallest prime factor of n, respectively.
$\mathbb{P}(\ldots)$ means a probability of the event (in some probability space) indicated within parenthesis.
The CM of ζ_1 and that of ζ_2 are said to be equivalent (in notation: $Q_{\xi_1} \asymp Q_{\xi_2}$), if

$$c_1 < \frac{Q_{\zeta_1}(h)}{Q_{\zeta_2}(h)} < c_2$$

holds for every $h > 0$.
Let $e(x) := e^{ix}$.

3. PRELIMINARIES AND LEMMATA

3.1 (Erdös and Wintner, 1939). The function $g \in \mathcal{A}$ has the limit distribution F, iff the series

$$\sum_{|g(p)|<1} \frac{g(p)}{p}, \qquad \sum_{|g(p)|<1} \frac{g^2(p)}{p}, \qquad \sum_{|g(p)|\geqslant 1} \frac{1}{p} \tag{3.1)–(3.3}$$

are convergent. Furthermore,

$$\varphi_F(\tau) = \prod_p \left(1 - \frac{1}{p}\right)\left(1 + \sum_{k=1}^{\infty} p^{-k} e\left(\tau g(p^k)\right)\right).$$

The distribution F can be interpreted as the DF of the RV $\eta = \sum \xi_p$, where ξ_p are IRV with purely discrete distributions, and

$$\varphi_{\xi_p}(\tau) = \left(1 - \frac{1}{p}\right)\left(1 + \sum_{k=1}^{\infty} p^{-k} e\left(\tau g(p^k)\right)\right). \tag{3.5}$$

3.2 (Lévy, 1931). If η is the convergent sum of the IRV ξ_p, $\eta = \sum \xi_p$, then F_η is continuous (everywhere) iff

$$\sum \mathbb{P}(\xi_p \neq 0) = \infty. \tag{3.6}$$

3.3. If (3.6) holds then F_η is of pure type, either absolutely continuous or singular (see (Lukács, 1960)).
3.4. We have

$$(\log t)^{-1} \ll Q_F(1/t) \ll (\log t)^{-1},$$

when one of the following conditions is valid:

a) $g(n) = \log \dfrac{\phi(n)}{n}$ (Tjan, 1966);

b) $g(n) = \log \dfrac{\sigma(n)}{n}$ (Erdös, 1974);

c) $g \in \mathcal{A}_s$ and

$$\sum_{p > t^A} \frac{|g(p)|}{p} < \frac{1}{t},$$

$$|g(p_1) - g(p_2)| > \frac{1}{t} \quad \text{if} \quad p_1 \neq p_2 < t^\delta$$

hold with suitable positive constants A and δ for every large t (Erdös and Kátai, 1979).
3.5. If $g \in \mathcal{A}_s$, $g(p) = (\log p)^{-\gamma}$, then

(a) $t^{-1/\gamma} \ll Q_F(1/t) \ll (\log\log t) t^{-1/\gamma}$ $(\gamma > 1)$,

(b) $t^{-1} \ll Q_F(1/t) \ll \dfrac{(\log\log t)^2 \log t}{t}$

(see (Erdös and Kátai, 1979)),

(c) F is absolutely continuous if $\gamma < 2$ (Babu, 1974/75).

3.6. Let γ_1, γ_2 be IRV, $\gamma = \gamma_1 + \gamma_2$. Then

$$Q_\gamma(h) \leqslant \min\left(Q_{\gamma_1}(h), Q_{\gamma_2}(h)\right).$$

If the distribution of γ_1 is purely discrete, then Q_{γ_2} and Q_γ are equivalent.
3.7. Let φ and ψ be CHF, such that

$$\kappa(\tau) = \frac{\varphi(\tau)}{\psi(\tau)} \neq 0, \qquad |\kappa(\tau)| > c > 0,$$

and κ is an almost periodic function with absolutely convergent series of Fourier-coefficients. Then the CM of F_φ and F_ψ are equivalent.
This is an immediate consequence of Wiener's theorem.
3.8. Let $P_1, \ldots, P_k$ be distinct irreducible polynomials with integer coefficients such that $P_j(n) > 0$ for every large n. Let $r_j(p^\alpha)$ be the number of solutions of

the congruence $P_j(n) \equiv 0 \pmod{p^\alpha}$, and $l_j(p^\alpha)$ be the number of those solutions $n \pmod{p^\alpha}$ of the above congruence for which $(n,p)=1$. Let $g_j \in \mathcal{A}$ be such that

$$g_j(p^\alpha)r_j(p^\alpha) \to 0 \qquad \text{as} \quad p^\alpha \to \infty,$$

for every $j = 1, \ldots k$. Furthermore let the series

$$\sum \frac{g_j(p)r_j(p)}{p}, \qquad \sum \frac{g_j^2(p)r_j(p)}{p}$$

be convergent for $j = 1, \ldots, k$.
Then $L_n := g_1(P_1(n)) + \ldots + g_k(P_k(n))$ has a limit distribution F. Furthermore, F can be represented as $F = F_\eta$, where

$$\eta = \eta^0 + \sum_{p>c_1} \xi_p,$$

η^0 and all ξ_p are IRV with discrete distributions, and

$$\varphi_{\xi_p}(\tau) = 1 - \frac{\sum_{j=1}^{k} r_j(p)}{p} + \sum_{j=1}^{k} \sum_{l=1}^{\infty} r_j(p^l)p^{-l}\left(1 - \frac{1}{p}\right)e(\tau g_j(p^l)). \tag{3.7}$$

Assume, additionally, that $P_j(0) \neq 0$ $(j = 1, \ldots, k)$. Then L_p, $p \in \mathbf{P}$ has the limit distribution G. It can be represented as G_ζ, where

$$\zeta = \zeta^{(0)} + \sum_{p>c_1} \xi'_p,$$

$\zeta^{(0)}$ and all ξ'_p are IRV with discrete distributions, and

$$\varphi_{\xi'_p}(\tau) = 1 - \frac{\sum_{j=1}^{k} r_j(p)}{p-1} + \sum_{j=1}^{k} \sum_{l=1}^{\infty} r_j(p^l)p^{-l}e(\tau g_j(p^l)). \tag{3.8}$$

For the proof see (Kátai, 1969a).
3.9. Let $g \in \mathcal{A}$ and an integer $l \neq 0$. Then g has the limit distribution on $\mathbf{P}^{(l)}$, iff (3.1)–(3.3) are convergent.
The sufficiency part was published in (Kátai, 1969b), the necessity part is a recent result of A. Hildebrand.
3.10. Let

$$\eta = \sum \xi_n, \qquad \eta' = \sum \xi'_n$$

be convergent sums of IRV. Assume that

$$\xi_n \in \{0\} \cup \{x_{nk}\colon\ k = 1, 2, \ldots\},$$

$$\xi'_n \in \{0\} \cup \{x'_{nk}\colon k = 1, 2, \dots\},$$

with the probabilities

$$\mathbb{P}(\xi_n = 0) = 1 - Q_n, \qquad \mathbb{P}(\xi_{nk}) = q_{nk}, \qquad Q_n = \sum_k q_{nk},$$

$$\mathbb{P}(\xi'_n = 0) = 1 - Q'_n, \qquad \mathbb{P}(\xi'_n = x'_{nk}) = q'_{nk}, \qquad Q'_n = \sum_k q'_{nk}.$$

Let $T(z) = 0$ if $z = 0$, and $T(z) = 1$ if $z \neq 0$. Let $Q'_n < -1/2$, and

$$e_n := \frac{1}{1 - 2Q'_n}\left(|Q'_n - Q_n| + \sum_k \left\{|q_{nk} - q'_{nk}| + q'_{nk}T(x_{nk} - x'_{nk})\right\}\right).$$

Then

$$\sum e_n < \infty$$

implies that $Q_n \asymp Q'_n$.

This assertion is an easy consequence of 3.6 and 3.7.

3.11. Assume that the conditions stated for g_j in 3.8 hold. Let $g'_j \in \mathcal{A}_s$ be defined by the relations

$$g'_j(p) = g_j(p), \qquad p \in \mathbf{P}, \quad j = 1, \dots, k,$$

and

$$L'_n = g_1(P_1(n)) + \dots + g_k(P_k(n)).$$

Then the CM of the LDFs of L_n, L'_n, and L_p are equivalent. This follows immediately from 3.10 and 3.8.

4. THE MAIN RESULT

THEOREM 1. *Let $t_j(p)$, $j = 1, \dots, k$, be non-negative functions defined on* $\mathbf{P}$ *the values of which are taken from a finite set, and $f_j \in \mathcal{A}_s$ be functions for which there exist positive constants δ and A such that for every $t \geqslant 2$:*

(a) $\displaystyle\sum_{p > t^A} \frac{t_j(p)|f_j(p)|}{p} < \frac{1}{t}$,

(b) *if $p_1, p_2 < t^\delta$ and $p_1 \neq p_2$, then $|f_i(p_1) - f_j(p_2)| > 1/t$ holds for every i and j for which $t_i(p_1) \neq 0$ and $t_j(p_2) \neq 0$.*

Let ξ_p be IRV defined for every prime $p > c_2$, the CHFs of which have the form

$$\varphi_{\xi_p}(\tau) = 1 + \sum_{j=1}^{k} \frac{t_j(p)(e(\tau f_j(p)) - 1)}{p}.$$

Let

$$\eta = \sum_{p > c_2} \xi_p,$$

where c_2 is a constant for which $t(p) < p/4$ holds,

$$t(p) := t_1(p) + \ldots + t_k(p).$$

Then

$$c_3 \prod_{c_2<p<t} \left(1 - \frac{t(p)}{p}\right) < Q_\eta\left(\frac{1}{t}\right) < c_4 \prod_{c_2<p<t} \left(1 - \frac{t(p)}{p}\right). \tag{4.2}$$

COROLLARY. *Let $P_1, \ldots, P_k$ be polynomials for which the conditions stated in 3.8 hold. For every $j \in \{1, \ldots, k\}$ let $T_j(n)$ be one of the functions*

$$\frac{\sigma(n)}{n}, \qquad \frac{n}{\sigma(n)}, \qquad \frac{\phi(n)}{n}, \qquad \frac{n}{\phi(n)}.$$

Let

$$A_n = T_1(P_1(n)) \ldots T_k(P_k(n)).$$

Let $F_1(x)$ (respectively, $F_2(x)$) be the LDF *of the sequence* $\log A_n$ *(respectively,* $\log A_p$*).*
Then, for $t \geqslant 2$ and $i = 1, 2$, we have

$$c_1(\log t)^{-k} < Q_{F_i}(1/t) < c_2(\log t)^{-k}. \tag{4.3}$$

Deduction of the Corollary from Theorem 1. Let

$$f_j \in \mathcal{A}_s, \qquad f_j(p) := \log T_j(p).$$

Then $f_j(p)$ is one of the numbers

$$\pm \log(1 + 1/p), \qquad \pm \log(1 - 1/p).$$

The existence of F_1 and F_2 is clear. Taking into account 3.11, 3.8, 3.6, 3.10, and that

$$\sum_{p \leqslant x} \frac{r_j(p)}{p} = \log \log x + O(1),$$

by the substitution $t_j(p) = r_j(p)$, we have $Q_{F_i} \asymp Q_\eta$. Therefore, it is enough to observe that the conditions (a) and (b) of Theorem 1 are satisfied.
Proof of Theorem 1. Let

$$\alpha(y) = \prod_{c_2<p<y} \left(1 - \frac{t(p)}{p}\right).$$

Since $0 \leqslant t(p) \leqslant c$, the function $\alpha(y)$ is decreasing and $\alpha(y^2) > c_1\alpha(y)$.
Let

$$\eta_1 = \sum \xi_p,$$

where the summation is extended over p, $p \in (c_2, t^\delta)$. We assume that t is a large positive number and δ is a small one.

Let H_j be multiplicative functions defined for square-free integers by

$$H_j(p) = \begin{cases} t_j(p)\left(1 - \dfrac{t(p)}{p}\right)^{-1} & \text{if } p \in (c_2, t^\delta), \\ 0 & \text{otherwise.} \end{cases}$$

For some real number x let

$$\Sigma^{(x,t)} = \sum_{d_1, \ldots, d_k} \frac{H_1(d_1) \ldots H_k(d_k)}{d_1 \ldots d_k},$$

where we sum over all those square-free, pairwise coprime integers $d_1, \ldots, d_k$, for which

$$f_1(d_1) + \ldots + f_k(d_k) \in \left[x, x + \frac{2}{t}\right].$$

Since

$$\varphi_{\xi_p}(\tau) = \left(1 - \frac{t(p)}{p}\right)\left(1 + \sum_{j=1}^{k} \frac{H_j(p) e(\tau f_j(p))}{p}\right),$$

we have

$$\mathbf{P}\left(\eta_1 \in \left[x, x + \frac{2}{t}\right]\right) \leqslant \alpha(t^\delta)\Sigma^{(x,t)}.$$

Let $d_1, \ldots, d_k$ be a k-tuple of square-free numbers which gives a positive contribution to $\Sigma^{(x,t)}$. Let $P = P(d_1 \ldots d_k)$, and assume that $P \mid d_i$. We define d'_l by

$$d'_l = \begin{cases} d_l & \text{if } l \neq i, \\ d_i / P & \text{if } l = i. \end{cases}$$

Then, due to the assumption (b) there exist no l and no prime Q, for which

$$P(d'_i \ldots d'_k) < Q < t^\delta, \qquad P \neq Q,$$

such that

$$d'_i, \ldots, d'_{l-1}, Q_d, d'_{l+1}, \ldots, d'_k$$

(in the role of $d_1, \ldots, d_k$) would give a positive contribution to $\Sigma^{(x,t)}$. Since

$$\frac{H_1(d_1) \ldots H_k(d_k)}{d_1 \ldots d_k} \leqslant \frac{c_5}{P(d'_1 \ldots d'_k)} \frac{H_1(d'_1) \ldots H_k(d'_k)}{d'_1 \ldots d'_k},$$

$$H_1(d'_1) \ldots H_k(d'_k) \leqslant c^{\omega(d'_1 \ldots d'_k)},$$

and the number of solutions of the equation $D = d'_1 \ldots d'_k$ is at most $O(c^{\omega(D)})$, we have

$$\Sigma^{(x,t)} \leqslant c_6 \sum \frac{c^{\omega(D)}}{DP(D)} + c_7, \tag{$*$}$$

where D runs over the square-free numbers for which

$$c_2 < p(D) \leqslant P(D) < t^{\delta}.$$

The right-hand side of $(*)$ is bounded by a constant. Indeed, the contribution to this sum of the $D's$ for which the inequalities $2^{2^h} < P(D) < 2^{2^{h+1}}$ hold is less than

$$2^{-2^h} \prod_{p<2^{2^{h+1}}} \left(1+\frac{c}{p}\right) < \frac{e^{c_1 h}}{2^{2^h}},$$

and the sum of the right-hand side over all h is convergent.
Since

$$Q_\eta\left(\frac{1}{t}\right) \leqslant Q_{\eta_1}\left(\frac{1}{t}\right) = \sup_x \mathbb{P}\left(\eta_1 \in \left[x, \frac{x_1}{t}\right]\right),$$

from the above estimations the second inequality in (4.2) follows immediately.
Let t be a large number. Then

$$\mathbb{P}\left(\sum_{p>t^{2A}} \xi_p > \frac{1}{t}\right) < \frac{1}{2}$$

due to (a), thus

$$Q_\eta\left(\frac{1}{t}\right) \geqslant \mathbb{P}\left(\eta \in \left[0, \frac{1}{t}\right]\right) \geqslant \mathbb{P}\left(\xi_p = 0 \text{ for } p \in \left(c_2, t^{2A}\right) \text{ and } \sum_{p>t^{2A}} \xi_p < \frac{1}{t}\right)$$
$$\leqslant \frac{1}{2}\alpha\left(t^{2A}\right) \leqslant c_3\alpha(t).$$

The proof of Theorem 1 is completed.

5. APPLICATION

Let

$$\sigma_0(n) := \frac{\sigma(n)}{n}, \qquad \phi_0(n) := \frac{\phi(n)}{n},$$

furthermore, let

$$A_y(n) := \prod_{\substack{p^\alpha \| n \\ p \leqslant y}} p^\alpha, \qquad B_y(n) := \frac{n}{A_y(n)}.$$

Let

$$a_n := \sigma_0(n) + \sigma_0(n+1),$$
$$b_n := \phi_0(n) + \phi_0(n+1),$$
$$c_n := \sigma_0(n) + \phi_0(n).$$

Let $F_a(x)$, $F_b(x)$, $F_c(x)$ denote the LDFs of the sequences a_n, b_n, c_n. Their existence follows from the fact that $\sigma(n)$ and $\phi(n)$ can be approximated by the periodic functions $\sigma_0(A_y(n))$ and $\phi_0(A_y(n))$ (see (Postnikov, 1986)).

THEOREM 2. *In the above notations, we have*

$$Q_{F_c}\left(\frac{1}{t}\right) \asymp (\log t)^{-1}$$

$$Q_{F_a}\left(\frac{1}{t}\right) \asymp Q_{F_b}\left(\frac{1}{t}\right) \asymp (\log t)^{-2}.$$

Proof. Let $t \geqslant 2$ and let A be an appropriate large constant.
We have

$$\frac{1}{N}\sum_{n\leqslant N}\left(\sigma_0\left(B_{t^2}(n)\right) - 1\right) < \frac{c_1}{t^2}, \tag{5.1}$$

$$\frac{1}{N}\sum_{n\leqslant N}\left(1 - \left(\phi_0\left(B_{t^2}(n)\right)\right)\right) < \frac{c_1}{t^2}. \tag{5.2}$$

The asymptotic density of the integers n for which $p(n) > t^A$ is

$$\prod_{p<t^A}\left(1 - \frac{1}{p}\right) \gg (\log t)^{-1}.$$

Taking into account (5.1) and (5.2) we deduce that the density of the integers n for which

$$c_n \in \left(2 - \frac{1}{2t}, 2 + \frac{1}{2t}\right)$$

is $\gg (\log t)^{-1}$. This proves the inequality

$$Q_{F_c}\left(\frac{1}{t}\right) \gg (\log t)^{-1}.$$

Now let us consider the set of those integers n for which

$$A_{t^A}(n) = 1, \qquad A_{t^A}(n+1) = 2.$$

The density of this set is $\gg (\log t)^{-2}$. Hence, by (5.1) and (5.2) we obtain that the density of those integers n for which $a_n \in [2, 2+1/t)$ is $\gg (\log t)^{-2}$. The same is true for the density of those integers n for which $b_n \in [2 - 1/t, 2)$. This proves that

$$Q_{F_a}\left(\frac{1}{t}\right) \gg (\log t)^{-2}, \qquad Q_{F_b}\left(\frac{1}{t}\right) \gg (\log t)^{-2}.$$

Now we shall prove the upper estimation for Q_{F_b}. For an arbitrary real number x we consider the set of those integers n for which $b_n \in [x, x+1/t]$. We may assume that $0 \leqslant x \leqslant 2$. From (5.2) we obtain that the density of the integers n for which

$$1 - \phi_0\left(B_{t^2}(n)\right) > t^{-3/2}$$

is $O(1/\sqrt{t})$, therefore, it is enough to estimate the density of those integers n for which

$$\phi_0\left(A_{t^2}(n)\right) + \phi_0\left(A_{t^2}(n+1)\right) \in \left[x, x + \frac{2}{t}\right]. \tag{5.3}$$

Let $\mathcal{D}$ be the set of coprime couples (d_1, d_2) of square-free numbers such that $P(d_i) < t^2$, $i = 1, 2$, and

$$\phi_0(d_1) + \phi_0(d_2) \in \left[x, x + \frac{2}{t}\right]. \tag{5.4}$$

The density of the integers n for which

$$A_{t^2}(n) = d_1, \qquad A_{t^2}(n+1) = d_2$$

is bounded by

$$\frac{c}{d_1 d_2} \prod_{\substack{p \neq 2 \\ p < t^2}} \left(1 - \frac{2}{p}\right).$$

We shall prove that

$$\sum_{(d_1, d_2) \in \mathcal{D}} \frac{1}{d_1 d_2} < c_1, \tag{5.5}$$

which implies the desired inequality

$$Q_{F_b}\left(\frac{1}{t}\right) \ll (\log t)^{-2}.$$

To prove (5.5), we choose a small positive δ and write

$$d_i = D_i E_i, \qquad D_i = A_{t^\delta}(d_i), \qquad i = 1, 2.$$

Let $P = P(D_1 D_2)$, and assume that $P \mid D_1$. Let

$$D_1' = \frac{D_1}{P}, \qquad d_1' = \frac{d_1}{P}.$$

Let Q be an arbitrary prime such that $P \neq Q$ and

$$P(d_1' d_2) < Q < t^\delta.$$

If $(d_1, d_2) \in \mathcal{D}$ and $(Q d_1', d_2) \in \mathcal{D}$, then

$$\left(\frac{1}{P} - \frac{1}{Q}\right) \phi_0(d_1') \in \left[-\frac{4}{t}, \frac{4}{t}\right], \tag{5.6}$$

and so

$$\phi_0(d_1') < c t^{\delta - 1}. \tag{5.7}$$

Since

$$\frac{1}{N}\sum_{n\leqslant N}\frac{1}{\phi_0(n)}\ll 1, \qquad N\to\infty,$$

the density of the integers n for which there exists a divisor d_1 satisfying (5.7) is less than $Q(t^{-1/2})$, if $\delta\leqslant 1/4$, say; furthermore, when (5.7) holds,

$$\sum_{(d_1,d_2)\in\mathcal{D}}\frac{1}{d_1 d_2}<c_2.$$

If (5.7) does not hold, then P is uniquely determined by x, d_1', and d_2. The case $P \mid d_2$ can be treated similarly. Thus the sum (5.5) is

$$\ll 1+\sum\left(\frac{1}{D_1'D_2P(D_1'D_2)}+\frac{1}{D_1D_2'P(D_1D_2')}\right)\frac{1}{E_1E_2}.$$

Repeating the argument which was used in the proof of Theorem 1, we get that this sum is bounded.

The proof of the inequality

$$Q_{F_a}\left(\frac{1}{t}\right)\ll(\log t)^{-2}$$

is similar, and we omit it.

Finally we sketch the proof of the inequality

$$Q_{F_c}\left(\frac{1}{t}\right)\ll(\log t)^{-1}.$$

We can reduce the problem to showing that

$$\sum\frac{1}{d} \tag{5.8}$$

is bounded by a constant, if d runs over all integers with $P(d)\leqslant t^A$ and such that

$$\phi_0(d)+\sigma_0(d)\in\left[x,x+\frac{2}{t}\right]. \tag{5.9}$$

Let $d=DE$, where $P(D)\leqslant t^\delta$, $p(E)>t^\delta$, and $P=P(D)$. The contribution to (5.8) of those d, for which $P^2\mid d$ or for which $1-\phi_0(d)<1/\sqrt{t}$, is clearly bounded. Let

$$D'=\frac{D}{P}, \qquad d'=\frac{d}{P}$$

and let Q be an arbitrary prime such that

$$P(D')<Q<t^\delta, \qquad Q\neq P.$$

If (5.9) holds for d and for $d'Q$; then

$$\left(\frac{1}{P}-\frac{1}{Q}\right)(\sigma_0(d')-\phi_0(d'))\in\left[-\frac{2}{t},\frac{2}{t}\right],$$

which by

$$\left|\frac{1}{P}-\frac{1}{Q}\right|>t^{-2\delta}$$

implies that

$$1-\phi_0(d')<2t^{-1+2\delta}<t^{-3/4},$$

and so that

$$1-\phi_0(d)<1/\sqrt{t},$$

if P is larger than a constant. Thus (5.8) is

$$\ll c_1+\sum\frac{1}{D'P(D')E}\,. \tag{5.10}$$

Since

$$\sum_E\frac{1}{E}\leqslant\prod_{t^\delta\leqslant p\leqslant t^A}\left(1-\frac{1}{p}\right)^{-1}\leqslant\frac{c_2A}{\delta},$$

$$\sum\frac{1}{D'P(D')}\leqslant\sum_{p\leqslant t^\delta}\frac{1}{p^2}\sum_{p(m)\leqslant p}\frac{1}{m}\leqslant c_3,$$

the sum in (5.10) is clearly bounded.
The proof of Theorem 2 is finished.

REFERENCES

Babu, G. J. (1974/75). Absolutely continuous function $f(p)=(\log p)^{-a}$, $a>0$. *Acta Arithm.* **26**, 401–403.

Erdös, P. (1974). On the distribution of numbers of the form $\sigma(n)/n$ and on some related questions. *Pacific J. Math.* **52**, 59–65.

Erdös, P. and Kátai, I. (1979). On the concentration of additive funtions. *Acta Sci. Math.* **41**, 295–305.

Erdös, P. and Wintner, A. (1939). Additive arithmetical functions and statistical independence. *Amer. J. Math.* **61**, 713–721.

Kátai, I. (1969a). On the distribution of arithmetical functions. *Acta Math. Hung.* **20**, 69–87.

Kátai, I. (1969b). On distribution of arithmetical functions on the set of prime plus one. *Compos. Math.* 278–288.

Lévy, P. (1931). Sur les séries dont les terms son des variables éventuelles indépandantes. *Studia Math.* **3**, 119–155.

Lukács, E. (1960). *Characteristic Functions.* Griffin, London.

Postnikov, A. G. (1986). *Introduction to Analytic Number Theory.* American Mathem. Soc., Providence, RI.

Tjan, M. M. (1966). On the question of the distribution of values of the Euler function $\phi(n)$. *Liet. Matem. Rink.* **6**, 105–119 (Russian).

New Trends in Probab. and Statist., pp. 235–249
F. Schweiger and E. Manstavičius (Eds.)
1992 VSP/TEV

TOPICS ON ARITHMETIC FUNCTIONS WITH RATIONAL ARGUMENTS

Z. KRYŽIUS
Institute of Mathematics and Informatics,
Akademijos 4, 2600 Vilnius, Lithuania

ABSTRACT

Probabilistic models for additive arithmetic functions over rational numbers are constructed. These models play the important role in the study of distribution of values of additive functions. The method applies when arguments of additive functions are s-tuples of pair-wise relatively prime numbers.

0. INTRODUCTION

Let $\mathbf{N}$ be the set of natural numbers,

$$\mathbf{N}^s = \mathbf{N} \times \ldots \times \mathbf{N} \quad (s \text{ times}), \qquad \overline{m} = \langle m_1, \ldots, m_s \rangle,$$

and

$$\widetilde{\mathbf{N}}^s = \{\overline{m} \in \mathbf{N}^s \colon \ (m_i, m_j) = 1 \text{ when } i \neq j\}.$$

The set $\widetilde{\mathbf{N}}^2$ may be interpreted as the set of rational numbers.
Let $\overline{m}, \overline{n} \in \mathbf{N}^s$. We say that $\mathrm{GCD}\,(\overline{m}, \overline{n}) = \overline{1} = \langle 1, \ldots, 1 \rangle$ iff $\mathrm{GCD}\,(m_i, n_i) = 1$ for $i = 1, \ldots, s$. The product $\overline{m}\,\overline{n}$ is defined by $\overline{m}\,\overline{n} = \langle m_1 n_1, \ldots, m_s n_s \rangle$. The function $h\colon \widetilde{\mathbf{N}}^s \longrightarrow \mathbf{R}$ is called additive (multiplicative) if $\mathrm{GCD}\,(\overline{m}, \overline{n}) = \overline{1}$ and $\bar{m}\bar{n} \in \widetilde{\mathbf{N}}^s$ imply $h(\overline{m}\,\overline{n}) = h(\bar{m}) + h(\overline{n})$ (or, respectively, $h(\overline{m}\,\overline{n}) = h(\bar{m})h(\overline{n})$).
The study of distribution of values of additive functions over rational numbers (i.e. over $\widetilde{\mathbf{N}}^2$) and over $\widetilde{\mathbf{N}}^s$ for $s > 2$ was initiated in (Stakėnas, 1987, 1988; Stakėnas and Šiaulys, 1988; Stakėnas, 1989).
Here we present the approach to this problem which seems to be more transparent than the original one, and to allow some further steps in the investigation of the problem. The classical way to treat the distribution of values of additive arithmetic functions is to construct some probability spaces called probabilistic models. This way takes its origin in the works of J. Kubilius (see (Kubilius, 1964; Elliott, 1979)). We show how these models may be constructed for additive functions over $\widetilde{\mathbf{N}}^s$.
Some words about notations. Everywhere p is a prime number, μ is the Möbius function, $\omega(m)$ is the number of distinct primes dividing m, φ is the Euler function, $\|$ means 'strictly divides', $\operatorname{card} A$ is the number of elements of a set A. Constants implied in symbols O are absolute unless other indications are made by subscripts or comments. This rule however is violated in § 4. Constants are denoted by $c_1, c_2, \ldots$.

1. APPLICATION OF THE FUNDAMENTAL LEMMA

We shall use the following notations. For $\bar{d} \in \tilde{\mathbf{N}}^s$ the set of multiples $\bar{d}\tilde{\mathbf{N}}^s$ is defined by

$$\bar{d}\tilde{\mathbf{N}}^s = \{\overline{m} \in \tilde{\mathbf{N}}^s\colon\ d_i \mid m_i \text{ for } i = 1, \ldots, s\},$$

$\bar{p}(i,\alpha)$ stands for $\langle 1, \ldots, p^\alpha, \ldots, 1\rangle$ with p^α as i-th component, $\bar{p}(i,\alpha) \parallel \bar{l}$ means that $p^\alpha \parallel l_i$, $\bar{d}^* = (\bar{d})^* = d_1 \cdot \ldots \cdot d_s$ for $\bar{d} \in \mathbf{N}^s$.
The general situation may be described as follows. Suppose that we have the fixed sequence $\{W_x\colon\ x \geqslant 1\}$ of regions $W_x \subset \mathbf{N}^s$ and the fixed sequence $\{f_x\colon\ x \geqslant 1\}$ of maps $f_x\colon \mathbf{N}^s \longrightarrow \mathbf{N}^s$ such that $f_x(\mathbf{N}^s) \cap \tilde{\mathbf{N}}^s \neq \varnothing$. The sequence of measurable spaces $\{(\Omega_x, \mathfrak{F}_{x,r})\colon\ x \geqslant 1\}$ is constructed, where $\Omega_x = f_x^{-1}(f_x(\mathbf{N}^s) \cap \tilde{\mathbf{N}}^s)$ and $\mathfrak{F}_{x,r} = \mathfrak{F}^s_{x,r}$ is the σ-algebra generated by the family of sets

$$E(\bar{p}(i,\alpha)) = \{\overline{m} \in \mathbf{N}^s\colon\ f_x(\overline{m}) \in \tilde{\mathbf{N}}^s,\ \bar{p}(i,\alpha) \parallel f_x(\overline{m})\},$$

$$p \leqslant r, \qquad i = 1, \ldots, s, \qquad \alpha \in \mathbf{N} \cup \{0\}.$$

The sets

$$A(\bar{l}) = \{\overline{m} \in \mathbf{N}^s\colon\ f_x(\overline{m}) \in \tilde{\mathbf{N}}^s,\ \bar{p}(i,\alpha) \parallel f_x(\overline{m}) \iff \bar{p}(i,\alpha) \parallel \bar{l}\},$$

where

$$p \mid \bar{l}^* \Rightarrow p \leqslant r,$$

are atoms in $\mathfrak{F}_{x,r}$. Thus for each $A \in \mathfrak{F}_{x,r}$ we have the representation

$$A = \sum_{\bar{l} \in L_A} A(\bar{l}),$$

where L_A is some family of indices, individual for each set A.
The sequence of empirical measures (in particular, densities) on $(\Omega_x, \mathfrak{F}_{x,r})$ is defined by

$$\nu_x(A) = \operatorname{card}(A \cap W_x)/F(x), \qquad A \in \mathfrak{F}_{x,r},$$

where $F(x)$ is some normalising function. The key point in the probabilistic model approach is the construction of σ-additive measure (asymptotic density) $\mathbf{P}$ on $(\Omega_x, \mathfrak{F}_{x,r})$ such that

$$\sup_{A \in \mathfrak{F}_{x,r}} |\nu_x(A) - \mathbf{P}(A)| \longrightarrow 0 \qquad \text{as} \quad x \to \infty.$$

In this construction the so-called fundamental lemma is used. The first such lemma was proved by J. Kubilius (see (Kubilius, 1964; Elliott, 1979)).
Now we show how the fundamental lemma may be directly applied in our situation. Let

$$V_x(\bar{l}) = \{\overline{m} \in W_x\colon\ f_x(\overline{m}) \in \bar{l}\tilde{\mathbf{N}}^s\}. \tag{1.1}$$

Suppose that the relations

$$\operatorname{card} V_x(\bar{l}) = N(x)\varrho(\bar{l}) + R_1(W_x, \bar{l}) \tag{1.2}$$

hold with some multiplicative function $\varrho(\bar{l})$. Denote

$$B_x(d,\bar{l}) = \{\overline{m} \in W_x\colon\ f_x(\overline{m}) \in \bar{l}\,\widetilde{\mathbf{N}}^s,\ d \mid (f_x(\overline{m})/\bar{l})^*\},$$

where $\overline{n}/\bar{l} = \langle n_1/l_1, \ldots, n_s/l_s\rangle$. We may write that

$$A(\bar{l}) \cap W_x = \{\overline{m} \in W_x\colon\ f_x(\overline{m}) \in \bar{l}\,\widetilde{\mathbf{N}}^s, p \nmid (f_x(\overline{m})/\bar{l})^* \text{ for all } p \leqslant r\} \tag{1.3}$$

Only for the sake of simplicity we suppose that (with some abuse of notations)

$$\varrho(\bar{l}) = \varrho(\bar{l}^*), \tag{1.4}$$

$$R_1(W_x,\bar{l}) = R_1(W_x,\bar{l}^*). \tag{1.5}$$

Then (1.4) and (1.5) imply that

$$\operatorname{card} B_x(d,\bar{l}) = N(x)\varrho(\bar{l}^*)\tilde{\varrho}(d) + R_2(W_x,d,\bar{l}^*)$$

for square-free numbers d, where $\tilde{\varrho}(d)$ is the multiplicative function,

$$\tilde{\varrho}(p) = \begin{cases} \varrho(p\bar{l}^*)/\varrho(\bar{l}^*) & \text{when } \ p \mid \bar{l}^*, \\ s\varrho(p\bar{l}^*)/\varrho(\bar{l}^*) & \text{when } \ p \nmid \bar{l}^*. \end{cases}$$

The remainder $R_2(W_x,d,\bar{l}^*)$ may be easily estimated by

$$|R_2(W_x,d,\bar{l}^*)| \leqslant s^{\omega(d)}|R_1(W_x,d\bar{l}^*)|.$$

Assume that

$$0 \leqslant \tilde{\varrho}(p) < 1 \qquad \text{for all} \quad p \leqslant r. \tag{1.6}$$

Now, having in mind (1.3), we may apply the fundamental lemma to obtain the expression for card $(A(\bar{l}) \cap W_x)$. We use the excellent form of this lemma formulated as Lemma 2.1 in (Elliott, 1979). Thus

$$\begin{aligned} \operatorname{card}\big(A(\bar{l}) \cap W_x\big) &= N(x)\varrho(\bar{l}^*)\prod_{p\leqslant r}\big(1-\tilde{\varrho}(p)\big)\big(1+2\theta_1 H\big) \\ &\quad + 2\theta_2 \sum_{d\mid Q,\ d\leqslant z^3} 3^{\omega(d)}|R_2(W_x,d,\bar{l}^*)|, \end{aligned} \tag{1.7}$$

where $Q = \prod_{p\leqslant r} p$, $|\theta_1| \leqslant 1$, $|\theta_2| \leqslant 1$, $r \geqslant 2$,

$$\max\big(\log r, S\big) \leqslant \frac{1}{8}\log z, \tag{1.8}$$

$$S = \sum_{p\leqslant r} \frac{\tilde{\varrho}(p)}{1-\tilde{\varrho}(p)}\log p,$$

$$H = \exp\left\{-\frac{\log z}{\log r}\left(\log\left(\frac{\log z}{\log S}\right) - \log\log\left(\frac{\log z}{S}\right) - \frac{2S}{\log z}\right)\right\}.$$

2. PRELIMINARY LEMMAS

LEMMA 2.1. *Let D be some natural number, $\mu(D) \neq 0$. Then*

$$\sum_{\substack{N_1 < m \leqslant N_2 \\ (m,D)=1}} 1 = (N_2 - N_1) \prod_{p|D} \left(1 - \frac{1}{p}\right) + O(2^{\omega(D)}).$$

Proof is quite easy and may be found in (Stakėnas, 1988).

The following lemma is the generalization of Theorem from (Postnikov, 1988, § 4.2.).

LEMMA 2.2. *Let the multiplicative function $\phi(m)$: $\mathbf{N} \longrightarrow \mathbf{C}$ have the form*

$$\phi(m) = m^r \prod_{p|m} \left(1 - \frac{\gamma(p)}{p}\right), \tag{2.1}$$

where $r \geqslant 0$, $|\gamma(p)| \leqslant c$ for all primes p. Then

$$\sum_{\substack{m \leqslant N \\ m \equiv b \,(\mathrm{mod}\, D)}} \phi(m) = \frac{N^{r+1}}{r+1} \frac{\mathfrak{S}(\gamma)}{D} \Pi(\delta, \gamma) \big(\mathfrak{S}(D, \gamma)\big)^{-1} + R_3(N, D), \tag{2.2}$$

where $\delta = (b, D)$,

$$\mathfrak{S}(\gamma) = \prod_{p} \left(1 - \frac{\gamma(p)}{p^2}\right), \qquad \mathfrak{S}(D, \gamma) = \prod_{p|D} \left(1 - \frac{\gamma(p)}{p^2}\right),$$

$$\Pi(\delta, \gamma) = \prod_{p|\delta} \left(1 - \frac{\gamma(p)}{p}\right),$$

$$R_3(N, D) = \begin{cases} O(N^r \log N) & \text{if } \ c = 1, \\ O_{c,\varepsilon}(N^{r+\varepsilon}) & \text{if } \ c > 1. \end{cases}$$

Proof. Let

$$S_N = \sum_{\substack{m \leqslant N \\ m \equiv b \,(\mathrm{mod}\, D)}} \phi(m).$$

Define the multiplicative function $\gamma(d)$ by

$$\gamma(d) = \prod_{p^\alpha \| d} (\gamma(p))^\alpha.$$

Then

$$\begin{aligned} S_n &= \sum_{\substack{m\leqslant N\\ m\equiv b\,(\mathrm{mod}\,D)}} m^r \sum_{d\mid m} \frac{\mu(d)\gamma(d)}{d} \\ &= \sum_{d\leqslant N} \mu(d)\gamma(d)d^{r-1} \sum_{\substack{t\leqslant N/d\\ dt\equiv b\,(\mathrm{mod}\,D)}} t^r \\ &= \sum_{\tau\mid D} \sum_{\substack{d\leqslant N\\ (d,D)=\tau}} \mu(d)\gamma(d)d^{r-1} \sum_{\substack{t\leqslant N/d\\ dt\equiv b\,(\mathrm{mod}\,D)}} t^r. \end{aligned} \tag{2.3}$$

The inner sum in the last part of (2.3) equals to zero if $\tau \nmid \delta$. If $\tau \mid (b, D)$ then $dt \equiv b \pmod D$ is equivalent to $t \equiv t_0 \pmod{D/\tau}$ for some t_0. Thus

$$S_N = \sum_{\tau\mid\delta} \sum_{\substack{d\leqslant N\\ (d,D)=\tau}} \mu(d)\gamma(d)d^{r-1} \sum_{\substack{t\leqslant N/d\\ t\equiv t_0\,(\mathrm{mod}\,D/\tau)}} t^r.$$

Applying the simple equality

$$\sum_{\substack{t\leqslant y\\ t\equiv t_0\,(\mathrm{mod}\,l)}} t^r = \frac{1}{r+1}\frac{y^{r+1}}{l} + O(y^r)$$

we obtain that

$$S_N = \frac{N^{r+1}}{r+1}\frac{1}{D}\Sigma_1 + O\Big(N^r\Sigma_2\Big), \tag{2.4}$$

where

$$\Sigma_1 = \sum_{\tau\mid\delta} \tau \sum_{\substack{d\leqslant N\\ (d,D)=\tau}} \frac{\mu(d)\gamma(d)}{d^2}, \tag{2.5}$$

$$\Sigma_2 = \sum_{\tau\mid\delta} \sum_{\substack{d\leqslant N\\ (d,D)=\tau}} \frac{\mu^2(d)|\gamma(d)|}{d}. \tag{2.6}$$

We easily get the estimate

$$\Sigma_2 \leqslant \prod_{p\leqslant N}\left(1+\frac{|\gamma(p)|}{p}\right) \leqslant \exp\left\{\sum_{p\leqslant N}\frac{|\gamma(p)|}{p}\right\} = O_c((\log N)^c). \tag{2.7}$$

Because $d = \tau l$ in the sum Σ_1, we have the equality

$$\Sigma_1 = \sum_{\tau\mid\delta} \tau \sum_{\substack{l\leqslant N/\tau\\ (l,D/\tau)=1}} \frac{\mu(\tau l)\gamma(\tau l)}{\tau^2 l^2}.$$

If $(\tau, l) > 1$ then $\mu(\tau l) = 0$, and thus

$$\Sigma_1 = \sum_{\tau|\delta} \frac{\mu(\tau)\gamma(\tau)}{\tau} \sum_{\substack{l \leqslant N/\tau \\ (l,D)=1}} \frac{\mu(l)\gamma(l)}{l^2}. \tag{2.8}$$

Evidently,

$$\sum_{\tau|\delta} \frac{\mu(\tau)\gamma(\tau)}{\tau} = \Pi(\delta, \gamma), \tag{2.9}$$

$$\sum_{\substack{l \leqslant N/\tau \\ (l,D)=1}} \frac{\mu(l)\gamma(l)}{l^2} = \prod_{p \nmid D} \left(1 - \frac{\gamma(p)}{p^2}\right) - S(N/\tau), \tag{2.10}$$

where

$$S(x) = \sum_{\substack{l > x \\ (l,D)=1}} \frac{\mu(l)\gamma(l)}{l^2}. \tag{2.11}$$

If $c = 1$, we easily obtain

$$S(N/\tau) = O\left(\sum_{l > N/\tau} \frac{1}{l^2}\right) = O(\tau N^{-1}). \tag{2.12}$$

If $c > 1$ then, for every $\varepsilon' \in (0, 1)$,

$$\begin{aligned} S(N/\tau) &= O\left((\tau/N)^{\varepsilon'} \prod_p \left(1 + |\gamma(p)| p^{-2+\varepsilon'}\right)\right) \\ &= O_{c,\varepsilon'}\left((\tau/N)^{\varepsilon'}\right). \end{aligned} \tag{2.13}$$

Equalities (2.4)–(2.8) and (2.10)–(2.13) imply that

$$\begin{aligned} S_N = {} & \frac{N^{r+1}}{r+1} \frac{1}{D} \sum_{\tau|\delta} \frac{\mu(\tau)\gamma(\tau)}{\tau} \left(\prod_{p \nmid D} \left(1 - \frac{\gamma(p)}{p^2}\right) + R_4(N, \tau)\right) \\ & + O_c\left(N^r (\log N)^c\right), \end{aligned} \tag{2.14}$$

where

$$R_4(N, \tau) = \begin{cases} O(\tau N^{-1}) & \text{if} \quad c = 1, \\ O_{c,\varepsilon'}\left((\tau N^{-1})^{\varepsilon'}\right) & \text{if} \quad c > 1. \end{cases}$$

If $c = 1$,

$$\sum_{\tau|\delta} \frac{\mu(\tau)\gamma(\tau)}{\tau} R_4(N, \tau) = O\left(\varphi(D) N^{-1}\right), \tag{2.15}$$

and if $c > 1$,

$$\begin{aligned}\sum_{\tau|\delta}\frac{\mu(\tau)\gamma(\tau)}{\tau}R_4(N,\tau) &= O\left(\frac{D^{\varepsilon'}}{N^{\varepsilon'}}\sum_{\tau|\delta}\frac{\mu^2(\tau)|\gamma(\tau)|}{\tau}\right)\\ &= O\left(\frac{D^{\varepsilon'}}{N^{\varepsilon'}}\prod_{p|\delta}\left(1+\frac{|\gamma(p)|}{p}\right)\right)\\ &= O\left(\frac{D^{\varepsilon'}}{N^{\varepsilon'}}\exp\left\{\sum_{p\leqslant D}\frac{c}{p}\right\}\right) = O_c\left(\frac{D^{\varepsilon'}(\log D)^c}{N^{\varepsilon'}}\right).\end{aligned} \tag{2.16}$$

Substitution of (2.9), (2.15), and (2.16) to (2.14) gives (2.2) with $\varepsilon = 1-\varepsilon'$.

Remark 2.3. One can easily see that in case $c > 1$ the estimate

$$R_3(N,D) = O_{c,\varepsilon}\left(N^{r+\varepsilon}\frac{(\log D)^c}{D^\varepsilon}\right), \qquad 0<\varepsilon<1,$$

is true.

COROLLARY 2.4. *If the multiplicative function $\phi(m)$ is defined by (2.1), and $D = D_1D_2$, $(D_1, D_2) = 1$, then*

$$\sum_{\substack{m\leqslant N\\ (m,D)=D_1}}\phi(m) = \frac{N^{r+1}}{r+1}\frac{\varphi(D_2)}{D}\mathfrak{S}(\gamma)\Pi(D_1,\gamma)\big(\mathfrak{S}(D,\gamma)\big)^{-1}+O\big(|R_3(N,D)|\varphi(D_2)\big).$$

LEMMA 2.5. *Let the multiplicative function $\pi_k(m)$ be defined by*

$$\pi_k(p^\alpha) = \begin{cases} 1-\dfrac{k}{p+k-1} & \text{if } \alpha = 0,\\ \dfrac{1}{p^\alpha}\left(1-\dfrac{k}{p+k-1}\right) & \text{if } \alpha \geqslant 1.\end{cases} \tag{2.17}$$

Let $L_r = \{\bar{l}\in\widetilde{\mathbb{N}}^k\colon\ p \mid \bar{l}^* \Rightarrow p\leqslant r\}$ and $r \geqslant 8$. Then

$$\sum_{\bar{l}\in L_r,\ \bar{l}^*\geqslant y}\pi_k(\bar{l}^*) = O_k\left(\exp\left\{-\frac{\log y}{\log r}\right\}\right). \tag{2.18}$$

Proof. Evidently,

$$\sum_{\alpha=0}^{\infty}\ \sum_{\bar{l}:\ \bar{l}^*=p^\alpha}\pi_k(\bar{l}^*) = 1$$

for each prime p. For every $\lambda \in (0, 1/2)$ we have

$$\begin{aligned}
\sum_{\bar{l} \in L_r,\ \bar{l}^* \geqslant y} \pi_k(\bar{l}^*) &\leqslant y^{-\lambda} \sum_{\bar{l} \in L_r} (\bar{l}^*)^{\lambda} \pi_k(\bar{l}^*) \\
&= y^{-\lambda} \prod_{p \leqslant r} \left(1 + k \sum_{\alpha=0}^{\infty} \sum_{\bar{l}:\ \bar{l}^* = p^{\alpha}} ((\bar{l}^*)^{\lambda} - 1) \pi_k(\bar{l}^*)\right) \\
&= y^{-\lambda} \prod_{p \leqslant r} \left(1 + k \sum_{\alpha=1}^{\infty} (p^{\alpha\lambda} - 1) \pi_k(p^{\alpha})\right) \\
&\leqslant y^{-\lambda} \prod_{p \leqslant r} (1 + c_1 k \lambda p^{-(1-\lambda)} \log p) \\
&\leqslant y^{-\lambda} \exp\left\{ c_1 k \lambda r^{\lambda} \sum_{p \leqslant r} p^{-1} \log p \right\} \\
&\leqslant \exp\{ -\lambda \log y + c_2 k \lambda r^{\lambda} \log r \},
\end{aligned}$$

and the lemma follows when $\lambda = (\log r)^{-1}$.
An inequality slightly sharper than (2.18) may be proved following (Stakėnas and Šiaulys, 1988, Lemma 3).

3. COUNTING OF MULTIPLES

a) Define $W_x \subset \mathbf{N}^2$ by

$$W_x = \{\bar{m}:\ m_1 \leqslant x,\ \xi m_1 < m_2 \leqslant \eta m_1\}, \tag{3.1}$$

where ξ and η may be some functions of x, and let

$$f_x(\langle m_1, m_2 \rangle) = \langle m_1 + a,\ m_1 + m_2 \rangle, \tag{3.2}$$

where a is a natural number possibly depending on x. Let

$$S'_x(\bar{d}) = \operatorname{card} V_x(\bar{d}), \qquad \bar{d} \in \widetilde{\mathbf{N}}^2,$$

where $V_x(\bar{d})$ is defined by (1.1). Then

$$S'_x(\bar{d}) = \sum_{k_1 \in J} \sum_{k_2 \in K} 1,$$

where

$$\begin{aligned}
J &= \{k_1:\ m_1 + a = k_1 d_1,\ m_1 \leqslant x,\ (k_1, d_2) = 1\}, \\
K &= \{k_2:\ m_1 + m_2 = k_2 d_2,\ \xi m_1 < m_2 \leqslant \eta m_1,\ (k_2, k_1 d_1) = 1\}.
\end{aligned}$$

By Lemma 2.1 we have that

$$\sum_{k_2 \in K} 1 = \frac{\eta - \xi}{d_2}(k_1 d_1 - a) \prod_{p | k_1 d_1} \left(1 - \frac{1}{p}\right) + O(2^{\omega(k_1 d_1)}), \tag{3.3}$$

because $m_1 = k_1 d_1 - a$. Further, applying Corollary 2.4 we have that

$$\begin{aligned}\sum_{k_1\in J} k_1 d_1 \prod_{p\mid k_1 d_1}\left(1-\frac{1}{p}\right) &= \sum_{\substack{a<m\leqslant x+a\\ (m,d_1 d_2)=d_1}} \varphi(m)\\ &= \frac{x^2+2ax}{2}\,\frac{\varphi(d_2)}{d_1 d_2}\mathfrak{S}(1)\Pi(d_1,1)\big(\mathfrak{S}(d_1 d_2,1)\big)^{-1}\\ &\quad + O\big(\varphi(d_2)(x+a)\log(x+a)\big).\end{aligned} \tag{3.4}$$

It is known (see (Kubilius, 1964, Lemma 9.2)) that

$$\sum_{m\leqslant y} 2^{\omega(m)} = O(y\log y). \tag{3.5}$$

Hence

$$\sum_{k_1\in J} 2^{\omega(k_1 d_1)} = O\big((x+a)\log(x+a)\big). \tag{3.6}$$

From (3.3), (3.4), and (3.6) we obtain the equality

$$\begin{aligned}S'_x(\bar d) &= \frac{1}{2}(x^2+2ax)(\eta-\xi)\mathfrak{S}(1)\frac{1}{d_1 d_2}\prod_{p\mid d_1 d_2}\left(1-\frac{1}{p+1}\right)\\ &\quad + O\Big(\big(1+(\eta-\xi)\big)(x+a)\log(x+a)\Big).\end{aligned} \tag{3.7}$$

b) Define $W_x \subset \mathbf{N}^3$ by

$$W_x = \{\overline{m}\colon\ m_3\leqslant x,\ \xi_2 m_3^3 < m_2 \leqslant \eta_2 m_3^3,\ \xi_1 m_3 m_2 < m_1 \leqslant \eta_1 m_3 m_2\}, \tag{3.8}$$

where $\eta_i,\ \xi_i,\ i=1,2$, may be some functions of x, and let

$$f_x(\overline{m}) = \overline{m}, \qquad \overline{m}\in\mathbf{N}^3. \tag{3.9}$$

Let

$$S''_x(\bar d) = \operatorname{card} V_x(\bar d), \qquad \bar d \in \widetilde{\mathbf{N}}^3,$$

where $V_x(\bar d)$ is defined by (1.1). Then

$$S''_x(\bar d) = \sum_{k_3\in J}\ \sum_{k_2\in K}\ \sum_{k_1\in L} 1, \tag{3.10}$$

where

$$\begin{aligned}J &= \{k_3\colon\ k_3 d_3 = m_3 \leqslant x,\ (k_3, d_2 d_1) = 1\},\\ K &= \{k_2\colon\ \xi_2 m_3^3 < k_2 d_2 = m_2 \leqslant \eta_2 m_3^3,\ (k_2, k_3 d_3 d_1) = 1\},\\ L &= \{k_1\colon\ \xi_1 m_3 m_2 < k_1 d_1 = m_1 \leqslant \eta_1 m_3 m_2,\ (k_1, k_3 k_2 d_3 d_2) = 1\}.\end{aligned}$$

Again by Lemma 2.1 we have that

$$\sum_{k_1 \in L} 1 = \frac{\eta_1 - \xi_1}{d_1} m_3 k_2 d_2 \prod_{p \mid k_2 d_2 k_3 d_3} \left(1 - \frac{1}{p}\right) + O\left(2^{\omega(m_3 k_2 d_2)}\right). \tag{3.11}$$

Using Corollary 2.4 and (3.5) we have that

$$\begin{aligned}\sum_{k_2 \in K} k_2 d_2 \prod_{p \mid k_2 d_2} \left(1 - \frac{1}{p}\right) &= \sum_{\substack{\xi_2 m_3^3 < m \leqslant \eta_2 m_3^3 \\ (m, k_3 d_3 d_2 d_1) = d_2}} \varphi(m) \\ &= \frac{\eta_2^2 - \xi_2^2}{2} (k_3 d_3)^6 \frac{\varphi(k_3 d_3 d_1)}{k_3 d_3 d_2 d_1} \mathfrak{S}(1) \prod_{p \mid d_2} \left(1 - \frac{1}{p}\right) \prod_{p \mid k_3 d_3 d_2 d_1} \left(1 - \frac{1}{p^2}\right)^{-1} \\ &+ O\left(\eta_2 k_3^3 d_3^3 \varphi(k_3 d_3 d_1) \log\left(\eta_2 k_3^3 d_3^3\right)\right),\end{aligned} \tag{3.12}$$

$$\sum_{k_2 \in K} 2^{\omega(m_3 k_2 d_2)} = O\left(2^{\omega(m_3)} \eta_2 m_3^3 \log\left(\eta_2 m_3^3\right)\right). \tag{3.13}$$

Now (3.11)–(3.13) give

$$\begin{aligned}\sum_{k_2 \in K} \sum_{k_1 \in L} 1 = \frac{1}{2}&(\eta_2^2 - \xi_2^2)(\eta_1 - \xi_1)(k_3 d_3)^7 \mathfrak{S}(1) \\ &\times \frac{1}{d_2 d_1} \prod_{p \mid d_2 d_1} \left(1 - \frac{1}{p+1}\right) \prod_{p \mid k_3 d_3} \left(1 - \frac{2}{p+1}\right) \\ &+ O\left(\eta_2 (\eta_1 - \xi_1) k_3^4 d_3^4 \varphi(k_3 d_3 d_1) d_1^{-1} \log\left(\eta_2 k_3^3 d_3^3\right)\right) \\ &+ O\left(2^{\omega(m_3)} \eta_2 m_3^3 \log\left(\eta_2 m_3^3\right)\right).\end{aligned} \tag{3.14}$$

Applying Corollary 2.4 once more (with $\gamma(p) = 2p/(p+1)$) we obtain

$$\begin{aligned}\sum_{k_3 \in J} (k_3 d_3)^7 \prod_{p \mid k_3 d_3} \left(1 - \frac{2}{p+1}\right) &= \sum_{\substack{m \leqslant x \\ (m, d_3 d_2 d_1) = d_3}} m^7 \prod_{p \mid m} \left(1 - \frac{2}{p+1}\right) \\ &= \frac{x^8}{8} \frac{1}{d_3} \mathfrak{S}\left(\frac{2p}{p+1}\right) \prod_{p \mid d_2 d_1} \left(1 - \frac{1}{p}\right) \prod_{p \mid d_3} \left(1 - \frac{2}{p+1}\right) \\ &\times \prod_{p \mid d_3 d_2 d_1} \left(1 - \frac{2}{p(p+1)}\right)^{-1} + O_\varepsilon\left(\varphi(d_2 d_1) x^{7+\varepsilon}\right).\end{aligned} \tag{3.15}$$

Further,

$$\sum_{m \leqslant x} m^4 \varphi(m) \log\left(\eta_2 m^3\right) = O\left(x^6 \log\left(\eta_2 x\right)\right), \tag{3.16}$$

$$\sum_{m \leqslant x} 2^{\omega(m)} \eta_2 m^3 \log\left(\eta_2 m^3\right) = O\left(\eta_2 x^4 \log\left(\eta_2 x\right) \log x\right). \tag{3.17}$$

From (3.10) and (3.14)–(3.17) we conclude that

$$S_x''(\bar{d}) = \frac{x^8}{16}(\eta_2^2 - \xi_2^2)(\eta_1 - \xi_1)\mathfrak{S}(1)\mathfrak{S}\Big(\frac{2p}{p+1}\Big)\frac{1}{\bar{d}^*}\prod_{p|\bar{d}^*}\Big(1 - \frac{2}{p+2}\Big) \\ + O_\varepsilon(R_5 + R_6 + R_7), \tag{3.18}$$

where

$$R_5 = (\eta_2^2 - \xi_2^2)(\eta_1 - \xi_1)x^{7+\varepsilon},$$
$$R_6 = \eta_2(\eta_1 - \xi_1)x^6 \log(\eta_2 x),$$
$$R_7 = \eta_2 x^4 \log(\eta_2 x) \log x.$$

c) Here we show how the above considerations may be extended for $s > 3$. Let $W_x \subset \mathbf{N}^s$ be some region of the type

$$W_x = \{\bar{m}\colon\ m_i \leqslant y_i,\ i = 1, \ldots, s\},$$

where

$$y_i = \eta_i(x) \prod_{j=i+1}^{s} m_j^{\alpha_{ij}}, \qquad \alpha_{ij} \geqslant 0 \qquad \text{for all} \quad i,\ j.$$

Let $V_x(\bar{d})$ be defined by (1.1), $\bar{d} \in \widetilde{\mathbf{N}}^s$. For some s-tuple $\bar{k} = \langle k_1, \ldots, k_s\rangle \in \mathbf{N}^s$ define $\bar{k}_i^* = k_i k_{i+1} \cdot \ldots \cdot k_s$. Thus $\bar{k}_1^* = \bar{k}^*$. Then

$$\operatorname{card} V_x(\bar{d}) = \sum_{k_s \in K_s} \cdots \sum_{k_1 \in K_1} 1,$$

where

$$K_i = \{k_i\colon\ m_i = k_i d_i \leqslant y_i,\ (k_i, \bar{d}^* d_i^{-1} \bar{k}_{i+1}^*) = 1\}.$$

Suppose that for some $t \geqslant 2$

$$\sum_{k_t \in K_t} \cdots \sum_{k_1 \in K_1} 1 = Z_t m_{t+1}^{\alpha} \prod_{j=1}^{t} \frac{1}{d_j} \prod_{r=1}^{t-1} \mathfrak{S}\Big(\frac{rp}{p+r-1}\Big) \prod_{p|\bar{k}_{t+1}^* d_{t+1}^*} \Big(1 - \frac{1}{p}\Big) \\ \times \prod_{p|\bar{d}^* \bar{k}_{t+1}^*} \Big(1 - \frac{t-1}{p+t-1}\Big) + O(R_t^*) \tag{3.19}$$

with some unspecified functions

$$Z_t = Z_t(x, m_{t+2}, \ldots, m_s) \quad \text{and} \quad R_t^* = R_t^*(x, m_{t+1}, \ldots, m_s).$$

Then we may apply Corollary 2.4 to find the expression for

$$\sum_{k_{t+1} \in K_{t+1}} \sum_{k_t \in K_t} \cdots \sum_{k_1 \in K_1} 1. \tag{3.20}$$

Indeed, Corollary 2.4 with $\gamma(p) = tp(p+t-1)^{-1}$ gives the equality

$$\sum_{k_{t+1}\in K_{t+1}} (k_{t+1}d_{t+1})^{\alpha} \prod_{p\mid k_{t+1}d_{t+1}} \Big(1-\frac{1}{p}\Big)\Big(1-\frac{t-1}{p+t-1}\Big)$$
$$= \sum_{\substack{m\leqslant y_{t+1}\\ (m,\bar{d}^*\bar{k}^*_{t+2})=d_{t+1}}} m^{\alpha}\prod_{p\mid m}\Big(1-\frac{t}{p+t-1}\Big)$$
$$= \frac{(y_{t+1})^{\alpha+1}}{\alpha+1}\,\frac{\varphi(\bar{d}^*d^{-1}_{t+1}\bar{k}^*_{t+2})}{\bar{d}^*\bar{k}^*_{t+2}}\,\mathfrak{S}\Big(\frac{tp}{p+t-1}\Big)\prod_{p\mid d_{t+1}}\Big(1-\frac{t}{p+t-1}\Big) \tag{3.21}$$
$$\times \prod_{p\mid \bar{d}^*\bar{k}^*_{t+2}}\Big(1-\frac{t}{p+t-1}\Big)^{-1} + O_{\varepsilon}\big(y^{\alpha+\varepsilon}_{t+1}\varphi(\bar{d}^*d^{-1}_{t+1}\bar{k}^*_{t+2})\big).$$

From (3.19) and (3.21) we obtain that (3.20) equals to

$$Z'_{t+1}\prod_{j=1}^{t+1}\frac{1}{d_j}\prod_{r=1}^{t}\mathfrak{S}\Big(\frac{rp}{p+r-1}\Big)\prod_{p\mid d^*_{t+2}\bar{k}^*_{t+2}}\Big(1-\frac{1}{p}\Big)\prod_{p\mid \bar{d}^*\bar{k}^*_{t+2}}\Big(1-\frac{t}{p+t}\Big)+O(R^*_{t+1})$$

with again unspecified functions

$$Z'_{t+1} = Z'_{t+1}(x, m_{t+2},\dots,m_s) \quad\text{and}\quad R^*_{t+1} = R^*_{t+1}(x,m_{t+2},\dots,m_s).$$

The equality (3.19) is really true for $t = 2$ when y_i, $1\leqslant i\leqslant s$, are chosen appropriately (see the formula (3.14)). Thus we may prove by induction that

$$\operatorname{card} V_x(\bar{d}) = F_1(x)\prod_{r=1}^{s-1}\mathfrak{S}\Big(\frac{rp}{p+r-1}\Big)\frac{1}{\bar{d}^*}\prod_{p\mid\bar{d}^*}\Big(1-\frac{s-1}{p+s-1}\Big)+O(F_2(x)),$$

and this expression may occur to be the true asymptotic expression with respect to x when the functions y_i, $1\leqslant i\leqslant s$, are chosen appropriately.

4. PARTICULAR PROBABILISTIC MODELS

Here we use notations, definitions, and results of §§ 1–3.

THEOREM 4.1. *Let $W_x\subset \mathbf{N}^2$ be defined by (3.1), $\eta-\xi\geqslant x^{-\lambda}$ for some $\lambda>0$. Let f_x be defined by (3.2), and $a\leqslant x^B$ for some $B>0$. Let $r=r(x)$ have the property $\log r(x) = o(\log x)$ as $x\to\infty$ and $\log r(x)\leqslant c(r)\log x$ for $x\geqslant 2$. Define the σ-additive measure $\mathbf{P}_2$ on $\mathfrak{F}^2_{x,r}$ by*

$$\mathbf{P}_2(A(\bar{l})) = \prod_{p^{\alpha}\|\bar{l}^*}\pi_2(p^{\alpha})$$

where $\pi_2(p^\alpha)$ is defined by (2.17). Let

$$F(x) = \frac{1}{2}(x^2 + 2ax)(\eta - \xi)\mathfrak{S}(1).$$

Then

$$\sup_{A \in \mathfrak{F}^2_{x,r}} |\nu_x(A) - \mathbb{P}_2(A)| = O\left(\exp\left\{ -\frac{1-\lambda}{10} \frac{\log x}{\log r} \right\} + x^{-(1-\lambda)/2} \right). \tag{4.1}$$

The constant implied in the symbol O may depend on λ, B, and $c(r)$.

Proof. Let A be an arbitrary set from $\mathfrak{F}^2_{x,r}$. Then $A = A_1 + A_2$ where

$$A_1 = \sum_{\substack{\bar{l} \in L_A \\ \bar{l}^* \leqslant x^\beta}} A(\bar{l}), \qquad A_2 = \sum_{\substack{\bar{l} \in L_A \\ \bar{l}^* > x^\beta}} A(\bar{l}).$$

The results of § 3 show that the assumptions (1.2), (1.4), and (1.5) hold. The equality (3.7) implies that

$$\operatorname{card} B_x(d, \bar{l}) = N(x)\varrho(\bar{l}^*)\tilde{\varrho}(d) + O(R'(x))$$

with

$$N(x) = F(x) = \frac{1}{2}(x^2 + 2ax)(\eta - \xi)\mathfrak{S}(1), \tag{4.2}$$

$$\varrho(\bar{l}^*) = \frac{1}{\bar{l}^*} \prod_{p \mid \bar{l}^*} \left(1 - \frac{1}{p+1}\right),$$

$$\tilde{\varrho} = \begin{cases} \dfrac{1}{p} & \text{when } p \mid \bar{l}^*, \\ \dfrac{2}{p}\left(1 - \dfrac{1}{p+1}\right) & \text{when } p \nmid \bar{l}^*, \end{cases}$$

$$R'(x) \leqslant 2^{\omega(d)} R(x), \tag{4.3}$$

$$R(x) = \big(1 + (\eta - \xi)\big)(x + a)\log(x + a). \tag{4.4}$$

The function $\tilde{\varrho}(p)$ satisfies (1.6). We easily obtain the estimate $S \leqslant c_3 \log r$. We choose $z = x^\delta$ with some $\delta \in (0,1)$. Then (1.8) is satisfied for $x \geqslant x_0$. The estimate

$$H = O\left(\exp\left\{ -\delta \frac{\log x}{\log r} \right\} \right)$$

is evident. Further,

$$\prod_{p \leqslant r} \left(1 - \frac{1}{p+1}\right) = O((\log r)^{c_4}), \tag{4.5}$$

$$\sum_{d|Q,\ d\leqslant x^{3\delta}} 6^{\omega(d)} = O\big(x^{3\delta}(\log r)^{c_8}\big) \tag{4.6}$$

(see, for example, (Stakėnas and Šiaulys, 1988)).
Now from (1.7) and (4.2)–(4.6) we obtain that

$$\big|\nu_x(A_1) - \mathbf{P}_2(A_1)\big| = O(H) + O\big(x^{4\delta+\beta+\lambda-1}\big) = O(H) + O\big(x^{-(1-\lambda)/2}\big) \tag{4.7}$$

when $\beta = \delta = (1-\lambda)/10$.
According to Lemma 2.5 (with $y = x^{\beta}$)

$$\mathbf{P}_2(A_2) = O\left(\exp\left\{-\beta\frac{\log x}{\log r}\right\}\right). \tag{4.8}$$

The estimate

$$\nu_x(A_2) = O\left(\exp\left\{-\beta\frac{\log x}{\log r}\right\}\right) \tag{4.9}$$

may be proved as in (Stakėnas and Šiaulys, 1988).
Estimates (4.7)–(4.9) are uniform in A_1 and A_2. Thus (4.1) follows.

THEOREM 4.2. *Let $W_x \subset \mathbf{N}^3$ be defined by* (3.8) *and f_x be defined by* (3.9). *Let*

$$(\eta_2^2 - \xi_2^2)(\eta_1 - \xi_1) \geqslant x^{-\lambda_2}, \qquad \lambda_2 > 0,$$

$$\eta_2 \leqslant x, \qquad (\eta_2^2 - \xi_2^2) \geqslant x^{-\lambda_1}, \qquad \lambda_1 > 0.$$

Let $\log r(x) = o(\log x)$ as $x \to \infty$ and $\log r(x) \leqslant c(r)\log x$ for $x \geqslant 2$. Define the σ-additive measure $\mathbf{P}_3$ on $\mathcal{F}^3_{x,r}$ by

$$\mathbf{P}_3\big(A(\bar{l})\big) = \prod_{p^\alpha \| \bar{l}^{\bullet}} \pi_3(p^\alpha)$$

where $\pi_3(p^\alpha)$ is defined by (2.17). *Let*

$$F(x) = \frac{1}{16}x^8(\eta_2^2 - \xi_2^2)(\eta_1 - \xi_1)\mathfrak{S}(1)\mathfrak{S}\left(\frac{2p}{p+1}\right).$$

Then

$$\sup_{A\in\mathcal{F}^3_{x,r}} \big|\nu_x(A) - \mathbf{P}_3(A)\big| = O\left(\exp\left\{-\frac{1}{7}\frac{\log x}{\log r}\right\} + x^{-1/7} + x^{-1+\lambda_1} + x^{-3+\lambda_2+\lambda_1}\right).$$

The constant implied in the symbol O may depend on λ_1, λ_2 and $c(r)$.

Proof is similar to the proof of Theorem 4.1 and is based on (3.18) instead of (3.7).

Acknowledgment. The author is grateful to V. Stakėnas for helpful remarks.

REFERENCES

Elliott, P. D. T. A. (1979). *Probabilistic Number Theory*, I. Springer-Verlag, Berlin, Heidelberg, New York.

Kubilius, J. (1964). *Probabilistic Methods in the Theory of Numbers.* American Math. Soc., Providence, RI.

Postnikov, A. G. (1988). *Introduction to Analytic Number Theory.* American Math. Soc., Providence, RI.

Stakėnas, V. (1987). Additive functions and approximation of real numbers by rational ones. In: *28th Conf. of Lithuanian Math. Soc. Abstr. of Comm.* IMC, Vilnius, pp. 318–319 (in Russian).

Stakėnas, V. (1988). The additive functions and rational approximations. *Lithuanian Math. J.* **28**, 3, 260–272.

Stakėnas, V. and Šiaulys, J. (1988). Distribution of values of additive functions of a rational argument. *Lithuanian Math. J.* **28**, 3, 273–284.

Stakėnas, V. (1989). The additive functions and simultaneous rational approximations. *Lithuanian Math. J.* **29**, 1, 62–74.

New Trends in Probab. and Statist., pp. 251–267
F. Schweiger and E. Manstavičius (Eds.)
1992 VSP/TEV

MEASURE THEORY AND ARITHMETICAL FUNCTIONS

J.-L. MAUCLAIRE
C.N.R.S., U R A 212, Théories géométriques,
Université Paris-VII, Tour n° 45–55, 5e étage,
2, Place Jussieu, 75251 Paris Cedex 05, France

ABSTRACT

The main purpose of this article is to clarify some aspects of the relationship between statistical properties of arithmetical functions and properties of associated sequences of functions defined on a measurable space.

I. INTRODUCTION

a/ *Notations.* **N** (respectively, $\mathbf{N}^*$) is the set of ordinary integers (respectively, positive integers), **P** is the set of prime numbers, p is a generic element of **P**, **C** (respectively, **R**) is the set of complex (respectively, real) numbers, $v_p(a)$ is the exponent of p in a, and a.s. means almost surely.

b/ *Some formalism.* Let

$$\mathcal{E}_p = \{1, p, p^2, \ldots\}, \qquad \mathcal{E} = \prod_p \mathcal{E}_p.$$

On $\mathcal{E}$ we have the natural filtration

$$\mathcal{E}_y = \prod_{p \leqslant y} \mathcal{E}_p,$$

and the family of measures

$$\mu_y = \bigotimes_{p \leqslant y} \mu_p$$

where μ_p is defined on $\mathcal{E}_p$ by

$$\mu_p(\{p^k\}) = \left(1 - \frac{1}{p}\right)\frac{1}{p^k}, \qquad k \geqslant 0.$$

is a generic element of $\mathcal{E}$, then $t_p = p^n$ for a unique $n \in \mathbf{N}$. We denote this integer n by $v_p(t)$. For a given $t \in \mathcal{E}$ and a real number y, the finite sequence

$$t_y = \{t_p \colon\ p \in \mathbf{P},\ p \leqslant y\}$$

defines an integer in a unique way.
An additive (respectively, multiplicative) arithmetical function is a function

$$f\colon \mathbf{N}^* \longrightarrow \mathbf{C}$$

such that, if $(m,n) = 1$ then $f(mn) = f(m) + f(n)$ (respectively, $f(mn) = f(m)f(n)$, $f(1) = 1$). If f is an additive arithmetical function and y any real positive number, we define f_y by

$$f_y(t) = f(t_y) = \sum_{p \leqslant y} f\big(p^{v_p(t)}\big).$$

c/ *A remark on the formalism.* During the last fifty years different formalisms have been explicitly introduced to treat problems related to functions defined on the multiplicative set of the positive integers. The first one, introduced in 1939, is due to (Kac, 1949) and has been widely developed by J. Kubilius (1964). E. V. Novoselov (1964), W. Schwarz and J. Spilker (1971), and the present author (Mauclaire, 1981b) have considered other approaches. In all the approaches the algebra $\mathcal{B}$ generated by the sets $a\mathbf{N}^*$, $a \in \mathbf{N}^*$, is considered, and a classical result of Stone gives that there exists a compact space $\mathcal{K}$ such that $\mathcal{B}$ can be identified to a subalgebra of $\mathfrak{p}(\mathcal{K})$, where $\mathfrak{p}(\mathcal{K})$ is the set of the all parts of $\mathcal{K}$, and every element of this subalgebra is open and closed. $\mathcal{K}$ is, in fact, a compactification of $\mathcal{E}$. So, from a topological viewpoint, there is no choice. Since it is also required that for any $a \in \mathbf{N}^*$ the measure of the set associated to $a\mathbf{N}^*$ would be $1/a$, from a measure-theoretic viewpoint there is also no choice: the measure on $\mathcal{K}$ is uniquely defined. So, all these viewpoints will lead to measurable spaces equivalent to $(\mathcal{E}, \mu)$. But the filtration $\mathcal{E}_y$ will not arise naturally in this context. Something more is needed; for instance, a truncation arising from the consideration of the arithmetical mean value, which induces an order on $\mathbf{P}$, see (Mauclaire, 1987a).

d/ *A comment on the history of Probabilistic Number Theory.* It is well known that the arithmetical mean value is also called the Cesàro mean. But in general, the fact that this kind of mean values has been used by Cesàro himself to obtain results in Probabilistic Number Theory is not mentioned. We may refer to the titles of some of his articles like "*Probabilité de certains faits arithmétiques*", or give some of his results in their original forms as it is possible to find, for instance, in "*Excursions arithmétiques à l'infini*" (Paris, 1885) and "*Sur diverses questions*

d'arithmétique" (1883). We recall that, by definition, an integer is a triangular number if it may be written as $n(n+1)/2$.
Cesàro writes:

> *la probabilité que la racine du plus grand diviseur carré d'un nombre soit triangulaire est égale à* $8(1-9/\pi^2)$;
> *la probabilité que deux nombres triangulaires soient premiers entre eux est*
> $$\frac{3}{4}\prod_{p\geqslant 3}\left(1-\frac{4}{p^2}\right).$$

and also:

> *dans une division quelconque, la probabilité que le k-ième chiffre décimal soit r est*
> $$\frac{1}{20}+\frac{10^k}{2}\int_0^1 \frac{1-\varphi}{1-\varphi^{10}}\varphi^{10^k-1+r}\,d\varphi.$$

These results, which can be found in the Complete Works of Cesàro (see (Cesàro, 1965)), had been formulated before 1885, and in such a context, it seems not so clear that the commonly expressed opinion that "Probabilistic Number Theory may be viewed as a Twentieth Century sport", found in (Elliott, 1980, p. 25), corresponds to the historical reality exactly.

II. REAL VALUED ADDITIVE ARITHMETICAL FUNCTIONS REVISITED

a/ *Introduction.* In this part a proof of the key result, which allows to give a complete description of the distribution function defined by an additive function with values in **R** considered as an additive group, will be sketched, directly issued of the proof of the Erdös–Wintner theorem given in (Mauclaire, 1986). Relying heavily on straightforward applications of measure-theoretic concepts, this approach works in the case of any locally compact metrizable abelian group. For more details, see (Mauclaire, 1991a) and a forthcoming detailed paper of the same author.

b/ *Another way to the key result.*

THEOREM. *Let f: $\mathbf{N}^* \to \mathbf{R}$ be an additive arithmetical function. Consider the set A of real numbers a such that*

$$\limsup_{x\to+\infty}\frac{1}{x}\left|\sum_{\substack{n\leqslant x\\(n,2)=1}} e^{iaf(n)}\right|>0. \qquad (\star)$$

The assertion

$$\operatorname{meas}(A)>0$$

is equivalent to the assertion

there exist a real b *and a sequence* $r = (r_p)$ *of real numbers such that*

$$\sum_{p=3}^{[x]} \left(f(p^{v_p(t)}) - bv_p(t)\log p - r_p\right) \quad \textit{converges a.s.} \tag{$*$}$$

Remark. It is easy to check that the classical formulations of the results on additive arithmetical functions with values in the additive group of the real numbers (see, for instance, (Elliott, 1979, Chap. 5 and 7)) are immediate applications of standard theorems of probability theory and harmonic analysis. The conditions for the continuity of the measures which appear in this context may be obtained quite easily, using Bohr compactification of the reals, an argument which works also in the case of any locally compact metrizable abelian group.

Sketch of the proof. Consider the set A of real numbers satisfying ($\star$). Then, by Halász Theorem (see (Elliott, 1979, Chap. 6)), if $a \in A$, there exists a real number $u(a)$ such that

$$\frac{1}{x}\sum_{\substack{n\leqslant x\\(n,2)=1}} e^{iaf(n)}$$

$$= \frac{x^{iu(a)}}{1+iu(a)}\prod_{3\leqslant p<x}\left(1-\frac{1}{p}\right)\left(\sum_{k=0}^{+\infty}\frac{e^{iaf(p^k)}p^{-iku(a)}}{p^k}\right) + o(1), \qquad x\to+\infty,$$

and

$$\lim_{x\to+\infty}\prod_{3\leqslant p\leqslant x}\left(1-\frac{1}{p}\right)\left|\sum_{k=0}^{+\infty}\frac{e^{iaf(p^k)}p^{-iku(a)}}{p^k}\right| > 0.$$

It can be easily proved that, if meas$(A) > 0$, then:

A is an open subgroup of $\mathbf{R}$, and therefore, $A = \mathbf{R}$;
the mapping $a \mapsto u(a)$ is a continuous group homomorphism $\mathbf{R} \to \mathbf{R}$, i.e. there exists b such that $u(a) = ba$;
for any real a the sequence F_x^* defined by

$$F_x^*(t) = \left(\prod_{3\leqslant p\leqslant x} e^{iaf(p^{v_p(t)})}p^{-ibav_p(t)}\right)\left(\prod_{3\leqslant p\leqslant x}\left(1-\frac{1}{p}\right)\sum_{k=0}^{+\infty}\frac{e^{iaf(p^k)}p^{-ikba}}{p^k}\right)^{-1}$$

is a regular martingale (apply Fatou theorem).

Now, since

$$\left|\prod_{3\leqslant p\leqslant x}\left(1-\frac{1}{p}\right)\sum_{k=0}^{+\infty}\frac{e^{iaf(p^k)}p^{-ikba}}{p^k}\right|^2$$

is convergent, we get that for any real a,

$$\prod_{3\leqslant p\leqslant x} e^{iaf(p^{v_p(t)})}p^{-ibav_p(t)} \cdot \overline{e^{iaf(p^{v_p(u)})}p^{-ibav_p(u)}}$$

$$= \exp\left\{ia\sum_{p=3}^{[x]}\left(\left(f\left(p^{v_p(t)}\right) - bv_p(t)\log p\right) - \left(f\left(p^{v_p(u)}\right) - bv_p(u)\log p\right)\right)\right\}$$

converges a.s. But this implies that for almost all (u,t) the sequence

$$\exp\left\{ia\sum_{p=3}^{[x]}\left(\left(f\left(p^{v_p(t)}\right) - bv_p(t)\log p\right) - \left(f\left(p^{v_p(u)}\right) - bv_p(u)\log p\right)\right)\right\}$$

converges for almost all a, and as a consequence of the structure of locally compact abelian group $\mathbf{R}$, we get that for almost all (u,t),

$$\sum_{p=2}^{[x]}\left(\left(f\left(p^{v_p(t)}\right) - bv_p(t)\log p\right) - \left(f\left(p^{v_p(u)}\right) - bv_p(u)\log p\right)\right)$$

converges. Now, this implies that for almost all u

$$\sum_{p=3}^{[x]}\left(\left(f\left(p^{v_p(t)}\right) - bv_p(t)\log p\right) - \left(f\left(p^{v_p(u)}\right) - bv_p(u)\log p\right)\right)$$

converges for almost all t, and selecting such u, we define a sequence $r = (r_p)$ by

$$r_p = f\left(p^{v_p(u)}\right) - bv_p(u)\log p.$$

This is the sequence $r = (r_p)$ for which $(*)$ holds. The converse is immediate: if $(*)$ holds and A is defined as above, the measure of A is positive.

c/ *A question on the history of Probability Theory.* Underlying the above development is the following method in Probability Theory: if σ_n is a sequence of probability measures such that for any u

$$\lim_{n\to+\infty}\int_{-\infty}^{+\infty} e^{itu}\,d\sigma_n(t)$$

exists and if this limit is a function continuous at 0, then there exists a probability measure σ being the limit of the sequence σ_n. It would be interesting to have more information on the genesis of this so important method, since well-known authors have not the same opinion about it; for instance, in (Elliott, 1988, p. 197) it is mentioned that the idea of this method is due to Lyapounov; but Kac writes that the idea of this method is due to Markof, and its justification to P. Lévy (see (Kac, 1959, p. 36)).

III. MEASURES ASSOCIATED TO MULTIPLICATIVE ARITHMETICAL FUNCTIONS

a/ *Introduction.* Some years ago I remarked that the existence of a nonzero mean value for a multiplicative function satisfying some condition limiting its size is equivalent to a pure measure-theoretic result (Mauclaire, 1981b). A typical example is the following

PROPOSITION. *A multiplicative arithmetical function f satisfies the following conditions*

i) $\lim_{x\to+\infty} \frac{1}{x}\sum_{n\leqslant x} f(n)$ *exists and is not zero,*

ii) $\sup_{x>0} \frac{1}{x}\sum_{n\leqslant x} |f(n)|^\alpha < \infty$ *for some* $\alpha > 1$

if and only if the sequence of functions F_y: $\mathcal{E} \mapsto \mathbf{C}$ *defined* μ*-almost everywhere by*

$$F_y(t) = \prod_{p\leqslant y} f(p^{v_p(t)})$$

converges a.s. and in $\mathcal{L}^\alpha(\mathcal{E},\mu)$ *to some F such that*

$$\int_{\mathcal{E}} F(t)\mu(dt) \neq 0.$$

It was not a fortuitous remark, but the fact appeared naturally in the method for getting the solution of the problem of the characterization of multiplicative functions satisfying conditions i) and ii) given in (Mauclaire, 1981a) with a sketch of proof. Problems of this kind are basically related to the existence and the regularity of the martingale

$$F_y(t)\left(\prod_{p\leqslant y}\left(1-\frac{1}{p}\right)\sum_{k\geqslant 0}\frac{f(p^k)}{p^k}\right)^{-1} = H_y(t).$$

Now we shall restrict ourselves to non-negative functions. In such case, $H_y(t)$ defines a non-negative Borel measure on $\mathcal{E}$, and since $\mu(H_y) = 1$, the probability theory allows only two possibilities.

1. H_y is a regular martingale. In such a case, many results are available, concerning the distribution of values of f, for statistical properties of the sequence $(f(n))$ turn to be almost direct translation of measure-theoretic properties of H_y.

2. H_y defines a probability measure m orthogonal to μ, given by

$$m\left(t_y\cdot\prod_{p>y}\mathcal{E}_p\right) = F_y(t_y)\left(\prod_{p\leqslant y}p^{v_p(t)}\right)^{-1}\left(\prod_{p\leqslant y}\sum_{k=0}^{\infty}\frac{f(p^k)}{p^k}\right)^{-1}.$$

An example of a "natural" function leading to an orthogonal measure is $\tau(n)^2n^{-11}$, where τ is the well-known Ramanujan function. Now the question arises: is there any relation between orthogonality of the measure defined by a function like $|\tau(n)|n^{-11/2}$ which has the zero asymptotic mean value (a result of (Elliott *et al.*, 1984)) and the fact that its mean value is zero? This question is also related to a statement of (Elliott, 1988, p. 205) concerning the possibility to apply the method developed in (Mauclaire, 1986) to multiplicative arithmetical functions with the zero asymptotic mean value. We have the following result.

THEOREM. *Let f be a non-negative multiplicative function satisfying*

i) *there exists a constant C such that*

$$\sum_{p^v \leqslant x} f^2(p^v) \log p^v \leqslant Cx \qquad \text{for all} \quad x,$$

ii) *for any $p \in \mathbf{P}$*

$$\sum_{v \geqslant 0} \frac{f^2(p^v)}{p^v} < \infty,$$

iii) *the products*

$$\prod_{p \leqslant y} \left(1 - \frac{1}{p}\right) \sum_{v \geqslant 0} \frac{f^2(p^v)}{p^v}$$

are bounded independently of y.

Then there exists a constant B such that for any δ, $0 \leqslant \delta \leqslant 2$, and $x \geqslant 0$

$$\frac{1}{x} \sum_{n \leqslant x} f^\delta(n) \leqslant BS^{h(\delta)}(x)$$

where

$$h(\delta) = \begin{cases} \delta & \text{if} \quad 0 \leqslant \delta \leqslant 1, \\ 2 - \delta & \text{if} \quad 1 \leqslant \delta \leqslant 2, \end{cases}$$

and $S(x)$ is defined by

$$S(x) = \prod_{p \leqslant x} \left(1 - \frac{1}{p}\right) \left(\sum_{k=0}^{\infty} \frac{f(p^k)}{p^k} \right) \left(\left(1 - \frac{1}{p}\right) \sum_{k=0}^{\infty} \frac{f^2(p^k)}{p^k} \right)^{-1/2}.$$

As a consequence, the orthogonality of m implies the fact that the arithmetical mean value of $f(n)$ is zero.

N.B. i) This result has been announced in (Mauclaire, 1990), without proof.

ii) Using standard techniques, it is not difficult to show that $|\tau(n)|n^{-11/2}$ satisfies the hypothesis i) and ii) of the above theorem.

b/ *Proof of the Theorem.* The last assertion is a consequence of a very classical result of Kakutani (see, for instance, (Bourbaki, 1967, Exer. 17b, p. 133)).
First of all, we apply the method which consists of bounding of an arithmetical mean value by an integral on $\mathcal{E}$.

LEMMA. *There exists a constant D such that, for $0 \leqslant r \leqslant 2$,*

$$\frac{1}{x}\sum_{n\leqslant 0} f^r(n) \leqslant D \prod_{p\leqslant x}\left(1-\frac{1}{p}\right)\sum_{k\geqslant 0}\frac{f^r(p^k)}{p^k}.$$

Proof. We have

$$\begin{aligned}
\sum_{x^{1/2}\leqslant n\leqslant x} f^r(n)\log n &\leqslant \sum_{n\leqslant x} f^r(n)\log n \\
&= \sum_{p^v\leqslant x}\log p^v \sum_{\substack{p^v \| n\\ n\leqslant x}} f^r(n) \leqslant \sum_{p^v\leqslant x}(\log p^v) f^r(p^v) \sum_{n\leqslant x/p^v} f^r(n) \\
&= \sum_{n\leqslant x} f^r(n) \sum_{p^v\leqslant x/n} f^r(p^v)\log p^v.
\end{aligned}$$

Now, since $f^r(p^v) \leqslant 1 + f^2(p^v)$, we majorize the last sum by

$$\sum_{n\leqslant x} f^r(n)\left(\sum_{p^v\leqslant x/n}\log p^v + \sum_{p^v\leqslant x/n} f^2(p^v)\log p^v\right),$$

and by the Chebyschev inequality and i), there exists a constant D_1 such that, uniformly in r,

$$\sum_{x^{1/2}\leqslant n\leqslant x} f^r(n)\log n \leqslant D_1 \sum_{n\leqslant x} f^r(n)\frac{x}{n}.$$

A fortiori, we have

$$\sum_{x^{1/2}\leqslant n\leqslant x} f^r(n) = D_1\frac{2}{\log x}\sum_{n\leqslant x}\frac{f^r(n)}{n} \leqslant D_2 x \prod_{p\leqslant x}\left(1-\frac{1}{p}\right)\left(\sum_{k=0}^{+\infty}\frac{f^r(p^k)}{p^k}\right)$$

by the Mertens formula, with D_2 not depending on r.
Now, since

$$\begin{aligned}
\frac{1}{x}\sum_{n\leqslant x^{1/2}} f^r(n) &= \sum_{n\leqslant x^{1/2}}\frac{f^r(n)}{n}\frac{n}{x} \leqslant \frac{1}{x^{1/2}}\sum_{n\leqslant x^{1/2}}\frac{f^r(n)}{n} \\
&\leqslant x^{-1/2}\prod_{p\leqslant x}\left(1-\frac{1}{p}\right)^{-1}\prod_{p\leqslant x}\left(1-\frac{1}{p}\right)\sum_{v=0}^{\infty}\frac{f^r(p^v)}{p^v},
\end{aligned}$$

the lemma is proved.
We finish the proof of the theorem. Denote

$$A = \sup_{y>0}\prod_{p\leqslant y}\left(1-\frac{1}{p}\right)\sum_{v\geqslant 0}\frac{f^2(p^v)}{p^v}.$$

If $r \leqslant 1$, we have

$$\sum_{n\leqslant x} f^r(n) \leqslant \left(\sum_{n\leqslant x} f(n)\right)^r \left(\sum_{n\leqslant x} 1\right)^{1-r},$$

i.e.

$$\frac{1}{x}\sum_{n\leqslant x} f^r(n) \leqslant \left(\frac{1}{x}\sum_{n\leqslant x} f(n)\right)^r \leqslant \left(D\prod_{p\leqslant x}\left(1-\frac{1}{p}\right)\left(\sum_{v\geqslant 0}\frac{f(p^v)}{p^v}\right)\right)^r,$$

by Lemma (where $r = 1$), and majorizing by iii) we get

$$\frac{1}{x}\sum_{n\leqslant x} f^r(n) \leqslant D^r A^{r/2} S^r(x).$$

If $r \geqslant 1$, let $p = r - 1$.
We have

$$\sum_{n\leqslant x} f(n)^{1+p} = \sum_{\substack{n\leqslant x\\ f(n)<1/S(x)}} f^{1+p}(n) + \sum_{\substack{n\leqslant x\\ f(n)\geqslant 1/S(x)}} f^{1+p}(n).$$

The first term on the right-hand side is trivially less than

$$\left(\sum_{n\leqslant x} f(n)\right) S^{-p}(x),$$

and by Lemma and iii), this is bounded by

$$A^{1/2} D S(x)^{1-p}.$$

We treat the case of the second term, and first use the Chebyshev inequality. Since

$$\left(\sum_{\substack{f(n)\geqslant 1/S(x)\\ n\leqslant x}} 1\right)\frac{1}{S^2(x)} \leqslant \sum_{n\leqslant x} f^2(n) \leqslant Dx\prod_{p\leqslant x}\left(1-\frac{1}{p}\right)\sum\frac{f^2(p^v)}{p^v} \leqslant DAx,$$

we have

$$\sum_{\substack{f(n)\geqslant 1/S(x)\\ n\leqslant x}} 1 \leqslant DAxS(x)^2.$$

Now we remark that

$$\sum_{\substack{n\leqslant x\\ f(n)\geqslant 1/S(x)}} f^{1+p}(n) \leqslant \left(\sum_{n\leqslant x} f^2(n)\right)^{\frac{1+p}{2}} \left(\sum_{\substack{f(n)\geqslant 1/S(x)\\ n\leqslant x}} 1\right)^{1-\frac{1+p}{2}}$$
$$\leqslant (DAx)^{\frac{1+p}{2}} \left(DAxS^2(x)\right)^{1-\frac{1+p}{2}} \leqslant DAxS^{1-p}(x)$$

by Lemma and iii).
If $r \geqslant 1$, this gives

$$\frac{1}{x}\sum_{n\leqslant x} f^r(n) \leqslant (DA + DA^{1/2})xS^{1-p}(x),$$

and the theorem is proved with

$$B = \max\left((DA + DA^{1/2}), D^r A^{r/2}\right).$$

c/ *Remark.* It may be remarked that the proof of the singularity of the measure associated to the function $|\tau(n)|n^{-11/2}$ given in (Mauclaire, 1985) relies on the result of (Moreno and Shahidi, 1983) which seems not to be obtainable by ordinary methods of Probabilistic Number Theory since its proof needs the consideration of integrable functions on spaces much larger than the product of the ordinary p-adic integers. It would be interesting to extend the classical Probabilistic Number Theory to these spaces.

IV. REMARKS ON PROBABILISTIC NUMBER THEORY ON ARITHMETICAL SEMIGROUPS

a/ *Introduction.* We say that a semigroup $\mathcal{A}$ generated by a set of primes $\mathfrak{P}$ with a norm N is an arithmetical semigroup if the following hypothesis $H(L)$ is satisfied.

$H(L)$: *There exists a positive constant L such that*

$$\sum_{\substack{a\in\mathcal{A}\\ N(a)\leqslant x}} 1 = Lx + o(x).$$

Extensions of the notions of additive or multiplicative functions and classical problems of Probabilistic Number Theory have been considered on arithmetical semigroups by many authors, and it has been remarked that although there are strong analogies with the case of ordinary arithmetical functions, as it appears in (Knopfmacher, 1975; Kryžius, 1985a, 1985b; Mauclaire, 1986, 1987b, 1987c), there are also differences. For instance, under some additional hypothesis, it may happen that the classical Chebyshev inequalities, or the equivalence between the existence of the Prime Number Theorem and the fact that the Möbius function has the zero mean value, do not hold (see (Hall, 1973; Zhang, 1987)). The same formalism as in I-b may be defined without difficulty, and the analogues of the classical definitions and results of the ordinary Probabilistic Number Theory may be formulated. Since among the tools mainly used in the classical Probabilistic Number Theory, we find the famous Turán–Kubilius inequality (see, for instance, (Elliott, 1979, Chap. 4)), it seems to be of interest to consider the problem of its existence on general arithmetical semigroups, and an answer to this question must be provided, since for some authors, "the existence of any direct sum or product structure practically guarantees the existence of an inequality of Turán–Kubilius type" (Elliott,

1988, p. 207). We shall prove a result which shows that the actual situation is not so simple.

b/ *Result.*

THEOREM. *Any arithmetical semigroup $\mathcal{A}$ is a subsemigroup of an arithmetical semigroup $\mathfrak{A} = \mathcal{A} \times \mathcal{A}'$, $\mathcal{A}'$ being a semigroup in which the Turán–Kubilius inequality fails. Moreover, if f is a multiplicative function on $\mathfrak{A}$, such that $|f| \leqslant 1$, and if*

$$M(f) = \lim_{x \to +\infty} \frac{1}{x} \sum_{\substack{a \in \mathcal{A} \\ N(a) \leqslant x}} f(a) \quad \textit{exists},$$

then we have that

$$\lim_{x \to +\infty} \frac{1}{x} \sum_{\substack{u \in \mathfrak{A} \\ N(u) \leqslant x}} f(u) \quad \textit{exists}$$

and is equal to

$$M(f) \sum_{a' \in \mathcal{A}'} \frac{f(a')}{a'}.$$

N.B. This result is mentioned in (Mauclaire, 1991b) without proof.

Consequences of this result.

1. There is no equivalence between the existence of a Turán–Kubilius type inequality and the existence of an Erdös–Wintner type theorem since if the Erdös–Wintner theorem holds on $\mathcal{A}$, the Erdös–Wintner theorem holds on $\mathfrak{A}$.
2. The equivalence of the Prime Number Theorem and the zero mean value of the Möbius function is not a general fact in arithmetical semigroups since if the Möbius function on $\mathcal{A}$ has the zero mean value (on $\mathcal{A}$), then the Möbius function on $\mathfrak{A}$ has the zero mean value; but there is no Prime Number Theorem on $\mathfrak{A}$, for a consequence of this theorem would be the existence of a Turán–Kubilius type inequality.

c/ *Proof of the Theorem.*

i/ For $v \in \mathbf{N}^*$ we set

$$M(v) = \exp\exp v^4, \qquad k(v) = \frac{M(v)}{v^3},$$

and we define a set $\mathfrak{P}_v$ of $k(v)$ prime elements p' by choosing a set of $k(v)$ real numbers in $]M(v), M(v)+1]$.

Now we denote

$$\mathfrak{P}' = \bigcup_{v \in \mathbf{N}^*} \mathfrak{P}_v,$$

and $\mathcal{A}'$ is the semigroup generated by $\mathfrak{P}'$, with the norm defined by "norm of p' equals p'". We remark that

$$\sum_{a'\in\mathcal{A}'}\frac{1}{a'}<+\infty,$$

since

$$\sum_{a'\in\mathcal{A}'}\frac{1}{a'}=\prod_{p'\in\mathfrak{P}'}\left(1-\frac{1}{p'}\right)^{-1}\leqslant\prod_{v\in\mathbf{N}^{*}}\left(1-\frac{1}{M(v)}\right)^{-k(v)}<+\infty,$$

and

$$\log\prod_{v\in\mathbf{N}^{*}}\left(1-\frac{1}{M(v)}\right)^{-k(v)}=O\left(\sum_{v\in\mathbf{N}^{*}}\frac{k(v)}{M(v)}\right)=O\left(\sum_{v\in\mathbf{N}^{*}}\frac{1}{v^{3}}\right)=O(1).$$

We denote

$$D=\sum_{a'\in\mathcal{A}'}\frac{1}{a'}.$$

ii/ We consider now $\mathfrak{A}$, the semigroup generated by $\mathfrak{P}\cup\mathfrak{P}'$, where $\mathfrak{P}$ is the set of the primes of $\mathcal{A}$, with the norm N defined by the product of the norms on $\mathcal{A}$ and $\mathcal{A}'$.

PROPOSITION. *If f is a multiplicative function defined on $\mathfrak{A}$, $|f|\leqslant 1$, and*

$$\sum_{\substack{a\in\mathcal{A}\\N(a)\leqslant x}}f(a)=M(f)x+o(x),\qquad x\to+\infty,$$

then

$$\lim_{x\to+\infty}\frac{1}{x}\sum_{\substack{u\in\mathfrak{A}\\N(u)\leqslant x}}f(u)\quad\textit{exists}$$

and is equal to

$$M(f)\sum_{a'\in\mathcal{A}'}\frac{f(a')}{a'}.$$

If we set $f(u)=1$, we get that $\mathfrak{A}$ satisfies the $H(L)$ hypothesis, and so we obtain the following

COROLLARY. *$\mathfrak{A}$ is an arithmetical semigroup and*

$$\lim_{x\to+\infty}\frac{1}{x}\sum_{\substack{u\in\mathfrak{A}\\N(u)\leqslant x}}1=LD.$$

Proof of the Proposition. First of all, we remark that the series

$$\sum_{a'\in\mathcal{A}}\frac{f(a')}{a'}$$

is absolutely convergent. We shall denote this sum by $P(f)$. We have

$$\frac{1}{x}\sum_{\substack{u\in\mathfrak{A}\\ N(u)\leqslant x}} f(u)=\frac{1}{x}\sum_{\substack{a\cdot a'\in\mathfrak{A}\\ N(aa')\leqslant x\\ a\in\mathcal{A}\\ a'\in\mathcal{A}'}} f(aa')=\frac{1}{x}\sum_{\substack{a'\in\mathcal{A}'\\ N(a')\leqslant x}} f(a')\sum_{\substack{a\in\mathcal{A}\\ N(a)\leqslant x/N(a')}} f(a).$$

Now, since $\mathcal{A}$ satisfies the $H(L)$ hypothesis, we know that there exists K such that

$$\sum_{N(a)\leqslant y} 1\leqslant Ky \qquad \text{for all} \quad y\geqslant 0,$$

and therefore,

$$\left|\sum_{N(a)\leqslant y} f(a)\right|\leqslant Ky \qquad \text{for all} \quad y\geqslant 0.$$

Moreover, for any $\varepsilon>0$ there exists $y(\varepsilon)$ such that, if $y\geqslant y(\varepsilon)$ then

$$\left|\sum_{N(a)\leqslant y} f(a)-M(f)y\right|\leqslant \varepsilon y.$$

So, we write, for a given $\varepsilon>0$:

$$\begin{aligned}\frac{1}{x}\sum_{\substack{a'\in\mathcal{A}'\\ N(a')\leqslant x}} f(a')\sum_{\substack{a\in\mathcal{A}\\ N(a)\leqslant x/N(a')}} f(a)&=\frac{1}{x}\sum_{\substack{a'\in\mathcal{A}'\\ a'\leqslant x/y(\varepsilon)}} f(a')\sum_{\substack{a\in\mathcal{A}\\ N(a)\leqslant x/a'}} f(a)\\ &\quad+\frac{1}{x}\sum_{\substack{a'\in\mathcal{A}'\\ x/y(\varepsilon)\leqslant a'\leqslant x}} f(a')\sum_{\substack{a\in\mathcal{A}\\ N(a)\leqslant x/a'}} f(a),\end{aligned}$$

for, if $a'\in\mathcal{A}'$ then $N(a')=a'$.
The second term is bounded by

$$\frac{1}{x}\sum_{x/y(\varepsilon)\leqslant a'\leqslant x}\frac{x}{a'}K,$$

which tends to zero when x tends to $+\infty$ for $\sum 1/a'$ is convergent.
For the first one we have

$$\begin{aligned}&\left|\sum_{\substack{a'\in\mathcal{A}'\\ a'\leqslant x/y(\varepsilon)}} f(a')\left(\sum_{\substack{a\in\mathcal{A}\\ N(a)\leqslant x/a'}} f(a)-M(f)\frac{x}{a'}\right)\right|\\ &\qquad\leqslant\sum_{\substack{a'\in\mathcal{A}'\\ a'\leqslant x/y(\varepsilon)}}\left|\sum_{\substack{a\in\mathcal{A}\\ N(a)\leqslant x/a'}} f(a)-M(f)\frac{x}{a'}\right|\leqslant\sum_{\substack{a'\in\mathcal{A}'\\ a'\leqslant x/y(\varepsilon)}}\varepsilon\frac{x}{a'},\end{aligned}$$

and we get

$$\left| \sum_{\substack{a' \in \mathcal{A}' \\ a' \leqslant x/y(\varepsilon)}} f(a') \left(\sum_{\substack{a \in \mathcal{A} \\ N(a) \leqslant x/a'}} f(a) - M(f) \frac{x}{a'} \right) \right| \leqslant \varepsilon D x,$$

and so

$$\left| \sum_{\substack{a' \in \mathcal{A}' \\ a' \leqslant x/y(\varepsilon)}} f(a') \sum_{\substack{a \in \mathcal{A} \\ N(a) \leqslant x/a'}} f(a) - x M(f) \sum_{a' \in \mathcal{A}'} \frac{f(a')}{a'} \right| \leqslant \varepsilon D x + L x \sum_{x/y(\varepsilon) \leqslant a'} \frac{1}{a'}.$$

Hence we get

$$\frac{1}{x} \sum_{\substack{u \in \mathfrak{A} \\ N(u) \leqslant x}} f(u) = P(f) M(f) + o(1), \qquad x \to +\infty.$$

iii/ Now we prove that on $\mathfrak{A}$ the Turán–Kubilius inequality does not hold. It is sufficient to show that there exists an additive function f on $\mathfrak{A}$ such that,

$$\limsup_{x \to +\infty} \left(\sum_{\substack{N(u) \leqslant x \\ u \in \mathfrak{A}}} \left| f(u) - \sum_{\substack{\pi \in \mathfrak{P} \cup \mathfrak{P}' \\ N(\pi) \leqslant x}} \frac{f(\pi)}{N(\pi)} \right|^2 \right) \left(x L D \sum_{\substack{\pi \in \mathfrak{P} \cup \mathfrak{P}' \\ N(\pi) \leqslant x}} \frac{f^2(\pi)}{N(\pi)} \right)^{-1} = +\infty.$$

We suppose that the Turán–Kubilius inequality holds on $\mathfrak{A}$. We define an additive function ω on $\mathfrak{A}$ by

$$\omega(u) = \sum_{\substack{\pi \mid u \\ \pi \in \mathfrak{P} \cup \mathfrak{P}'}} 1.$$

Since the Turán–Kubilius inequality holds, there exists an absolute constant C such that for any x

$$\left(\sum_{\substack{N(u) \leqslant x \\ u \in \mathfrak{A}}} \left| \omega(u) - \sum_{\substack{\pi \in \mathfrak{P} \cup \mathfrak{P}' \\ N(\pi) \leqslant x}} \frac{\omega(\pi)}{N(\pi)} \right|^2 \right) \leqslant C x \sum_{\substack{\pi \in \mathfrak{P} \cup \mathfrak{P}' \\ N(\pi) \leqslant x}} \frac{\omega(\pi)^2}{N(\pi)}.$$

This may be written as

$$\sum_{\substack{N(u) \leqslant x \\ u \in \mathfrak{A}}} \left| \omega(u) - \sum_{\substack{\pi \in \mathfrak{P} \cup \mathfrak{P}' \\ N(\pi) \leqslant x}} \frac{1}{N(\pi)} \right|^2 \leqslant C x \sum_{\substack{\pi \in \mathfrak{P} \cup \mathfrak{P}' \\ N(\pi) \leqslant x}} \frac{1}{N(\pi)}.$$

By the Proposition C_3 in (Mauclaire, 1986, p. 112), since $\mathfrak{A}$ satisfies the $H(L)$ hypothesis, we have

$$\sum_{\substack{\pi \in \mathfrak{P} \cup \mathfrak{P}' \\ N(\pi) \leqslant x}} \frac{1}{N(\pi)} = \log \log x + O(1).$$

Remark. In (Elliott, 1988, p. 206) there is a quite solemn warning about this Proposition C_3: the validity of its proof requires necessarily the assumption that the set of the values of the norms of the prime elements of the arithmetical semigroup must have a minimum greater than 1. But there is no problem: this assumption is satisfied in any arithmetical semigroup since it is an evident consequence of the condition $H(L)$.

The Turán–Kubilius inequality for the additive function ω may be written as

$$\sum_{\substack{N(u)\leqslant x \\ u\in\mathfrak{A}}} |\omega(u) - \log\log x|^2 \leqslant Hx \log\log x,$$

where H is a constant, and $x \geqslant e^e$.
Now, for $x = M(v) + 1$, we have

$$\sum_{N(u)\leqslant M(v)+1} |\omega(u) - \log\log(M(v)+1)|^2 \leqslant H(M(v)+1)\log\log(M(v)+1),$$

i.e. since $M(v) = \exp\exp v^4$,

$$\sum_{N(u)\leqslant M(v)+1} \left|\omega(u) - v^4\right|^2 = O\big(v^4 M(v)\big), \qquad v \in \mathbf{N}^*.$$

A fortiori, we have

$$\sum_{M(v)<N(u)\leqslant M(v)+1} \left|\omega(u) - v^4\right|^2 = O\big(v^4 M(v)\big),$$

and, still a fortiori,

$$\sum_{\substack{M(v)\leqslant N(\pi)\leqslant M(v)+1 \\ \pi\in\mathfrak{P}\cup\mathfrak{P}'}} \left|1 - v^4\right|^2 = O\big(v^4 M(v)\big),$$

and, once more a fortiori,

$$\sum_{p'\in\mathfrak{P}_v} \left|1 - v^4\right|^2 = O\big(v^4 M(v)\big).$$

But this gives that

$$v^8 \operatorname{card}\mathfrak{P}_v = O\big(v^4 M(v)\big),$$

i.e. by definition of $\mathfrak{P}_v$,

$$v^8 k(v) = O\big(v^4 M(v)\big),$$

and since $k(v) = M(v)/v^3$, we get

$$v^8 \frac{M(v)}{v^3} = O\big(v^4 M(v)\big),$$

i.e.

$$v = O(1), \qquad v \in \mathbf{N}^*,$$

which is not true.
Hence we get that the Turán–Kubilius inequality fails on $\mathfrak{A}$.

d/ *Last remarks.* It would be interesting to know what kind of conditions are necessary and sufficient to allow the validity of a theorem of the Erdös–Wintner type or the Turán–Kubilius inequality on an arithmetical semigroup. In (Mauclaire, 1986, p. 141) it is mentioned that the analogue of the Erdös–Wintner theorem holds in any arithmetical semigroup satisfying the condition

$$\sum_{N(p)\leqslant x} \log N(p) = O(x), \qquad x \to +\infty,$$

and it is asked to weaken this requirement (ibid., p.130). In this direction, we have the following result.

THEOREM. *If an arithmetical semigroup $\mathcal{A}$ has the property that the set $\mathfrak{P}$ of its prime elements satisfies the conditions*

1) *there exist $\mathfrak{P}_1$ and $\mathfrak{P}_2$ such that*

$$\mathfrak{P} = \mathfrak{P}_1 \cup \mathfrak{P}_2, \qquad \sum_{p\in\mathfrak{P}_1} \frac{1}{N(p)} < +\infty,$$

2)

$$\sum_{\substack{p\in\mathfrak{P}_2\\ N(p)\leqslant x}} \log N(p) = O(x), \qquad x \to +\infty,$$

then a theorem of the Erdös–Wintner type holds on $\mathcal{A}$.

REFERENCES

Bourbaki, N. (1967). *Intégration. Chapitre V.* Hermann, Paris.

Cesàro, E. (1965). *Opere Scelte*, Vol.1, P. I, II. Edizioni Cremonese, Roma.

Elliott, P. D. T. A. (1979). *Probabilistic Number Theory*, I. Springer-Verlag, Berlin, Heidelberg, New York.

Elliott, P. D. T. A. (1980). *Probabilistic Number Theory*, II. Springer-Verlag, Berlin, Heidelberg, New York.

Elliott P. D. T. A. (1988). Review of "Intégration et Théorie des Nombres". *Bull. Amer. Math. Soc.* **18**, 193–209.

Elliott, P. D. T. A., Moreno, C. J. and Shahidi, F. (1984). On the absolute value of Ramanujan's τ-function. *Math. Ann.* **266**, 507–511.

Hall, R. (1973). Beurling generalized prime number systems in which Chebyshev inequalities fail. *Proc. Amer. Math. Soc.* **40**, 79–83.

Kac, M. (1949). Probability methods in some problems of Analysis and Number Theory. *Bull. Amer. Math. Soc.* **55**, 641–665.

Kac, M. (1959). *Statistical Independence in Probability, Analysis and Number Theory.* Math. Association of America, New York.

Knopfmacher, J. (1975). *Abstract Analytic Number Theory.* North-Holland, Amsterdam, Oxford.

Kryžius, Z. (1985a) Additive arithmetic functions on semigroups and the preservation of weak convergence of measures. *Lithuanian Math. J.* **25**, 35–43.

Kryžius, Z. (1985b) Almost-even arithmetic functions on semigroups. *Lithuanian Math. J.* **25**, 128–136.

Kubilius, J. (1964). *Probabilistic Methods in the Theory of Numbers.* American Mathem. Soc., Providence, RI.

Mauclaire, J.-L. (1981a). Suites limite-périodiques et théorie des nombres. II. *Proc. Japan. Acad. Ser. A* **56**, 5, 223–224.

Mauclaire, J.-L. (1981b). Fonctions arithmétiques et analyse harmonique. In: *Analytic Number Theory (Proc. Symp., Tokyo, 1980).* Maison Franco-Japon., Tokyo, pp. 83–94.

Mauclaire, J.-L. (1985). On multiplicative arithmetical functions. In: *Number Theory and Combinatorics, Japan 1984.* World Sci. Publ., Singapour, pp. 297–304.

Mauclaire, J.-L. (1986). *Intégration et Théorie des Nombres.* Hermann, Paris.

Mauclaire, J.-L. (1987a). Integration and number theory. In: *Prospects in Mathematical Science.* World Sci. Publ., Singapour, pp. 97–125.

Mauclaire, J.-L. (1987b). Sur la notion de fonction multiplicative et quelques problèmes qui lui sont associés. II. *Proc. Japan Acad., Ser. A* **63**, 5, 157–161.

Mauclaire, J.-L. (1987c) Multiplicative number theory in probability spaces: an example. In: *Proc. Fifth Soviet–Japan Symposium in Probab. Theory and Mathem. Statist.* (*Lecture Notes in Math.* **1299**). Springer-Verlag, Berlin, Heidelberg, New York, pp. 278–285.

Mauclaire, J.-L. (1990). Deux résultats de théorie probabiliste des nombres. *C.R. Acad. Sci. Paris, Sér. I* **311**, 69–72.

Mauclaire, J.-L. (1991a) Distribution des valeurs d'une fonction arithmétique additive à valeurs dans un groupe abélien localement compact métrisable. *C. R. Acad. Sci. Paris, Sér. I* **313**, 345–348.

Mauclaire, J.-L. (1991b). Remarques sur quelques problèmes de théorie des nombres. In: *Séminaire d'Arithmétique de Saint-Etienne, 1989–1990, Exposé n°5*, pp. 58–72.

Moreno, C. J. and Shahidi, F. (1983). The fourth moment of Ramunujan τ-function. *Math. Ann.* **266**, 233–239.

Novoselov, E. V. (1964). A new method in probabilistic number theory. *Izv. Akad. Nauk SSSR, Ser. Mat.* **28**, 307–364 (in Russian).

Schwarz, W. and Spilker, J. (1971). Eine Anwendung des Approximationssatzes von Weierstrass–Stone auf Ramanujan-Summen. *Niew Archief vor Wiskunde* (3) **19**, 198–209.

Zhang, W. (1987). A generalization of Halász's theorem to Beurling's generalized integers and its application. *Ill. J. Math.* **31**, 645–664.

New Trends in Probab. and Statist., pp. 269–283
F. Schweiger and E. Manstavičius (Eds.)
1992 VSP/TEV

GENERALIZATIONS OF KUBILIUS' CLASS OF ADDITIVE FUNCTIONS. I

I. Z. RUZSA*
Mathematical Institute of the Hungarian Academy of Sciences,
Budapest, Pf. 127, H-1364, Hungary.

ABSTRACT

The Kubilius approach in the theory of additive arithmetic functions is extended to deal with their distributions in general topological groups. New concepts of separated and controlled functions are proposed. Asymptotic independence of two parts of an additive function defined on small prime numbers and large ones is established. This leads to the probabilistic proofs of limit theorems available earlier only by analytical methods.

1. INTRODUCTION

A real-valued additive arithmetic function f belongs to Kubilius' class H if with the notation

$$B_N^2 = \sum_{p \leqslant N} \frac{f(p)^2}{p}$$

it satisfies

$$\sum_{N^t \leqslant p \leqslant N} \frac{f(p)^2}{p} = o(B_N^2) \tag{1.1}$$

for every $t \in (0,1)$ as $N \to \infty$. A considerable part of Kubilius' famous book (Kubilius, 1959, 1962, 1964) is devoted to the study of this class. A function from H often behaves like a sum of independent random variables; for an exact description of this phenomenon we refer to (Kubilius, 1964) or (Elliott, 1979, 1980). This method of comparing an additive function to a sum of independent random variables made it possible to apply the techniques of probability theory to additive functions and obtain many results in a unified way. Later limiting laws were established for wider classes of additive functions, typically by analytic methods. (See, for instance, (Levin and Timofeev, 1975, 1977)). Our aim is to show that still there are some hidden reserves in the Kubilius approach, and sufficient conditions comprising the majority of known cases (as well as some not yet investigated ones) can be formulated and proved in similar terms.

We shall work in the framework of additive functions mapping into a commutative topological group. The reason for this is twofold. First, it gives a unified way of formulating results that work for real, complex additive functions, distribution

*Supported by Hungarian National Foundation for Scientific Research, Grant No. 1901.

modulo one, nonzero multiplicative functions etc. equally. (Some recent investigations, where additive functions are connected with stochastic process, can be regarded as the examples of dealing with additive functions with values in a space of functions, a less obvious example of a group; see, for instance, (Manstavičius, 1984, 1985, 1986).) However, the reader not interested in abstract structures will not lose much by supposing that the functions are real. Second, it imposes a certain puritanity of through: we have to avoid the vast arsenal of moments, integral transforms and other analytical means that offer themselves in the real case. Some of the results are probably true even for semigroups, but the proofs seem to present technical difficulties.

Among others, the square mean to form B_N and division by it as a way of norming are both unavailable in an abstract setting. To avoid the problem of normalization, we shall work with a *sequence of functions* instead. We say that a sequence (f_N) of additive functions belongs to class H if

$$\sum_{N^t \leqslant p \leqslant N} \frac{f_N(p)^2}{p} \longrightarrow 0 \tag{1.3}$$

as $N \to \infty$, for all $t \in (0,1)$. Then a single function f belongs to H if and only if so does the sequence defined by $f_N(n) = f(n)/B_N$.

Next, observe that the meaning of (1.3) is that "the contribution of large primes is negligible", and a consequence of (1.3) is that

$$\sum_{\substack{N^t \leqslant p \leqslant N \\ |f_N(p)| > \delta}} \frac{1}{p} \longrightarrow 0 \tag{1.4}$$

for every positive t and δ. Let us call a sequence satisfying (1.4) to be of the *Kubilius type*. This is slightly more general than the class H, and a part of results that hold for H (those involving only limiting distribution and not variance or other moments) work equally for this class. This definition can be transferred to general groups without difficulty, and this is our starting point.

Let us fix first some notations and terminology. We shall work with a commutative topological group G, the operation in which will be denoted by $+$ and the neutral element (zero) by $\mathbf{0}$. A G-valued arithmetical function f: $\mathbf{N} \longrightarrow G$ is called *additive* if it satisfies $f(mn) = f(m) + f(n)$ whenever the $\gcd(m,n) = 1$. By an additive function we shall always mean a G-valued additive function and ν_N denotes the uniform measure on the set of numbers $\{1, 2, \ldots, N\}$.

The distributions occurring will be measures in our group G; we always assume that they are tight. By their convergence we always mean weak convergence, which will be denoted by $\Rightarrow$. It will be more convenient to work with convolution of measures rather than sums of independent variables. The convolution of two measures ϑ and λ will be denoted by $\vartheta\lambda$, without the usual $*$, and for convolution of several terms we use the $\prod$ notation.

DEFINITION 1.1. A sequence (f_N) of additive functions is of the *Kubilius type* if for every $t \in (0,1)$ and every neighbourhood U of $\mathbf{0}$ we have

$$\sum_{\substack{N^t \leqslant p \leqslant N \\ f_N(p) \notin U}} \frac{1}{p} \longrightarrow 0. \tag{1.5}$$

We use δ_x or $\delta(x)$ to denote the *point mass* at a point $x \in G$, that is,

$$\delta_x(X) = \begin{cases} 1 & \text{if} \quad x \in X, \\ 0 & \text{otherwise.} \end{cases}$$

We write $\delta = \delta(\mathbf{0})$, the unit mass at $\mathbf{0}$, which is the unity of the convolution semigroup. Define the measures ξ_{Np} and ζ_N by

$$\xi_{Np} = \left(1 - \frac{1}{p}\right) \sum_{j=0}^{\infty} \frac{\delta(f_N(p^j))}{p^j}, \qquad \zeta_N = \prod_{p \leqslant N} \xi_{Np}. \tag{1.6}$$

DEFINITION 1.2. We say that the sequence (f_N) has a *limiting distribution* ϑ *with centering* A_N if the probability measures ϑ_N defined by

$$\vartheta_N(X) = \nu_N(f_N(n) - A_N \in X)$$

tend weakly to the probability measure ϑ as $N \to \infty$.
The following result corresponds to Theorem 4.1 of (Kubilius, 1964).

THEOREM 1. *Assume that the sequence (f_N) of additive functions is of the Kubilius type. It has a limiting distribution with centering A_N if and only if the sequence $\zeta_N \delta(-A_N)$ of measures converges weakly, and in this case the limiting distributions are the same.*

We do not give a separate proof of this theorem. It can be proved by a minimal modification of Kubilius' proof; it is also a particular case of Theorem 3 below, except for the identity of the limiting distributions.
Now we define a wider class of functions.

DEFINITION 1.3. A sequence (f_N) of additive functions is *controlled* if for every t, $0 < t < 1$, the measures

$$\pi_N(t) = \sum_{N^t \leqslant p \leqslant N} \frac{1}{p} \delta(f_N(p)) \tag{1.7}$$

tend weakly to some measure π_t as $N \to \infty$.
For a Kubilius-type sequence π_t is a degenerate measure with mass $\log 1/t$ at $\mathbf{0}$.
The idea here is that large primes may not be negligible, just we want to know their contribution exactly. Condition (1.7) is in the real case equivalent to condition (7) of (Levin and Timoveev, 1975) for the convergence of

$$\sum_{p \leqslant N,\ f(p) \leqslant u B_N} \frac{1}{p} \left(\frac{\log p}{\log N}\right)^k$$

for every real u and positive integer k.
For a controlled sequence of functions the limiting distributions of f_N and ζ_N may not be the same, but their existences are related.

THEOREM 2. *Let (f_N) be a controlled sequence of additive functions, and assume that G does not have any compact subgroup except $\{\mathbf{0}\}$. If for some $A_N \in G$ the sequence $\zeta_N \delta(-A_N)$ converges weakly, then (f_N) has a limiting distribution with the same centering A_N (the limiting distributions may be different).*

I can show that the corresponding result does not hold for the circle group (group of complex numbers with $|z| = 1$ and the operation of multiplication). I conjecture that the restriction about compact subgroups can be weakened to that G does not contain a subgroup isomorphic to the circle group.
The restriction on G can be lifted by imposing an extra condition on f_N.

DEFINITION 1.4. A sequence (f_N) of additive functions is *separated* if for every $t \in (0,1)$, $\varepsilon > 0$, and every neighbourhood U of $\mathbf{0}$ there is a number α,

$$0 < \alpha = \alpha(U,t) \leqslant 1,$$

such that

$$\limsup \sum_{\substack{N^{\alpha t} \leqslant p \leqslant N^{\alpha} \\ f_N(p) \notin U}} \frac{1}{p} \leqslant \varepsilon. \tag{1.8}$$

For a Kubilius-type sequence $\alpha = 1$ is such a choice.

THEOREM 3. *Let (f_N) be a controlled, separated sequence of additive functions. If the sequence $\zeta_N \delta(-A_N)$ converges weakly for some A_N, then (f_N) has a limiting distribution with centering A_N.*

In the second part of the paper we plan to present results where a "logarithmic" type summand may also occur.

Acknowledgement. I profited much from the correspondence with E. Manstavičius.

2. LARGE PRIMES ONLY

The total variation of a signed measure ν will be denoted by $|\nu|$, and the *norm* of ν is defined by

$$\|\nu\| = |\nu|(G).$$

In an equation involving measures, $O(c)$ will mean a signed measure whose norm is $O(c)$.

LEMMA 2.1. *Let ν_N be a sequence of measures on G. Assume that for every $\varepsilon > 0$ there is a convergent sequence λ_N of measures such that*

$$\limsup \|\nu_N - \lambda_N\| < \varepsilon.$$

The the sequence ν_N is convergent as well.

Real analogs of this lemma are familiar. We leave the easy proof to the reader.

Now we treat the case of functions that depend on large primes only.

STATEMENT 2.2. *Let (f_N) be a controlled sequence of additive functions, and assume that there is a constant α, $0 < \alpha < 1$, such that $f_N(p^j) = 0$ for all primes $p \leqslant N^\alpha$ and every j. The functions f_N have a limiting distribution.*

Proof. Every integer has the unique decomposition of the form $n = aw$ where the prime divisors of a are not greater than N^α and those of w are greater than N^α, and we have $f(n) = f(w)$. (We shall omit the subscript N when we can do it without confusion). Thus if ϑ_N denotes the distribution of f on the set of integers $\{1, \ldots, N\}$, A and W are the sets of these numbers a and w, then we have

$$\vartheta_N = \sum_{w \in W} \delta(f(w)) A\Big(\frac{N}{w}\Big), \tag{2.1}$$

where we use the same letter A for the counting function of the set A. The function $A(x)$ is a particular case of de Bruijn's function: we have

$$A(x) = \psi(x, N^\alpha) \sim x\varrho\Big(\frac{\log x}{\alpha \log N}\Big) \tag{2.2}$$

with a certain continuous function ϱ, and this relation is uniform in the range $N^\alpha < x \leqslant N$ (see (de Bruijn, 1950)). For $x \leqslant N^\alpha$ we have $A(x) = [x]$, hence by putting $\varrho(t) = 1$ for $0 < t \leqslant 1$, (2.2) will hold as long as $x \to \infty$.
To approximate ϑ_N, we shall divide W into classes. Select a positive integer k and define r by $\alpha r^k = 1$. We consider the intervals

$$I_j = \big(N^{\alpha r^{j-1}}, N^{\alpha r^j}\big], \qquad j = 1, \ldots, k,$$

whose union is the interval $(N^\alpha, N]$.
For a sequence $\underline{\varepsilon} = (\varepsilon_1, \ldots, \varepsilon_k)$ of zeroes and units, let $B(\underline{\varepsilon})$ be the set of those integers n which are divisible by exactly ε_j prime number (counted with multiplicity) from each interval I_j. Let $W(\underline{\varepsilon})$ be the set of those numbers from $B(\underline{\varepsilon})$ which are free of primes $p \leqslant N^\alpha$. With the above decomposition $n = aw$ we have $n \in B(\underline{\varepsilon})$ if and only if $w \in W(\underline{\varepsilon})$. We shall group the integers according to their class $B(\underline{\varepsilon})$.
First we estimate the number of those $n \leqslant N$ that do not lie in any $B(\underline{\varepsilon})$. These numbers are divisible by at least two primes from some interval I_j. For an I_j their number is

$$\leqslant \sum_{p,q \in I_j} \Big[\frac{N}{pq}\Big] \leqslant N\Big(\sum_{p \in I_j} \frac{1}{p}\Big)^2.$$

Since (for fixed k and $N \to \infty$) we have

$$\sum_{p \in I_j} \frac{1}{p} \to \log r = \frac{\log 1/\alpha}{k},$$

the total number of these numbers is

$$\ll Nk(\log r)^2 \ll N/k.$$

Write

$$\sigma(\underline{\varepsilon}) = \alpha \sum \varepsilon_j r^{j-1}.$$

This quantity is related to the size of numbers in $W(\underline{\varepsilon})$; we have obviously

$$W(\underline{\varepsilon}) \subset \left(N^{\sigma(\underline{\varepsilon})}, N^{r\sigma(\underline{\varepsilon})}\right]. \tag{2.3}$$

We call $\underline{\varepsilon}$ *small*, if $\sigma(\underline{\varepsilon}) \leqslant 1/r^2$, and *large* otherwise. We estimate the number of n's belonging to a large $\underline{\varepsilon}$. In the decomposition $n = aw$ we have

$$w > N^{1/r^2}, \qquad a < N^{1-1/r^2},$$

hence n is free of prime divisors from the interval

$$\left(N^{1-1/r^2}, N^{\alpha}\right].$$

By standard sieve estimates the number of such integers $n \leqslant N$ is

$$\ll N \frac{\log N^{1-1/r^2}}{\log N^{\alpha}} \ll N/k.$$

Consequently we have

$$\begin{aligned} \vartheta_N &= \frac{1}{N} \sum_{\underline{\varepsilon} \text{ small}} \sum_{w \in W(\underline{\varepsilon})} \delta(f(w)) A\left(\frac{N}{w}\right) + O(1/k) \\ &= \sum_{\underline{\varepsilon} \text{ small}} \sum_{w \in W(\underline{\varepsilon})} \delta(f(w)) \frac{1}{w} \varrho\left(\frac{\log N/w}{\alpha \log N}\right) + O(1/k). \end{aligned} \tag{2.4}$$

In the second line of (2.4) we used the fact that $N/w > N^{1-1/r} \to \infty$ for $\underline{\varepsilon}$ small, hence the asymptotic formula (2.2) applies.
For $w \in W(\underline{\varepsilon})$ we have

$$\frac{\log N/w}{\alpha \log N} = \frac{1 - \sigma(\underline{\varepsilon})}{\alpha} + O(1/k),$$

thus (2.4) yields (using the differentiability of ϱ except at 1)

$$\vartheta_N = \sum_{\underline{\varepsilon} \text{ small}} \varrho\left(\frac{1 - \sigma(\underline{\varepsilon})}{\alpha}\right) \sum_{w \in W(\underline{\varepsilon})} \frac{\delta(f(w))}{w} + O(1/k).$$

Finally we observe that for a fixed $\underline{\varepsilon}$ we have

$$\begin{aligned} \sum_{w \in W(\underline{\varepsilon})} \frac{\delta(f(w))}{w} &= \prod_{j=1}^{k} \left(\sum_{p \in I_j} \frac{1}{p} \delta(f(p)) \right)^{\varepsilon_j} = \prod_{j=1}^{k} \left(\pi_N(\alpha r^{j-1}) - \pi_N(\alpha r^j)\right)^{\varepsilon_j} \\ &\Rightarrow \prod_{j=1}^{k} \left(\pi_{\alpha r^{j-1}} - \pi_{\alpha r^j}\right)^{\varepsilon_j}. \end{aligned}$$

Now an application of Lemma 2.1 concludes the proof.

3. SEPARATION AND INDEPENDENCE

Here we show that for a separated sequence the contributions of large and small primes are almost independent.
Assume $1 \leqslant y < z \leqslant N$, and let f be an additive function such that $f(p^k) = 0$ whenever $y < p \leqslant z$. Define the additive function g by

$$g(p^k) = \begin{cases} 0 & \text{for } p < z, \\ f(p^k) & \text{for } p \geqslant z. \end{cases}$$

Let ϑ and λ be the distributions of f and g on the set of integers $n \leqslant N$, that is,

$$\vartheta(X) = \nu_N(f(n) \in X), \qquad \lambda(X) = \nu_N(g(n) \in X).$$

Write

$$\xi_p = \left(1 - \frac{1}{p}\right) \sum_{j=0}^{\infty} \frac{\delta(f(p^j))}{p^j}, \qquad \zeta = \prod_{p \leqslant y} \xi_p.$$

Finally put

$$h = \frac{\log z}{\log y} > 1.$$

THEOREM 4. *With the above notations and assumptions we have*

$$\|\vartheta - \zeta\lambda\| \ll h^{-1/3} \tag{3.1}$$

with an absolute constant.

The case $z = N$ is the classical Kubilius model. In this case the estimate in (3.1) may be improved to $\exp\{-ch \log h\}$. In the general case I can improve the $h^{-1/3}$ to $(\log h)/h$, but I can show that an improvement over $1/h$ is in general impossible. Let U, V, and W denote the sets of integers all prime factors of which are not greater than y, belong to $(y, z]$, or are greater than z, respectively. Generic elements of these sets will be denoted by u, v, and w. Put $UV = R$, $VW = S$ (sets of all pairwise products). The counting functions of these sets will be denoted by $U(t)$, etc.

LEMMA 3.1. *For $t \geqslant y^K$, $K \geqslant 1$, we have*

$$S(t) = \gamma t(1 + O(e^{-cK})) \tag{3.2}$$

with a positive absolute constant c and

$$\gamma = \prod_{p \leqslant y} \left(1 - \frac{1}{p}\right).$$

Proof. S is the set of numbers free of prime factors not greater than y, thus (3.2) is a standard sieve estimate. (As a general reference for sieve results, (see (Halberstam and Richert, 1974).)

LEMMA 3.2. *For $t \geqslant z$ we have*

$$W(t) \ll \frac{t}{\log z}. \tag{3.3}$$

Proof. W is the set of integers free of primes less than z, so this is a standard sieve estimate as well.

LEMMA 3.3. *For $t > 1$, $T \geqslant 2t$, we have*

$$\sum_{t \leqslant w < T} \frac{1}{w} \ll \frac{\log (T/t)}{\log z}. \tag{3.4}$$

Proof. Since W has no elements between 1 and z, we may assume $t \geqslant z$, apply the previous lemma and partial summation. (For $t \leqslant 1$ this lemma may fail because $1 \in W$.)

Proof of Theorem 4. For every set $X \subset G$ we have

$$\vartheta(X) = \frac{1}{N} \sum_{f(u)+f(w) \in X} V\left(\frac{N}{uw}\right),$$

$$\lambda(X) = \frac{1}{N} \sum_{f(w) \in X} R\left(\frac{N}{w}\right),$$

$$\zeta(X) = \gamma \sum_{f(u) \in X} \frac{1}{u}.$$

Consequently

$$(\zeta\lambda)(X) = \frac{1}{N} \sum_{f(u)+f(w) \in X} \frac{\gamma}{u} R\left(\frac{N}{w}\right),$$

and we obtain

$$\|\vartheta - \zeta\lambda\| \leqslant \frac{1}{N} \sum_{u,w} \left| V\left(\frac{N}{uw}\right) - \frac{\gamma}{u} R\left(\frac{N}{w}\right) \right| = A. \tag{3.5}$$

We want to estimate this sum, a typical term of which is

$$B_{uw} = V\left(\frac{N}{uw}\right) - \frac{\gamma}{u} R\left(\frac{N}{w}\right).$$

LEMMA 3.4. *For $uw \leqslant N/y$ we have*

$$B_{uw} \ll \frac{N}{uw} \left(\frac{1}{\log \min (z, N/(uw))} + \frac{\log u}{\log y \log z} \right). \tag{3.6}$$

Proof. Introduce temporarily the notations $M = N/(uw) \geqslant y$,

$$B = B_{uw} = V(M) - \frac{\gamma}{u} R(Mu).$$

By an obvious Möbius inversion we have

$$V(t) = \sum_{w} \mu(w) S\Big(\frac{t}{w}\Big),$$

$$R(t) = \sum_{w} \mu(w) \Big[\frac{t}{w}\Big],$$

hence

$$B \leqslant \sum_{w} \Big| S\Big(\frac{M}{w}\Big) - \frac{\gamma}{u}\Big[\frac{Mu}{w}\Big]\Big|.$$

(This variable w is different from the previous one, which disappeared by the introduction of M.)

Consider first the contribution of $w = 1$. This is

$$S(M) - \frac{\gamma}{u}[Mu] = S(M) - \gamma M + O(\gamma) \ll \gamma M e^{-ck}$$

by Lemma 3.1, where $k = (\log M)/(\log y)$. Since $e^{-x} < 1/x$ and $\gamma \ll 1/\log y$, this error term is smaller than that in (3.6).

Consider the numbers $w \geqslant z$ satisfying $y^k \leqslant M/w < y^{k+1}$ with an integer $k \geqslant 1$. By Lemma 3.1 each term is $O(\gamma e^{-ck} M/w)$, the sum over w is by Lemma 3.3 $O(\gamma e^{-ck} M/\log z)$, finally the sum over k yields $O(\gamma M/\log z)$ as wanted.

Finally we take those w's for which $M/w < y$, that is, $M/y < w \leqslant Mu$. Here $S(M/w) = 1$ for $w \leqslant M$ and 0 for $w > M$, the total contributions is $O(M/\log z)$ by Lemma 3.2. The second summand is $(\gamma/u)[Mu/w] \leqslant \gamma M/w$, thus their contribution is not greater than

$$\frac{\gamma M \log(uy)}{\log z} \ll \frac{M}{\log z} + \frac{M \log u}{\log y \log z}.$$

The first term is swallowed by the first error term.

We continue the proof of Theorem 4.

The signed sum of the numbers B_{uw} is

$$\sum B_{uw} = N\big(\vartheta(G) - (\zeta\lambda)(G)\big) = 0,$$

hence

$$\sum |B_{uw}| = 2 \sum B_{uw}^{+},$$

where $x^{+} = \max(x, 0)$. To estimate this, we use Lemma 3.4 for "typical" u and w, and other arguments for atypical values. We call them typical if $w > z^{-L} N$ and $uw < y^{-K} N$, where the parameters K and L will be specified later. We assume that $K < h$, that is, $y^K < z$.

For typical values, Lemma 3.4 yields

$$|B_{uw}| \ll \frac{N}{w \log y}\Big(\frac{1}{uK} + \frac{\log u}{u \log z}\Big).$$

Since

$$\sum \frac{1}{u} = \frac{1}{\gamma} \ll \log y,$$

and the decomposition

$$\log u = \sum_{d|u} \Lambda(d)$$

easily yields that

$$\sum \frac{\log u}{u} = \gamma^{-1} \sum_{p \leqslant y} \frac{\log p}{p-1} \ll (\log y)^2,$$

we obtain

$$\sum_u |B_{uw}| \ll \frac{N}{wK}. \tag{3.7}$$

From Lemma 3.3 we obtain that

$$\sum_{z^{-L} N < w \leqslant N} \frac{1}{w} \ll L; \tag{3.8}$$

the range of summation may include $w = 1$, but (3.8) still holds if $L > 1$. Formulae (3.7) and (3.8) together yield that

$$\sum_{u,w \text{ typical}} |B_{uw}| \ll \frac{NL}{K}. \tag{3.9}$$

For atypical values we use

$$B_{uw}^{+} \leqslant V\big(N/(uw)\big).$$

Consequently,

$$\sum_{u,w \text{ typical}} B_{uw}^{+}$$

is not greater than the number C of those integers $n = uvw \leqslant N$ in whose decomposition either $w \leqslant z^{-L}N$ (first kind) or $uw \geqslant y^{-K}N$ (second kind). (The first class may be empty.)
If n is of the first kind, then either $n \leqslant N/z$ or $uv > z^{L-1}$. Since

$$\log(uv) = \sum_{p|n,\ p \leqslant z} \log p$$

has a mean value $O(\log z)$, the second possibility gives $O(N/L)$ solutions; the first gives $O(N/z)$, which is swallowed by the second.
If n is of the second kind, then $v \leqslant y^K$, consequently, n is free of prime divisors from $(y^K, z]$. For the cardinality of these numbers a standard sieve estimate gives

$$\ll N \frac{K \log y}{\log z} = N \frac{K}{h}.$$

Combining our estimates we obtain

$$\sum_{u,w} B_{uw}^{+} \ll N\Big(\frac{L}{K} + \frac{1}{L} + \frac{K}{h}\Big).$$

We obtain Theorem 4 with $L = h^{1/3}$, $K = h^{2/3}$.

4. PROOF OF THEOREM 3

DEFINITION 4.1. By the *Poisson measure with spectrum* μ, where μ is a (totally finite) signed measure, we mean the signed measure

$$\Pi(\mu) = e^{-\mu(G)}\left(\delta + \sum_{j=1}^{\infty} \frac{\mu^j}{j!}\right). \tag{4.1}$$

If μ is positive, $\Pi(\mu)$ is a probability measure. Otherwise we can claim only that

$$\Pi(\mu)(G) = 1, \qquad \|\Pi(\mu)\| \leqslant e^{\|\mu\|-\mu(G)}. \tag{4.2}$$

This operation behaves like an exponentiation in the sense that

$$\Pi(\mu + \lambda) = \Pi(\mu)\Pi(\lambda). \tag{4.3}$$

LEMMA 4.2. *Let (f_N) be a controlled sequence of functions. Then for every α and β, $0 < \alpha \leqslant \beta \leqslant 1$, the measures*

$$\zeta_N(\alpha, \beta) = \prod_{N^\alpha < p \leqslant N^\beta} \xi_{Np} \tag{4.4}$$

converge weakly to

$$\zeta(\alpha, \beta) = \Pi\big(\pi_\alpha - \pi_\beta\big). \tag{4.5}$$

Proof. We have

$$\xi_{Np} = \Pi\Big(\frac{1}{p}\delta(f_N(p))\Big) + O\Big(\frac{1}{p^2}\Big)$$

by an easy calculation. Multiplying these equations and taking into account that

$$\sum_{N^\alpha < p \leqslant N^\beta} \frac{1}{p^2} \longrightarrow 0$$

we obtain the desired convergence.

LEMMA 4.3. *Let (f_N) be a controlled sequence of functions and assume that the measures $\zeta_N\delta(-A_N)$ converge weakly to ζ. Then for every α, $0 \leqslant \alpha \leqslant 1$, we have (with the notation (4.4))*

$$\zeta_N(0,\alpha)\delta(-A_N) \Longrightarrow \zeta(0,\alpha) = \zeta\Pi(-\pi_\alpha). \tag{4.6}$$

Proof. We know that

$$\zeta_N(0,\alpha)\zeta_N(\alpha,1)\delta(-A_N) = \zeta_N\delta(-A_N) \Longrightarrow \zeta,$$

and by multiplying this by $\Pi(-\pi_\alpha)$ we get

$$\zeta_N(0,\alpha)\delta(-A_N)\zeta_N(\alpha,1)\Pi(-\pi_\alpha) \Longrightarrow \zeta\Pi(-\pi_\alpha). \tag{4.7}$$

On the other hand, a calculation similar to the previous lemma shows

$$\zeta_N(\alpha,1)\Pi(-\pi_\alpha) = \delta(\mathbf{0}) + \eta_N, \qquad \|\eta_N\| \longrightarrow 0,$$

and substituting this into (4.7) we obtain (4.6).

LEMMA 4.4. *A sequence μ_N of probability measures converges weakly if and only if the sequence $\int F d\mu_N$ converges for every bounded uniformly continuous function F: $G \longrightarrow \mathbf{R}$.*

See (Topsøe, 1970). The difference from the usual definition of weak convergence is that only *uniformly* continuous functions are considered.
Proof of Theorem 3. Take a uniformly continuous bounded function F. We have to show the existence of the limit

$$\lim \frac{1}{N} \sum_{n \leqslant N} F\big(f_N(n) - A_N\big). \tag{4.8}$$

Take a positive ε, $\varepsilon < 1/2$. Let U be such a neighbourhood of $\mathbf{0}$ that

$$|F(x) - F(y)| < \varepsilon$$

whenever $x - y \in U$. Let k be a positive integer satisfying

$$k > \frac{1}{\varepsilon \log 1/\varepsilon},$$

V be a neigbourhood of $\mathbf{0}$ satisfying $Vk = V + \ldots + V \subset U$. Finally put $t = \varepsilon^2$ and choose a number α, $0 < \alpha < 1$, such that

$$\sum_{\substack{N^{\alpha t} < p \leqslant N^{\alpha} \\ f_N(p) \notin V}} \frac{1}{p} < \varepsilon \tag{4.9}$$

for large N; such an α exists by the assumption of separation.
We split f_N into three parts

$$f_N = g_N + h_N + r_N = f'_N + r_N,$$

where g_N is the contribution of primes not greater than $N^{\alpha t}$, h_N is the contribution of primes greater than N^{α}, r_N is that of primes in $I = (N^{\alpha t}, N^{\alpha}]$.
First we show that r_N is small. Indeed, $r_N(n) \notin U$ is possible in three ways. First, if n is divisible by the square of a prime $p \in I$; this is in $o(n)$ cases. Next, n may be divisible by at least k primes from I. Since the average number of prime divisors from I is not greater than

$$\sum_{p \in I} \frac{1}{p} \ll \log(1/\varepsilon),$$

the number of such n's is not greater than

$$\ll \frac{\log 1/\varepsilon}{k} N \ll \varepsilon N.$$

Finally, if n is the product of less than k distinct primes from the interval I (and others outside I), then by $Vk \subset U$, $r_N(n) \notin U$ is possible only if $f_N(p) \notin V$ for some prime $p \mid n$. By (4.9), this gives $O(\varepsilon N)$ possibilities.
We learned that $r_N(n) \in U$ for $n \leqslant N$ with $O(\varepsilon N)$ exceptions. By the definition of U this means that

$$\frac{1}{N} \sum_{n \leqslant N} F(f_N(n)) = \frac{1}{N} \sum_{n \leqslant N} F(f'_N(n)) + O(\varepsilon). \tag{4.10}$$

Let ϑ_N be the distribution of f'_N, and λ_N be that of h_N, both on the set of integers $\{1, \dots, N\}$. Applying Theorem 4 to the function f'_N with the choice $y = N^{\alpha t}$, $z = N^{\alpha}$ we obtain that

$$\|\vartheta_N - \lambda_N \zeta_N(0, \alpha t)\| \ll \sqrt{t} \ll \varepsilon.$$

Consequently,

$$\frac{1}{N} \sum_{n \leqslant N} F(f'_N(n)) = \int F \, d\vartheta_N = \int F \, d\lambda_N \zeta_N(0, \alpha t) + O(\varepsilon). \tag{4.11}$$

Statement 2.2 implies that $\lambda_N \Rightarrow \lambda$ with some λ, and Lemma 4.3 shows that $\zeta_N(0, \alpha t) \Rightarrow \zeta(0, \alpha t)$. Consequently, the right hand side of (4.11) is equal to

$$\int F \, d\lambda \zeta(0, \alpha t) + O(\varepsilon) \tag{4.12}$$

for $N > N_0(\varepsilon)$. Equations (4.10)–(4.12) together show that the difference between the lim inf and lim sup in (4.8) is $O(\varepsilon)$. Since this holds for any ε, the existence of the limit is established.

5. PROOF OF THEOREM 2

Theorem 2 will be a consequence of Theorem 3 and the following statement.

STATEMENT 5.1. *Assume that G has no nontrivial compact subgroup. If (f_N) is a controlled sequence of additive functions for which $\zeta = \lim \zeta_N$ exists, then it is separated.*

For any $\varepsilon > 0$ and a neighbourhood U of $\mathbf{0}$ put

$$\mathcal{U} = \mathcal{U}(U, \varepsilon) = \{\mu \colon \ \mu(U) > 1 - \varepsilon\}.$$

This set $\mathcal{U}$ of measures is a neighbourhood of the measure $\delta = \delta(\mathbf{0})$ in the weak topology. Put

$$\mathcal{U}' = \{\mu\delta(x)\colon\ \mu \in \mathcal{U}, x \in G\}.$$

Let $\mathcal{D}(G)$ denote the topological semigroup of distributions on G with the operation of convolution. For $\mu, \lambda \in \mathcal{D}(G)$ write $\mu \mid \lambda$ if $\lambda = \mu\vartheta$ with some $\vartheta \in \mathcal{D}(G)$.

LEMMA 5.2. *For every $\lambda \in \mathcal{D}(G)$ and every neighbourhood $\mathcal{U}$ of δ there exists a positive integer m with the following property: whenever*

$$\mu_1 \dots \mu_m \mid \lambda,$$

then $\mu_j \in \mathcal{U}'$ for some j.

Proof. By a theorem of Choquet–Deny–Tortrat (see (Ruzsa and Székely, 1988, Sec. 3.4)), the idempotents of $\mathcal{D}(G)$ are the Haar measures on compact subgroups. By the assumption that G has no nontrivial compact subgroup, the only idempotent is δ. In particular, λ has no other idempotent factors than δ, or (with the terminology of (Ruzsa and Székely, 1988), it is *bald*.

We write $\mu \sim \lambda$ and call them *associates* if $\mu \mid \lambda$ and $\lambda \mid \mu$. This is equivalent to $\mu = \lambda\delta(x)$ with some $x \in G$ (ibid., Cor. 4.8, Chap. 3). The factor semigroup $\mathcal{D}(G)^* = \mathcal{D}(G)/\sim$ will be Hun (ibid., Lemma 7.1, Chap. 3). An application of Lemma 8.10 from (ibid., Chap. 2) to the semigroup $\mathcal{D}(G)^*$ and the image of λ in it yields Lemma 5.2.

Proof of the Statement. Assume that $\varepsilon < 1/2$ and $0 < t < 1$. Take a neighbourhood V of $\mathbf{0}$ such that the closure of $V - V$ lies in U and let $\mathcal{U} = \mathcal{U}(V, \varepsilon/2)$. Consider the decomposition

$$\zeta = \zeta(0, t^m) \prod_{j=1}^{m} \Pi(\pi_{t^j} - \pi_{t^{j-1}}) \tag{5.1}$$

which follows from Lemma 4.2, with the number m provided by Lemma 5.2 for $\lambda = \zeta$ and the $\mathcal{U}$ defined above. We conclude that there must be a j such that all factors of $\Pi(\pi_{t^j} - \pi_{t^{j-1}}) = \Pi(\vartheta)$ lie in $\mathcal{U}'$.

Write $W = V - V$. We prove that $\vartheta(G \setminus W) < \varepsilon$. Indeed, assume that $\vartheta(G \setminus W) \geqslant \varepsilon$. Take a decomposition $\vartheta = \vartheta_1 + \vartheta_2$ (into positive measures) such that $\vartheta_1(W) = 0$, $\vartheta_1(G \setminus W) = \varepsilon$.

$$\Pi(\vartheta_1) = e^{-\varepsilon}\delta + e^{-\varepsilon}\vartheta_1 + \dots \tag{5.2}$$

is a factor of $\Pi(\vartheta)$, hence $\Pi(\vartheta_1)(V + x) > 1 - \varepsilon/2$ for some $x \in G$. Since

$$\Pi(\vartheta_1)(G \setminus \{\mathbf{0}\}) \leqslant 1 - e^{-\varepsilon} < \varepsilon,$$

we must have $\mathbf{0} \in V + x$, hence $x \in -V$ and $V + x \subset W$. This implies

$$\Pi(\vartheta_1)(W) > 1 - \varepsilon/2, \qquad \Pi(\vartheta_1)(G \setminus W) < \varepsilon/2.$$

On the other hand (5.2) yields

$$\Pi(\vartheta_1)(G \setminus W) \geqslant e^{-\varepsilon}(\vartheta_1)(G \setminus W) = e^{-\varepsilon}\varepsilon > \varepsilon/2,$$

a contradiction.

Since

$$\vartheta = \pi_{t^j} - \pi_{t^{j-1}} = \lim \varrho_N$$

with

$$\varrho_N = \sum_{N^{\alpha t} < p \leqslant N^{\alpha}} \frac{1}{p} \delta(f_N(p)), \qquad \alpha = t^{j-1},$$

the facts that $(\vartheta_1)(G \setminus W) < \varepsilon$ and U is a neighbourhood of the closure of W imply that $\varrho_N(G \setminus U) < \varepsilon$ for large N, and this was to be proved.

REFERENCES

de Bruijn, N. G. (1950). On the number of uncancelled elements in the sieve of Eratosthenes. *Indag. Math.* **12**, 247–256.

Elliott, P. D. T. A. (1979). *Probabilistic Number Theory, I.* Springer-Verlag, Berlin, Heidelberg, New York.

Elliott, P. D. T. A. (1980). *Probabilistic Number Theory, II.* Springer-Verlag, Berlin, Heidelberg, New York.

Halberstam, H. and Richert, H.-E. (1974). *Sieve Methods.* Academic Press, London, New York.

Kubilius, J. (1959). *Probabilistic Methods in the Theory of Numbers.* Valst. Polit. Moksl. Lit. Leid., Vilnius (in Russian)

Kubilius, J. (1962). *Probabilistic Methods in the Theory of Numbers. 2nd ed.* Valst. Polit. Moksl. Lit. Leid., Vilnius (in Russian).

Kubilius, J. (1964). *Probabilistic Methods in the Theory of Numbers.* American Math. Soc., Providence, RI.

Levin, B. V. and Timofeev, N. M. (1975). On the distribution of values of additive functions. *Acta Arithm.* **26**, 333–364.

Levin, B. V. and Timofeev, N. M. (1977). On the distribution of values of additive functions. II. *Acta Arithm.* **33**, 327–351 (in Russian).

Manstavičius, E. (1984). On an arithmetical modelling of stochastic process. *Liet. Matem. Rink.* **24**, 148–161 (in Russian).

Manstavičius, E. (1985). Additive functions and stochastic processes. *Liet. Matem. Rink.* **25**, 97–109 (in Russian).

Manstavičius, E. (1986). Additive functions and stochastic processes. II. *Liet. Matem. Rink.* **26**, 478–481 (in Russian).

Ruzsa, I. Z. and Székely, G. J. (1988). *Algebraic Probability Theory.* John Wiley and Sons, London.

Topsøe, F. (1970). *Topology and Measure.* Springer-Verlag, Berlin, Heidelberg, New York.

New Trends in Probab. and Statist., pp. 285–291
F. Schweiger and E. Manstavičius (Eds.)
1992 VSP/TEV

THE MULTI-DIMENSIONAL LOCAL THEOREM FOR ADDITIVE ARITHMETIC FUNCTIONS

G. STEPANAUSKAS
Department of Mathematics, Vilnius University,
Naugarduko 24, 2006 Vilnius, Lithuania

ABSTRACT

The s-dimensional local limit theorem for integer-valued additive arithmetic functions is proved. The proof is based on the method of characteristic functions (the inversion formula) and the estimates for the mean values of multiplicative functions.

There are many papers dealing with the local limit theorems for additive arithmetic functions. The more exhaustive introduction to this field may be found in (Kubilius, 1964, 1968; Manstavičius, 1980).

This article is the direct continuation of the author's papers (Stepanauskas, 1980, 1986). Here we shall obtain a multi-dimensional result and give the more exact dependence relations for some terms involved into its formulation.

Let us introduce some notations. Let p be a prime number. Let $c_1, c_2, \ldots$ be positive constants. Throughout the paper we assume that subscripts i and j will take values from the set $\{1, 2, \ldots, s\}$. Also we shall use the following notations: $\mathbf{f}(n) = (f_1(n), \ldots, f_s(n))$ where each component is an additive integer-valued arithmetic function; $\mathbf{P}_x$ is a set of prime numbers not exceeding x and such that for all x

$$\sum_{\substack{p \leqslant x \\ p \notin \mathbf{P}_x}} \frac{1}{p} \leqslant c_1;$$

further we write $\mathbf{A} = (A_1, \ldots, A_s)$ with

$$A_i = \sum_{p \in \mathbf{P}_x} \frac{f_i(p)}{p}; \qquad B_{ij} = \sum_{p \in \mathbf{P}_x} \frac{f_i(p) f_j(p)}{p};$$

$$M_i = \max_{p \in \mathbf{P}_x} |f_i(p)|; \qquad M = \max_i M_i;$$

$$\mu_i = \max_{p^r \leqslant x} |f_i(p^r)| / \sqrt{B_{ii}}; \qquad \mu = \max_i \mu_i,$$

the inequality $\mu \leqslant c_2$ being satisfied for all x. We shall denote the determinant of the quadratic matrix (B_{ij}) by B. Let $\mathbf{f}(\mathbf{P}_x) = \{\mathbf{F}_1, \ldots, \mathbf{F}_k\}$ where $\mathbf{F}_k \subset \mathbf{Z}^s$. Now we put two assumptions on $\mathbf{P}_x$ and (B_{ij}).

(A) The set $\mathbf{P}_x$ may be decomposed into nonintersecting sets of prime numbers $\mathbf{P}_k$, $k = 1, \ldots, K$, such that

$$\mathbf{f}(\mathbf{P}_k) = \mathbf{F}_k, \qquad \alpha_k = \sum_{p \in \mathbf{P}_k} \frac{1}{p} \longrightarrow \infty \qquad \text{when} \quad x \to \infty.$$

(B) The matrix (B_{ij}) is positive definite and $B_{ii} \to \infty$ $(i = 1, \ldots, s)$ when $x \to \infty$. From the positive definiteness of the matrix (B_{ij}) it follows that among the vectors $\mathbf{F}_1, \ldots, \mathbf{F}_k$ there exist s linearly independent ones. Consider a system of $(s-1)$-dimensional hyperplanes in the s-dimensional space of real variables $\mathbf{u} = (u_1, \ldots, u_s)$ defined by

$$\mathbf{F}_{k_i}\mathbf{u} = 2\pi l_i, \qquad i = 1, \ldots, s, \quad 1 \leqslant k_1 \leqslant \ldots \leqslant k_s \leqslant K. \tag{1}$$

We select the integers l_i so that the hyperplanes (1) would intersect the s-dimensional cube

$$\Pi = \{\mathbf{u}\colon\ -\pi \leqslant u_i \leqslant \pi,\ i = 1, \ldots, s\}.$$

Let us regard (1) as a system of equations in respect of $\mathbf{u}$. In the case when the vectors $\mathbf{F}_{k_i}$ from (1) are independent, the solution of this system is a point in s-dimensional space. Varying the vectors $\mathbf{F}_{k_i}$ and the integers l_i we get other systems. Let us denote all different solutions of such systems falling into the cube Π by $\mathbf{u}_1, \ldots, \mathbf{u}_N$. It is clear that among these solutions there is a zero one. Therefore, we shall regard $\mathbf{u}_1 = \mathbf{0}$.
Now we impose one more assumption.

(C) In the case when $N \geqslant 2$, for every $\mathbf{u}_r$, $2 \leqslant r \leqslant N$, there exist the hyperplanes $\mathbf{F}_{k_r}\mathbf{u} = 2\pi l$, $l \in \mathbf{Z}$, such that

$$\mathbf{u}_r \notin \bigcup_{l \in \mathbf{Z}} \{\mathbf{u}\colon\ \mathbf{F}_{k_r}\mathbf{u} = 2\pi l\}.$$

Let us denote eigen-values of the symmetric quadratic matrix $(B_{ij}/\sqrt{B_{ij}B_{jj}})$ by $\lambda_1, \ldots, \lambda_s$, $\lambda = \min_i \lambda_i$. Finally, let

$$\alpha = \exp\Big\{ -\min_{\substack{1 \leqslant k \leqslant K \\ \mathbf{F}_k \neq \mathbf{0}}} \alpha_k \Big\},$$

and let $\mathbf{a} = (a_1, \ldots, a_s)$ be an integer-valued vector.
Now we can formulate our main result.

THEOREM. *Let $\mu \to 0$ as $x \to \infty$. Assume that the conditions* (A), (B), *and* (C) *are satisfied. Then there exist positive constants c_3, c_4, and c_5 such that for all $x \geqslant 2$*

$$\frac{\sqrt{B}}{x} \sum_{\substack{n \leqslant x \\ \mathbf{f}(n) = \mathbf{a}}} 1 - \frac{1}{(2\pi)^{s/2}} \exp\Big\{ -\frac{1}{2}(\mathbf{a} - \mathbf{A})(B_{ij})^{-1}(\mathbf{a} - \mathbf{A})' \Big\}$$
$$= O\big(\mu + c_3^s \exp\{ -c_4 \lambda s^{-2} \mu^{-2/3}\} + \sqrt{B}\alpha^{c_5 \varepsilon}\big) \tag{2}$$

with $\varepsilon = 1/(2^s M^{s(s+3)/2} 1! \dots s!(s!)^2)$. *Here the constants* c_3, c_4, c_5, *and the constant in the symbol* O *depend on* c_1 *and* c_2 *only.*

Proof. Denoting the left-hand side of (2) by R and applying the inversion formula we have

$$R = \frac{\sqrt{B}}{(2\pi)^s}\left(\int_{\Pi} e^{-i\mathbf{a}\mathbf{u}} \frac{1}{x}\sum_{n\leqslant x} e^{i\mathbf{u}\mathbf{f}(n)}\, d\mathbf{u} - \int_{\mathbf{R}^s} e^{-i(\mathbf{a}-\mathbf{A})\mathbf{u}-\frac{1}{2}r(\mathbf{u})}\, d\mathbf{u}\right),$$

where $r(\mathbf{u}) = \mathbf{u}(B_{ij})\mathbf{u}'$ is a quadratic form. Let us denote also

$$R_0 = \int_{T_0} e^{-i\mathbf{a}\mathbf{u}}\left(\frac{1}{x}\sum_{n\leqslant x} e^{i\mathbf{u}\mathbf{f}(n)} - e^{i\mathbf{A}\mathbf{u}-\frac{1}{2}r(\mathbf{u})}\right) d\mathbf{u}$$

with

$$T_0 = \{\mathbf{u}\colon\ |u_i| \leqslant c_6 s^{-1}\mu_i^{-1/3} B_{ii}^{-1/2},\ i = 1,\dots,s\}$$

(we shall choose the constant c_6 later),

$$R_k = \int_{T_k} e^{-i\mathbf{a}\mathbf{u}} \frac{1}{x}\sum_{n\leqslant x} e^{i\mathbf{u}\mathbf{f}(n)}\, d\mathbf{u}, \qquad k = 1, 2,$$

with

$$T_1 = \{\mathbf{u}\colon\ |u_i| \leqslant \pi/(sM_i),\ i = 1,\dots,s\} \setminus T_0 \quad \text{and} \quad T_2 = \Pi \setminus (T_0 \cup T_1),$$

$$R_3 = \int_{\Pi\setminus T_0} e^{-i(\mathbf{a}-\mathbf{A})\mathbf{u}-\frac{1}{2}r(\mathbf{u})}\, d\mathbf{u}.$$

Hence

$$R = \frac{\sqrt{B}}{(2\pi)^s}(R_0 + R_1 + R_2 + R_3). \tag{3}$$

At first we shall estimate the integral R_3. After an orthogonal change of variables reducing the quadratic form $r(\mathbf{u})$ to the canonical one, we obtain

$$R_3 \ll \frac{1}{\sqrt{B_{11}\dots B_{ss}}} \int_{u_1^2+\dots+u_s^2\geqslant \xi^2} \exp\left\{-\frac{1}{2}(\lambda_1 u_1^2 + \dots + \lambda_s u_s^2)\right\} d\mathbf{u}$$

where $\xi = c_6/(s\mu^{1/3})$. Now from the identity

$$\det(B_{ij}/\sqrt{B_{ii}B_{jj}}) = \lambda_1 \dots \lambda_s$$

we easily obtain the following estimate

$$\begin{aligned} R_3 &\ll \frac{1}{\sqrt{B}} \int_{u_1^2+\dots+u_s^2\geqslant \lambda\xi^2} \exp\left\{-\frac{1}{2}(u_1^2 + \dots + u_s^2)\right\} d\mathbf{u} \\ &\ll \frac{1}{\sqrt{B}} \exp\left\{-\frac{1}{4}\lambda\xi^2\right\} \int_{\mathbf{R}^s} \exp\left\{-\frac{1}{4}(u_1^2 + \dots + u_s^2)\right\} d\mathbf{u} \\ &= \frac{(2\sqrt{\pi})^s}{\sqrt{B}} \exp\left\{-\frac{1}{4}\lambda\xi^2\right\}. \end{aligned} \tag{4}$$

The estimation of the integral R_0 is based on the modification of G. Halász's lemma.

LEMMA 1 (Mačiulis, 1983). *Let $g(n)$ be a complex-valued multiplicative function, $|g(n)| = 1$, $|g(p) - 1| \leqslant \eta$. Then there exists an absolute constant $c_7 < 1$ such that if $0 < \eta \leqslant c_7$, then*

$$\begin{aligned}\frac{1}{x}\sum_{n\leqslant x} g(n) &= \prod_{p\leqslant x}\left(1-\frac{1}{p}\right)\sum_{r=0}^{\infty}\frac{g(p^r)}{p^r}\\ &\quad + O\left((\eta + \log^{-c_7} x)\exp\left\{\sum_{p\leqslant x}\frac{\operatorname{Re} g(p) - 1}{p}\right\}\right)\end{aligned} \tag{5}$$

with the absolute constant in the symbol O.

Let us assume

$$g(n) = e^{i\mathbf{uf}(n)}, \qquad \eta = |u_1|\sqrt{B_{11}}\mu_1 + \ldots + |u_s|\sqrt{B_{ss}}\mu_s$$

and choose $c_6 = c_7/c_2^{2/3}$. Now for the estimation of the integrand in the integral R_0 we use Lemma 1. The main term from the right-hand side of equality (5) is equal to

$$\prod_{p\leqslant x}\left(1-\frac{g(p)}{p}\right)\sum_{r=0}^{\infty}\frac{g(p^r)}{p^r}\exp\left\{\sum_{p\leqslant x}\frac{g(p)-1}{p} + O\left(\frac{|g(p)-1|}{p^2}\right)\right\}.$$

It is not difficult to understand that

$$\begin{aligned}&\prod_{p\leqslant x}\left(1-\frac{g(p)}{p}\right)\sum_{r=0}^{\infty}\frac{g(p^r)}{p^r}\\ &= \exp\left\{\sum_{p\leqslant x}\log\left(1+\frac{g(p^2)-g(p)g(p)}{p^2}+\frac{g(p^3)-g(p^2)g(p)}{p^3}+\ldots\right)\right\}\\ &= \exp\left\{O\left(\eta + \sum_{p\leqslant x}\ \sum_{r>\log x/\log p}\frac{1}{p^r}\right)\right\} = 1 + O\left(\eta + \frac{1}{\log x}\right).\end{aligned}$$

Taking into account that

$$|v_1 + \ldots + v_s|^3 \leqslant s^2\left(|v_1|^3 + \ldots + |v_s|^3\right)$$

for all complex v_i, we obtain

$$\begin{aligned}\sum_{p\leqslant x}\frac{g(p)-1}{p} &= i\mathbf{uA} - \frac{1}{2}r(\mathbf{u}) + O\left(\sum_{p\leqslant x}\frac{|\mathbf{uf}(p)|^3}{p}\right)\\ &= i\mathbf{uA} - \frac{1}{2}r(\mathbf{u}) + O\Big(\min\big(s^2\big(|u_1|^3 B_{11}^{3/2}\mu_1 + \ldots\\ &\quad + |u_s|^3 B_{ss}^{3/2}\mu_s\big), r(\mathbf{u})\eta\big)\Big).\end{aligned}$$

In addition, if

$$|u_i| \leqslant \frac{\pi}{sM_i}, \qquad i = 1, \ldots, s,$$

then

$$\sum_{p \leqslant x} \frac{\operatorname{Re} g(p) - 1}{p} = -2 \sum_{p \leqslant x} \frac{\sin^2(\mathbf{uf}(p)/2)}{p} \leqslant -\frac{2}{\pi^2} r(\mathbf{u}). \tag{6}$$

Hence

$$\frac{1}{x} \sum_{n \leqslant x} g(n) = \exp\left\{ i\mathbf{uA} - \frac{1}{2} r(\mathbf{u}) \right\} + O\Big(\exp\left\{ -\frac{2}{\pi^2} r(\mathbf{u}) \right\} \left(\eta + r(\mathbf{u})\eta + \log^{-c_7} x \right) \Big).$$

Estimating in a similar way as in (4) we have

$$\int_{T_0} \exp\left\{ -\frac{2}{\pi^2} r(\mathbf{u}) \right\} d\mathbf{u} \ll \frac{1}{\sqrt{B_{11} \ldots B_{ss}}} \int_{\mathbf{R}^s} \exp\left\{ -\frac{2}{\pi^2} (\lambda_1 u_1^2 + \ldots + \lambda_s u_s^2) \right\} d\mathbf{u} \ll \frac{1}{\sqrt{B}} \left(\frac{\pi^3}{2} \right)^{s/2},$$

$$\int_{T_0} \exp\left\{ -\frac{2}{\pi^2} r(\mathbf{u}) \right\} (|u_1| B_{11}^{1/2} + \ldots + |u_s| B_{ss}^{1/2}) \, d\mathbf{u} \ll \frac{\sqrt{s}}{\sqrt{B_{11} \ldots B_{ss}}} \int_{\mathbf{R}^s} \exp\left\{ -\frac{2}{\pi^2} (\lambda_1 u_1^2 + \ldots + \lambda_s u_s^2) \right\} (|u_1| + \ldots + |u_s|) \, d\mathbf{u}$$
$$\ll \frac{\sqrt{s}}{\sqrt{B}} \left(\frac{\pi^3}{2} \right)^{s/2} \left(\frac{1}{\sqrt{\lambda_1}} + \ldots + \frac{1}{\sqrt{\lambda_s}} \right),$$

$$\int_{T_0} \exp\left\{ -\frac{2}{\pi^2} r(\mathbf{u}) \right\} r(\mathbf{u}) (|u_1| B_{11}^{1/2} + \ldots + |u_s| B_{ss}^{1/2}) \, d\mathbf{u} \ll \frac{s\sqrt{s}}{\sqrt{B}} \left(\frac{\pi^3}{2} \right)^{s/2} \left(\frac{1}{\sqrt{\lambda_1}} + \ldots + \frac{1}{\sqrt{\lambda_s}} \right).$$

From these estimates it follows that

$$R_0 \ll \frac{1}{\sqrt{B}} \left(\frac{\pi^3}{2} \right)^{s/2} \frac{s^{5/2}}{\lambda^{1/2}} \mu.$$

The estimation of the integrals R_1 and R_2 is based on the modification of another G. Halasz's lemma.

LEMMA 2 (Stepanauskas, 1987). *Let $g(n)$ be a complex-valued multiplicative function, $|g(n)| = 1$, and let there exist the numbers θ and δ, $\delta > 0$, such that*

$$\left|e^{i \arg g(p)} - e^{i\theta}\right| \geqslant \delta$$

for all primes $p \in \mathbf{P}_x$. Then

$$\frac{1}{x}\sum_{n \leqslant x} g(n) = O\left(\exp\left\{ - c_8\delta^3 \sum_{p \leqslant x} \frac{1 - \operatorname{Re} g(p)}{p} \right\}\right)$$

where the constant c_8 and the constant in the symbol O depend on c_1 only.

For the estimation of the integrand in R_1 we apply Lemma 2 with $\theta = \pi$ and $\delta = 1$. In view of (6) it follows that

$$\left|\frac{1}{x}\sum_{n \leqslant x} e^{i\mathbf{uf}(n)}\right| \ll \exp\left\{ - c_8 \sum_{p \leqslant x} \frac{1 - \cos(\mathbf{uf}(p))}{p}\right\} \ll \exp\left\{ - c_9 r(\mathbf{u})\right\}.$$

Further, using the same ideas as in the estimation of the integral R_3, we obtain

$$\int_{T_1} e^{-c_9 r(\mathbf{u})}\, d\mathbf{u} \ll \frac{1}{\sqrt{B}} e^{-c_9\lambda\xi^2/2}\left(\frac{2\pi}{c_9}\right)^{s/2}$$

It remains to estimate the integral R_2. We shall divide the domain of integration T_2 into parts. Let $\mathbf{u} \in T_2$ satisfy the condition

$$\left|\mathbf{F}_k\mathbf{u} - 2\pi l\right| \geqslant \varepsilon_1 \tag{7}$$

for all $l \in \mathbf{Z}$ and all $\mathbf{F}_k$, $\mathbf{F}_k \neq \mathbf{0}$. The value of ε_1 we shall define later. For the estimation of the mean value of the multiplicative function $e^{i\mathbf{uf}(n)}$ we may apply Lemma 2 with $\theta = \varepsilon_1/2$ and $\delta = \varepsilon_1/4$. Because in this case $1 - \cos(\mathbf{uf}(p)) \geqslant c_{10}\varepsilon_1^2$, we have

$$\left|\frac{1}{x}\sum_{n \leqslant x} e^{i\mathbf{uf}(n)}\right| \ll \exp\left\{ - c_{11}\varepsilon_1^5 \sum_{\substack{p \leqslant x \\ \mathbf{f}(p) \neq \mathbf{0}}} \frac{1}{p}\right\}.$$

Further let $\mathbf{u} \in T_2$ satisfy the conditions

$$\left|\mathbf{F}_{k_1}\mathbf{u} - 2\pi l_1\right| \leqslant \varepsilon_j, \ \ldots, \ \left|\mathbf{F}_{k_1}\mathbf{u} - 2\pi l_j\right| \leqslant \varepsilon_j, \tag{8}$$

where j is one of the numbers $1, \ldots, s-1$ and the vectors $\mathbf{F}_{k_1}, \ldots, \mathbf{F}_{k_j}$ are linearly independent. Moreover, if $\mathbf{F}_k$, $\mathbf{F}_{k_1}, \ldots, \mathbf{F}_{k_j}$ are linearly independent, let

$$\left|\mathbf{F}_k\mathbf{u} - 2\pi l\right| \geqslant \varepsilon_{j+1} \tag{9}$$

for all possible l, $l \in \mathbf{Z}$. We shall assume in what follows that $\mathbf{F}_k \neq \mathbf{0}$. It is clear that in the case when $\mathbf{F}_k$ depends linearly on $\mathbf{F}_{k_1}, \ldots, \mathbf{F}_{k_j}$ but the $(s+1)$-dimensional vectors $(\mathbf{F}_k, l)$, $(\mathbf{F}_{k_1}, l_1), \ldots, (\mathbf{F}_{k_j}, l_j)$ are linearly independent, the inequality

$$\left|\mathbf{F}_k\mathbf{u} - 2\pi l\right| \geqslant \varepsilon_{j+1} \tag{10}$$

is valid. (We shall choose the values of ε_j later. They will be sufficiently small.) On the other hand, if the vector $(\mathbf{F}_k, l)$ depends linearly on $(\mathbf{F}_{k_1}, l_1), \ldots, (\mathbf{F}_{k_j}, l_j)$, then using this dependence and the inequalities (8) we can easily get that

$$|\mathbf{F}_k\mathbf{u} - 2\pi l| \leqslant j!jM^j\varepsilon_j.$$

Assume that $\mathbf{u}$ satisfies (8), (9) and

$$j!jM^j\varepsilon_j = \varepsilon_{j+1}/2.$$

It is evident now that we may apply Lemma 2 again with $\delta = \varepsilon_{j+1}/8$ and suitably chosen θ. Thus

$$\left|\frac{1}{x}\sum_{n\leqslant x} e^{i\mathbf{u}\mathbf{f}(n)}\right| \ll \exp\left\{-c_8\delta^3\sum_{p\leqslant x}\frac{1-\cos(\mathbf{u}\mathbf{f}(p))}{p}\right\}$$

$$\ll \exp\left\{-c_{12}\varepsilon_{j+1}^5\sum{}^{*}\frac{1}{p}\right\} \ll \alpha^{c_{12}\varepsilon_{j+1}^5}.$$

The superscript * at the summation sign means that the summation is expanded over all primes p from the sets $\mathbf{P}_k$, $k = 1, \ldots, K$, corresponding to those $\mathbf{F}_k$ which are linearly independent of the system of vectors $\mathbf{F}_{k_1}, \ldots, \mathbf{F}_{k_j}$.

In the case when $N = 1$ the set T_2 can be covered by the sets defined by the inequalities (7) or by the inequalities (8) and (9). In the case when $N \geqslant 2$ we must include into this covering the set defined by (8) with $j = s$. Let us notice that the distances between the points $\mathbf{u}_1, \ldots, \mathbf{u}_N$ are not smaller than $2\pi/(s!M_1\ldots M_s)$. Therefore, choosing $\varepsilon_{s+1} = 1/(s!M^s)$ we guarantee that only one of these points may belong to the set defined by the inequalities (8) with $j = s$. In addition, the inequality (10) is satisfied as well. And in this case the condition (C) allows the application of Lemma 2 with $\delta = \varepsilon_{s+1}/s$. Hence

$$R_2 \ll \int_{T_2} \alpha^{c_{13}\varepsilon_1^5}\,d\mathbf{u} \ll (2\pi)^s\alpha^{c_{13}\varepsilon_1^5}$$

with $\varepsilon_1 = 1/(2^sM^{s(s+3)/2}1!\ldots s!(s!)^2)$.

Finally, collecting all obtained estimates into (3) we complete the proof of our theorem.

REFERENCES

Kubilius, J. (1964). *Probabilistic Methods in the Theory of Numbers.* American Mathem. Soc., Providence, RI.

Kubilius, J. (1968). On local theorems for additive number-theoretic functions. In: *Abhandlungen aus Zahlentheorie und Analysis (zur Erinnerung an E. Landau).* VEB Deutscher Verlag der Wissenschaften, Berlin, pp. 175–191.

Mačiulis, A. (1983). Application of sums of multiplicative functions to the study of large deviations. *Lithuanian Math. J.* **23**, 196–206.

Manstavičius, E. (1980). Probabilistic number theory. *Lithuanian Math. J.* **20**, 206–216.

Stepanauskas, G. (1980). Local limit theorems for a class of arithmetic functions. *Lithuanian Math. J.* **20**, 165–175.

Stepanauskas, G. (1986). Local distribution law for additive functions on an arithmetic progression. *Lithuanian Math. J.* **26**, 383–389.

Stepanauskas, G. (1987). Remark on a theorem of G. Halász. *Lithuanian Math. J.* **27**, 159–163.

New Trends in Probab. and Statist., pp. 293–310
F. Schweiger and E. Manstavičius (Eds.)
1992 VSP/TEV

THE VON MISES THEOREM IN NUMBER THEORY

J. ŠIAULYS
Department of Mathematics, Vilnius University,
Naugarduko 24, 2006 Vilnius, Lithuania.

ABSTRACT

Let $\{f_x(m),\ x \geqslant 1\}$ be a set of additive functions,

$$\nu_x(f_x(m) < u) = x^{-1}\#\{m\colon\ f_x(m) < u\}.$$

The necessary and sufficient conditions for the weak convergence of the distribution functions $\nu_x(f_x(m) < u)$ to the Poisson law are obtained.

INTRODUCTION

Notations We always denote prime numbers by $p,\ p_1,\ p_2,\ \ldots$. The function $\varepsilon(x)$ always tends to zero, when x tends to infinity. The absolute constants will be denoted by $c,\ c_1,\ c_2,\ \ldots$. We use B to denote a quantity which is bounded by an absolute constant. The expression $a \ll b$ is equivalent to $|a| \leqslant cb$. If the bounding constant depends on a parameter a, we shall write $\varepsilon_a(x),\ c_a,\ B_a,\ \ll_a$. If the distribution functions $F_x(u)$ converge weakly to the limit distribution function $F(u)$, we shall write $F_x(u) \underset{x\to\infty}{\Longrightarrow} F(u)$.
The following statement is well known in the probability theory.

THEOREM M. *Let*

$$\xi_{nk}, \qquad n = 1,2,\ldots, \quad k = 1,2,\ldots,k_n, \quad \lim_{n\to\infty} k_n = \infty,$$

be the array of random variables where for each n the variables

$$\xi_{n1},\ \xi_{n2},\ \ldots,\ \xi_{nk_n}$$

are independent. Let ξ_{nk} take the values 0 *and* 1 *only.*
In order that the distribution laws of the sums

$$\xi_{n1} + \xi_{n2} + \ldots + \xi_{nk_n}$$

converge weakly to the Poisson law as $n \to \infty$, it is necessary and sufficient that

$$\lim_{n\to\infty} \max_{1\leqslant k\leqslant k_n} \mathbb{P}(\xi_{nk}=1)=0,$$

$$\lim_{n\to\infty} \sum_{k=1}^{k_n} \mathbb{P}(\xi_{nk}=1)=\lambda.$$

The proof of this fact is due to (von Mises, 1921). Another, much simpler, proof of this theorem was obtained in (Mačys, 1979).

Now let $\{f_x(m)\colon\ x\geqslant 1\}$ be a set of real valued additive functions,

$$\nu_x\big(f_x(m)<u\big)=\frac{1}{[x]}\sum_{\substack{m\leqslant x\\ f_x(m)<u}} 1.$$

If $f_x(m)=g(m)/B(x)$, where $g(m)$ is an additive function and

$$\lim_{x\to\infty} B(x)=\infty,$$

then the Poisson law cannot be a limit for the sequence of distribution functions $\nu_x(f_x(m)<u)$. P. D. T. A. Elliott suspected that the Poisson law cannot occur as the limit law for distributions $\nu_x(h(m)<u)$ where $h(m)$ is an additive function (see (Elliott, 1980, p. 173)).

Let $\Pi_\lambda(u)$ denote the distribution function of the Poisson law with the parameter $\lambda>0$. In general, the Poisson law can occur as the limit law for distributions $\nu_x(f_x(m)<u)$. For example,

$$\nu_x\big(f_x(m)<u\big)\underset{x\to\infty}{\Longrightarrow}\Pi_{\log 2}(u),$$

where $f_x(m)$ are the strongly additive functions defined by

$$f_x(p)=\begin{cases}1 & \text{if } \log x<p\leqslant \log^2 x,\\ 0 & \text{otherwise.}\end{cases}$$

The aim of this work is to prove the following assertion.

THEOREM 1. *Let $\{f_x(m)\colon\ x\geqslant 1\}$ be a set of strongly additive functions, $f_x(p)\in\{0,1\}$ for each prime number p. Then*

$$\nu_x\big(f_x(m)<u\big)\underset{x\to\infty}{\Longrightarrow}\Pi_\lambda(u) \tag{1}$$

if and only if

$$\lim_{x\to\infty} \max_{\substack{p\leqslant x\\ f_x(p)=1}} \frac{1}{p}=0, \tag{2}$$

$$\lim_{x\to\infty} \sum_{\substack{p\leqslant x\\ f_x(p)=1}} \frac{1}{p}=\lambda, \tag{3}$$

$$\lim_{x\to\infty} \frac{1}{\log x}\sum_{\substack{p\leqslant x\\ f_x(p)=1}} \frac{\log p}{p}=0. \tag{4}$$

PRELIMINARY RESULTS

LEMMA 1 (Halász, 1975). *Let $h(m)$ be an arbitrary real-valued additive function. There is an absolute constant c_1 such that*

$$\sum_{\substack{m\leqslant x\\ h(m)=a}} 1 \leqslant c_1 x\left(\sum_{\substack{p\leqslant x\\ h(p)\neq 0}} \frac{1}{p}\right)^{-1/2}.$$

LEMMA 2 (Levin and Timofeev, 1975). *Let $g(m,x,t)$ be a function, multiplicative in m and satisfying the conditions*

$$|g(m,x,t)| \leqslant 1,$$

$$\sum_{p\leqslant x} \frac{1-\operatorname{Re} g(p,x,t)}{p} \ll 1.$$

Then

$$\frac{1}{x}\sum_{m\leqslant x} g(m,x,t)$$

$$= \frac{1}{2\pi i}\int_{1-iK}^{1+iK} \frac{e^z}{z} \prod_{p\leqslant x^{1/u}} \left(1-\frac{1}{p}e^{-z\frac{\log p}{\log x}}\right)\left(1+\sum_{r=1}^{\infty}\frac{g(p^r,x,t)}{p^r}e^{-z\frac{\log p^r}{\log x}}\right)dz$$

$$+\frac{B}{K^{1/48}} + \varepsilon_K(x)$$

for all $a>0$ and $K>1$ uniformly in u, $0<a\leqslant u\leqslant 1$.

LEMMA 3 (Raikov, 1937). *Let $\varphi_1(t)$ and $\varphi_2(t)$ be characteristic functions such that*

$$\varphi_1(t)\varphi_2(t) = e^{\lambda(e^{it}-1)} \qquad \text{for all} \quad t\in\mathbf{R},$$

where $\lambda>0$. Then

$$\varphi_1(t) = e^{\lambda_1(e^{it}-1)}, \qquad \varphi_2(t) = e^{\lambda_2(e^{it}-1)},$$

with $\lambda_1,\lambda_2\geqslant 0$, $\lambda_1+\lambda_2=\lambda$.

LEMMA 4 (Rusza, 1984). *Let $h(m)$ be an arbitrary complex-valued additive function. If F is a non-negative increasing function satisfying*

$$F(2x)\leqslant c_2 F(x),$$

then we have

$$\frac{1}{x}\sum_{m\leqslant x} F\left(\left|h(m)-\frac{1}{x}\sum_{m\leqslant x} h(m)\right|\right)$$

$$\leqslant c_3\left(F\left(\left(\sum_{p^k\leqslant x}\frac{|h(p^k)|^2}{p^k}\right)^{1/2}\right)+\sum_{p^k\leqslant x}\frac{F(|h(p^k)|)}{p^k}\right),$$

where c_3 depends on c_2.

LEMMA 5 (Fréchet and Shohat, 1931). *Let $F_n(u)$, $n = 1, 2, \dots$, be a sequence of distribution functions. For each non-negative integer l let*

$$\alpha_l = \lim_{n\to\infty} \int_{-\infty}^{+\infty} u^l \, dF_n(u)$$

exist.
Then there is a subsequence F_{n_k}, $n_1 < n_2 < \dots$, which converges weakly to the limiting distribution $F(u)$ satisfying

$$\alpha_l = \int_{-\infty}^{+\infty} u^l \, dF(u), \qquad l = 0, 1, \dots .$$

Moreover, if the set of moments α_l determines $F(u)$ uniquely, then the distributions $F_n(u)$ converge weakly to $F(u)$ as $n \to \infty$.

LEMMA 6 (Elliott, 1980). *Let $\{f_x(m)\colon x \geqslant 1\}$ be a set of real-valued strongly additive arithmetic functions. Let a real function $\alpha(x)$ be defined for $x \geqslant 1$. Let ξ_{xp} be independent random variables, defined for each prime p in the range $2 \leqslant p \leqslant x$, and distributed according to*

$$\xi_{xp} = \begin{cases} f_x(p) & \text{with probability } \dfrac{1}{p}, \\ 0 & \text{with probability } 1 - \dfrac{1}{p}. \end{cases}$$

Then there is a positive absolute constant c_4 such that the inequality

$$\rho\Bigg(\nu_x(f_x(m) - \alpha(x) < u), \mathbf{P}\Big(\sum_{p\leqslant x} \xi_{xp} - \alpha(x) < u\Big)\Bigg)$$
$$\leqslant c_4 \left(\sum_{\substack{x^\varepsilon < p \leqslant x \\ |f_x(p)| > u}} \frac{1}{p} + \frac{u}{\varepsilon} + \exp\Big\{ -\frac{1}{8\varepsilon} \log \frac{1}{\varepsilon} \Big\} + x^{-1/15} \right)$$

holds uniformly for all $u > 0$, $0 < \varepsilon < 1$, $x \geqslant 1$, and functions $f_x(m)$. Here $\rho(F(u), G(u))$ denotes the Lévy distance between distribution functions $F(u)$ and $G(u)$.

THE PROOF OF NECESSITY

To make our formulae more tractable we introduce some notations. For any set of primes $\mathcal{R}(y)$, depending on a parameter y,

$$\sum_{p\in\mathcal{R}(y)}^{*} \quad \text{means} \quad \sum_{\substack{p\in\mathcal{R}(y)\\ f_y(p)=1}},$$

$$\prod_{p\in\mathcal{R}(y)}^{*} \quad \text{means} \quad \prod_{\substack{p\in\mathcal{R}(y)\\ f_y(p)=1}},$$

$$\max_{p\in\mathcal{R}(y)}{}^{*} \quad \text{means} \quad \max_{\substack{p\in\mathcal{R}(y)\\ f_y(p)=1}}.$$

Further, let

$$\varpi(x) = \exp\left\{\sqrt{\log x}\right\},$$

$$\psi_k(x) = 1 - \frac{1}{(\log x)^{1/k}}, \qquad \psi(x) = \psi_2(x),$$

$$\Psi_k(x) = x^{\psi_k(x)}, \qquad \Psi(x) = \Psi_2(x),$$

and

$$P_l = p_1 \cdot \ldots \cdot p_{l-1},$$

$$\Lambda_l(x, p_1, p_2, \ldots, p_{k-l+1}) := \log\left(\frac{1 - \dfrac{\log P_{k-l+2}}{\log x}}{\psi_l(x) - \dfrac{\log P_{k-l+2}}{\log x}}\right),$$

$$\Theta_l(x, p_1, p_2, \ldots, p_{k-l+1}) := \log\frac{\Psi_l(x)}{P_{k-l+2}},$$

for some primes p_1, ..., p_l, $l \geqslant 2$, and $k \geqslant 2$.

Step One. We shall prove the assertion (2) of the theorem. It is clear that

$$\lim_{x\to\infty} \nu_x\big(f_x(m) = 0\big) = e^{-\lambda}.$$

Applying Lemma 1 we see that

$$\sum_{p\leqslant x}^{*} \frac{1}{p} \leqslant c_\lambda \qquad \text{for} \quad x \geqslant 1. \tag{5}$$

Hence

$$\sum_{p\leqslant x} \frac{1 - \operatorname{Re} e^{itf_x(p)}}{p} \leqslant 2c_\lambda \qquad \text{for} \quad x \geqslant 1 \quad \text{and} \quad t \in \mathbf{R}.$$

We apply Lemma 2 with the function

$$g(m,x,t)=e^{itf_x(m)}.$$

In this case we obtain the equality

$$\frac{1}{[x]}\sum_{m\leqslant x}e^{itf_x(m)}=\frac{1}{2\pi i}\int_{1-iK}^{1+iK}\frac{e^z}{z}\prod_{p\leqslant x}\left(1+\frac{e^{itf_x(p)}}{p^{1+(z/\log x)}}\right)dz + \frac{B}{K^{1/48}}+\varepsilon_K(x), \tag{6}$$

where $K\geqslant 1$.
Let for each fixed $x\geqslant 1$

$$\xi_{xp},\qquad p\leqslant x,$$

be independent random variables. Let

$$\mathbb{P}(\xi_{xp}=f_x(p))=\frac{1}{p},\qquad \mathbb{P}(\xi_{xp}=0)=1-\frac{1}{p}.$$

For $u\geqslant c_\lambda$

$$\mathbb{P}\left(\Big|\sum_{p\leqslant\varpi(x)}\xi_{xp}\Big|>u\right)\leqslant\mathbb{P}\left(\Big|\sum_{p\leqslant\varpi(x)}(\xi_{xp}-\mathbf{E}\xi_{xp})\Big|>u-c_\lambda\right)$$
$$\leqslant\frac{1}{(u-c_\lambda)^2}\sum_{p\leqslant\varpi(x)}\mathbf{D}\xi_{xp}=\frac{1}{(u-c_\lambda)^2}\sum_{p\leqslant\varpi(x)}^{*}\frac{1}{p}\leqslant\frac{c_\lambda}{(u-c_\lambda)^2}.$$

Hence, the set of distribution functions

$$\mathbb{P}\left(\sum_{p\leqslant\varpi(x)}\xi_{xp}<u\right)$$

is compact.
Let $f_x^*(m)$ be a strongly additive function which satisfy

$$f_x^*(m)=\sum_{\substack{p|m\\ \varpi(x)<p\leqslant x}}^{*}1.$$

We deduce from our hypothesis (1) that

$$\lim_{u\to\infty}\limsup_{x\to\infty}\nu_x\big(|f_x^*(m)|>u\big)\leqslant\lim_{u\to\infty}\limsup_{x\to\infty}\nu_x\big(|f_x(m)|>u\big)=0.$$

Hence, the set of distribution functions

$$\nu_x\big(f_x^*(m)<u\big)$$

is compact, too.
Let $\{x_k\}$ be an arbitrary sequence of real numbers such that

$$\lim_{k\to\infty} x_k = \infty.$$

There exists a subsequence $\{x_k^*\}$, which satisfy the following conditions:

$$\mathbb{P}\left(\sum_{p\leqslant \varpi(x_k^*)} \xi_{x_k^* p} < u\right) \underset{k\to\infty}{\Longrightarrow} F_1(u), \tag{7}$$

$$\nu_{x_k^*}\left(f_{x_k^*}^*(m) < u\right) \underset{k\to\infty}{\Longrightarrow} F_2(u). \tag{8}$$

Here $F_1(u)$ and $F_2(u)$ are the limiting distribution functions.
If $z = 1 + i\tau$ and $|\tau| \leqslant K$, then

$$\left|\prod_{p\leqslant\varpi(x)}^{*}\left(1+\frac{e^{it}-1}{p^{1+(z/\log x)}}\right) - \prod_{p\leqslant\varpi(x)}^{*}\left(1+\frac{e^{it}-1}{p}\right)\right|$$

$$= \left|\sum_{p\leqslant\varpi(x)}^{*}\left(e^{-z\frac{\log p}{\log x}}-1\right)\frac{e^{it}-1}{p}\prod_{\substack{p_1<p\\ f_x(p_1)=1}}\left(1+\frac{e^{it}-1}{p_1^{1+(z/\log x)}}\right)\right.$$

$$\left.\times \prod_{p<p_1\leqslant x}^{*}\left(1+\frac{e^{it}-1}{p_1}\right)\right| \leqslant 4|z|\prod_{p\leqslant x}^{*}\left(1+\frac{2}{p}\right)\frac{1}{\log x}\sum_{p\leqslant\varpi(x)}^{*}\frac{\log p}{p}$$

$$\leqslant 4e^{2c_\lambda}|z|\frac{1}{\log x}\sum_{p\leqslant\varpi(x)}\frac{\log p}{p} \ll_\lambda \frac{K}{(\log x)^{1/2}}.$$

Now the application of the equality (6) shows that

$$\frac{1}{[x]}\sum_{m\leqslant x} e^{itf_x(m)} = \frac{B}{K^{1/48}} + \varepsilon_K(x)$$

$$+\left(\prod_{p\leqslant\varpi(x)}^{*}\left(1+\frac{e^{it}-1}{p}\right)\right)\frac{1}{2\pi i}\int_{1-iK}^{1+iK}\frac{e^z}{z}\prod_{\varpi(x)<p\leqslant x}^{*}\left(1+\frac{e^{it}-1}{p^{1+(z/\log x)}}\right)dz, \tag{9}$$

where $K \geqslant 1$ and $t \in \mathbf{R}$.
From (7) we obtain

$$\lim_{k\to\infty}\prod_{p\leqslant\varpi(x_k^*)}^{*}\left(1+\frac{e^{it}-1}{p}\right) = \varphi_1(t). \tag{10}$$

Here $\varphi_1(t)$ is the characteristic function of the distribution function $F_1(u)$.

Applying Lemma 2 to the function

$$g(m,x,t) = e^{itf_x^*(m)}$$

we have

$$\frac{1}{[x]}\sum_{m\leqslant x} e^{itf_x^*(m)}$$

$$= \frac{1}{2\pi i}\int_{1-iK}^{1+iK} \frac{e^z}{z} \prod^*_{\varpi(x)<p\leqslant x}\left(1+\frac{e^{it}-1}{p^{1+(z/\log x)}}\right)dz + \frac{B}{K^{1/48}} + \varepsilon_K(x),$$

where $K \geqslant 1$ and $t \in \mathbf{R}$.
The last equality and the condition (8) show that

$$\lim_{K\to\infty}\lim_{k\to\infty}\int_{1-iK}^{1+iK} \frac{e^z}{z} \prod^*_{\varpi(x_k^*)<p\leqslant x_k^*}\left(1+\frac{e^{it}-1}{p^{1+(z/\log x_k^*)}}\right)dz = \varphi_2(t), \tag{11}$$

where $\varphi_2(t)$ is the characteristic function of $F_2(u)$.
By virtue of assertions (9), (10), (11), and hypothesis (1) of the theorem, we obtain the equality

$$e^{\lambda(e^{it}-1)} = \varphi_1(t)\varphi_2(t)$$

for $t \in \mathbf{R}$.
By means of Lemma 3 we have that

$$\varphi_1(t) = e^{\lambda_1(e^{it}-1)}, \qquad \varphi_2(t) = e^{\lambda_2(e^{it}-1)},$$

with $\lambda_1, \lambda_2 \geqslant 0,\ \lambda_1 + \lambda_2 = \lambda$.
Hence, $F_1(u) = \Pi_{\lambda_1}(u)$, and

$$\mathbb{P}\left(\sum_{p\leqslant\varpi(x_k^*)} \xi_{x_k^*p} < u\right) \underset{k\to\infty}{\Longrightarrow} \Pi_{\lambda_1}(u).$$

Now the Theorem M gives that

$$\lim_{k\to\infty}\ \max^*_{p\leqslant\varpi(x_k^*)} \frac{1}{p} = 0.$$

Since $\{x_k\}$ is an arbitrary sequence of real numbers (for which $\lim_{k\to\infty} x_k = \infty$), then

$$\lim_{x\to\infty}\ \max^*_{p\leqslant\varpi(x)} \frac{1}{p} = 0.$$

Now the desired result follows from the last equality.
Step Two. Here we prove the assertion (3) of our theorem.

Applying the inequality (5) and the proposition (2) we have

$$\left|\sum_{p\leqslant x}^{*}\left(\log\left(1+\frac{e^{it}-1}{p^{1+(z/\log x)}}\right)-\frac{e^{it}-1}{p^{1+(z/\log x)}}\right)\right|$$
$$\leqslant \sum_{5<p\leqslant x}^{*}\frac{1}{p^2}+\varepsilon(x)\leqslant \max_{p\leqslant x}{}^{*}\frac{1}{p}\sum_{p\leqslant x}^{*}\frac{1}{p}+\varepsilon(x)=\varepsilon(x)$$

for $z=1+i\tau$ and $|\tau|\leqslant K$.
The last estimate, (1), and (6) show that

$$\begin{aligned}&\lim_{x\to\infty}\frac{1}{2\pi i}\int_{1-iK}^{1+iK}\frac{e^z}{z}\exp\left\{(e^{it}-1)\sum_{p\leqslant x}^{*}\frac{1}{p}e^{-z\frac{\log p}{\log x}}\right\}dz\\&=e^{\lambda(e^{it}-1)}+\frac{B}{K^{1/48}},\end{aligned}\tag{12}$$

where $K\geqslant 1$ and $t\in\mathbf{R}$.
Let $\{x_k\}$ be an arbitrary unbounded increasing sequence of real numbers. There is a subsequence $\{x_k^*\}$ such that the following limits exist:

$$\lim_{k\to\infty}\sum_{p\leqslant x_k^*}^{*}\frac{1}{p}=\mu_0,$$

$$\lim_{k\to\infty}\frac{l}{\log^l x_k^*}\sum_{p\leqslant x_k^*}^{*}\frac{\log^l p}{p}=\mu_l,\qquad l=1,2,\dots.$$

Let

$$\mu(z)=\mu_0+\sum_{l=1}^{\infty}\frac{(-1)^l}{l!l}\mu_l z^l.$$

It follows from (12) that

$$\frac{1}{2\pi i}\int_{1-iK}^{1+iK}\frac{e^z}{z}e^{(e^{it}-1)\mu(z)}\,dz=e^{\lambda(e^{it}-1)}+\frac{B}{K^{1/48}}.\tag{13}$$

Since $0\leqslant\mu_l\leqslant 1$ for $l=1,2,\dots$, we have that the function $\mu(z)$ is analytic on the plane of complex numbers $\mathbf{C}$. Now we shall estimate the function $\mu(z)$. Suppose that $z=\xi+i\eta$, $\xi\geqslant 0$. Since

$$\left|\log\prod_{p\leqslant x}^{*}\left(1+\frac{1}{p^{1+(z/\log x)}}\right)-\sum_{p\leqslant x}^{*}\frac{1}{p^{1+(z/\log x)}}\right|$$
$$\leqslant\sum_{p\leqslant x}^{*}\frac{1}{p^2}\leqslant\max_{p\leqslant x}{}^{*}\frac{1}{p}\sum_{p\leqslant x}^{*}\frac{1}{p},$$

from (2) and (5) we obtain the equality

$$\mu(z) = \lim_{k\to\infty}\left(\sum_{p\leqslant x_k^*}{}^{*}\frac{1}{p} + \sum_{l=1}^{\infty}\frac{(-z)^l}{l!l}\frac{l}{\log^l x_k^*}\sum_{p\leqslant x_k^*}{}^{*}\frac{\log^l p}{p}\right)$$

$$= \lim_{k\to\infty}\sum_{p\leqslant x_k^*}{}^{*}\frac{1}{p^{1+(z/\log x_k^*)}} = \lim_{k\to\infty}\log\prod_{p\leqslant x_k^*}{}^{*}\left(1+\frac{1}{p^{1+(z/\log x_k^*)}}\right).$$

Applying inequalities

$$1+u \leqslant e^u, \qquad u\in\mathbf{R},$$

and

$$1-u \geqslant e^{-eu} \qquad \text{for} \quad 0\leqslant u\leqslant 1/e,$$

we have that

$$\left|\prod_{p\leqslant x}{}^{*}\left(1+\frac{1}{p^{1+((\xi+i\eta)/\log x)}}\right)\right| \leqslant \exp\left\{\sum_{p\leqslant x}{}^{*}\frac{1}{p}\right\} \leqslant e^{c_\lambda},$$

$$\left|\prod_{p\leqslant x}{}^{*}\left(1+\frac{1}{p^{1+((\xi+i\eta)/\log x)}}\right)\right| \geqslant \prod_{p\leqslant x}{}^{*}\left(1-\frac{1}{p}\right) \geqslant \frac{1}{2}\exp\left\{-e\sum_{p\leqslant x}{}^{*}\frac{1}{p}\right\} \geqslant \frac{1}{2}e^{-ec_\lambda}.$$

From these inequalities it follows that

$$|\mu(z)| \ll_\lambda 1 \qquad \text{if} \quad \operatorname{Re} z\geqslant 0.$$

Now let $z=-\xi+i\eta$, $\xi>0$. From the equality

$$\mu(z) = \lim_{k\to\infty}\sum_{p\leqslant x_k^*}{}^{*}\frac{1}{p}e^{-z\log p/\log x_k^*}$$

it is not difficult to derive that

$$|\mu(z)| \leqslant \mu(-\xi).$$

If $\xi\leqslant 1$, then

$$\mu(-\xi) = \mu_0 + \sum_{l=1}^{\infty}\frac{\xi^l}{l!l}\mu_l \ll_\lambda 1 + \sum_{l=1}^{\infty}\frac{\xi^l}{l!l} \leqslant e.$$

If $\xi\geqslant 1$, then

$$\mu(-\xi) \ll_\lambda 1 + \sum_{l=1}^{\infty}\frac{\xi^l}{l!l} = 1 + \int_0^{\xi}\frac{e^u-1}{u}\,du$$

$$= 1 + \int_0^{1}\frac{e^u-1}{u}\,du + \int_1^{\sqrt{\xi}}\frac{e^u-1}{u}\,du + \int_{\sqrt{\xi}}^{\xi}\frac{e^u-1}{u}\,du$$

$$\leqslant e + \frac{e^{\sqrt{\xi}}-1}{2}\log\xi + \frac{e^{\xi}-e^{\sqrt{\xi}}-\xi+\sqrt{\xi}}{\sqrt{\xi}} \ll \frac{e^{\xi}}{\sqrt{\xi}}.$$

From the obtained relations we conclude that

$$|\mu(z)| \ll_\lambda \begin{cases} 1 & \text{if} \quad \operatorname{Re} z \geqslant -1, \\ \dfrac{e^{-\operatorname{Re} z}}{\sqrt{-\operatorname{Re} z}} & \text{if} \quad \operatorname{Re} z < -1. \end{cases} \tag{14}$$

Let $I(K,t)$ be the integral in the equality (13). Then we may write

$$e^{\lambda(e^{it}-1)} = I(K,t) + R(K,t), \qquad |R(K,t)| \ll \frac{1}{K^{1/48}}. \tag{15}$$

From the estimates (14) for function $\mu(z)$ we have

$$\{I(K,t)\}_t' = \frac{1}{2\pi} \int\limits_{1-iK}^{1+iK} \frac{e^z}{z} e^{(e^{it}-1)\mu(z)} e^{it} \mu(z)\, dz,$$

$$\{I(K,t)\}_t'' = \frac{-1}{2\pi i} \int\limits_{1-iK}^{1+iK} \frac{e^z}{z} e^{(e^{it}-1)\mu(z)} \left(\mu^2(z) e^{2it} + \mu(z) e^{it}\right) dz.$$

The equality (15) shows that

$$i\lambda e^{\lambda(e^{it}-1)} e^{it} = \{I(K,t)\}_t' + \{R(K,t)\}_t', \tag{16}$$

$$-e^{\lambda(e^{it}-1)}\left(\lambda^2 e^{2it} + \lambda e^{it}\right) = \{I(K,t)\}_t'' + \{R(K,t)\}_t''. \tag{17}$$

From the estimates (14) it follows that

$$\left|\{I(K,t)\}_t''\right| \ll_\lambda \int\limits_{-K}^{K} \frac{d\tau}{\sqrt{1+\tau^2}} \ll \int\limits_0^1 d\tau + \int\limits_1^K \frac{d\tau}{\tau} = 1 + \log K.$$

Now applying the equality (17) we see that

$$\left|\{R(K,t)\}_t''\right| \ll_\lambda 1 + \log K. \tag{18}$$

Suppose that u is a real number. Since the function $\operatorname{Re} R(K,t)$ has a derivative we may write

$$\operatorname{Re} R(K, t+u) - \operatorname{Re} R(K,t) = u\left(\operatorname{Re} \{R(K,v)\}_{v=t+\theta_1 u}'\right),$$

where $|\theta_1| \leqslant 1$.
From the last equality and the relation (15) it follows that

$$\left|\operatorname{Re} \{R(K,v)\}_{v=t+\theta_1 u}'\right| \ll_\lambda \frac{1}{|u| K^{1/48}}.$$

The function $\mathrm{Re}\,\{R(K,t)\}'_t$ is differentiable. Then applying estimation (18) we obtain that

$$\begin{aligned}&\left|\mathrm{Re}\left\{R(K,v)\right\}'_{v=t+\theta_1 u}-\mathrm{Re}\left\{R(K,t)\right\}'_t\right|\\&=\left|\theta_1 u\mathrm{Re}\left\{R(K,v)\right\}''_{v=t+\theta_1\theta_2 u}\right|\ll_\lambda |u|(1+\log K),\end{aligned}$$

where $|\theta_2|\leqslant 1$.
Let now $u=1/K^{1/96}$. Then we have

$$\begin{aligned}\left|\mathrm{Re}\left\{R(K,t)\right\}'_t\right|&\leqslant\left|\mathrm{Re}\left\{R(K,v)\right\}'_{v=t+\theta_1 u}\right|\\&\quad+\left|\mathrm{Re}\left\{R(K,v)\right\}'_{v=t+\theta_1 u}-\mathrm{Re}\left\{R(K,t)\right\}'_t\right|\\&\ll_\lambda\frac{1}{|u|K^{1/48}}+|u|(1+\log K)\ll\frac{1+\log K}{K^{1/96}}.\end{aligned}$$

It can be found similarly that

$$\left|\mathrm{Im}\left\{R(K,t)\right\}'_t\right|\ll_\lambda\frac{1+\log K}{K^{1/96}}.$$

We see from these inequalities and from (16) that

$$i\lambda e^{\lambda(e^{it}-1)}e^{it}=\left\{I(K,t)\right\}'_t+\frac{B(1+\log K)}{K^{1/96}}$$

and

$$\lambda e^{\lambda(e^{it}-1)}=\frac{1}{2\pi i}\int_{1-iK}^{1+iK}\frac{e^z}{z}e^{(e^{it}-1)\mu(z)}\mu(z)\,dz+\frac{B(1+\log K)}{K^{1/96}},$$

where $t\in\mathbf{R}$ and $K\geqslant 1$.
If $t=0$, we conclude that

$$\lambda=\frac{1}{2\pi i}\int_{1-iK}^{1+iK}\frac{e^z}{z}\mu(z)\,dz+\frac{B(1+\log K)}{K^{1/96}}. \tag{19}$$

Let γ_k be the rectangle with vertices

$$1-iK,\ 1+iK,\ -K+iK,\ -K-iK.$$

The relations (14) imply that the following estimates hold:

$$\begin{aligned}&\left|\frac{1}{2\pi i}\int_{-K+iK}^{1+iK}\frac{e^z}{z}\mu(z)\,dz\right|=\left|\frac{1}{2\pi i}\int_{-K}^{1}\frac{e^{t+iK}}{t+iK}\mu(t+iK)\,dt\right|\\&\ll\int_{-1}^{1}\frac{e^t}{\sqrt{t^2+K^2}}\left|\mu(t+iK)\right|dt+\int_{1}^{K}\frac{e^{-t}}{\sqrt{t^2+K^2}}\left|\mu(-t+iK)\right|dt\\&\ll\frac{1}{K}+\frac{1}{K}\int_1^K\frac{dt}{\sqrt{t}}\ll\frac{1}{\sqrt{K}},\end{aligned}$$

$$\left|\frac{1}{2\pi i}\int\limits_{-K-iK}^{1-iK}\frac{e^z}{z}\mu(z)\,dz\right| \ll \frac{1}{\sqrt{K}},$$

$$\left|\frac{1}{2\pi i}\int\limits_{-K-iK}^{-K+iK}\frac{e^z}{z}\mu(z)\,dz\right| = \left|\frac{1}{2\pi i}\int\limits_{-K}^{K}\frac{e^{-K+it}}{-K+it}\mu(-K+it)\,dt\right|$$

$$\ll \int\limits_{-K}^{K}\frac{e^{-K}}{\sqrt{K^2+t^2}}\frac{e^K}{\sqrt{K}}\,dt \ll \frac{1}{\sqrt{K}}.$$

From these inequalities and from (19), the equality

$$\frac{1}{2\pi i}\int\limits_{\gamma_K}\frac{e^z}{z}\mu(z)\,dz = \lambda + \frac{B(1+\log K)}{K^{1/96}}$$

follows. Because the function $e^z\mu(z)$ is analytic on the complex plane $\mathbf{C}$, we have

$$\frac{1}{2\pi i}\int\limits_{\gamma_K}\frac{e^z}{z}\mu(z)\,dz = \mu_0.$$

The two last equalities show that

$$\lim_{k\to\infty}\sum_{p\leqslant x_k^*}{}^{*}\,\frac{1}{p} = \lambda.$$

Since $\{x_k\}$ is an arbitrary unbounded increasing sequence, the assertion (3) of our theorem holds.

Step Three. Let α_l, $l=1,2,\ldots$, be the moments of the Poisson law with parameter λ. In this part we shall prove the equalities

$$\lim_{x\to\infty}\frac{1}{x}\sum_{m\leqslant x} f_x^l(m) = \alpha_l, \qquad l=1,2,\ldots. \tag{20}$$

If $l=1$, from (3) we have

$$\begin{aligned}\frac{1}{x}\sum_{m\leqslant x} f_x(m) &= \frac{1}{x}\sum_{m\leqslant x}\sum_{\substack{p\mid m\\ f_x(p)=1}} 1 = \frac{1}{x}\sum_{p\leqslant x}{}^{*}\sum_{\substack{m\leqslant x\\ p\mid m}} 1\\ &= \sum_{p\leqslant x}{}^{*}\frac{1}{p} + B\frac{1}{x}\sum_{p\leqslant x} 1 = \lambda + \varepsilon(x).\end{aligned} \tag{21}$$

Now let $l\geqslant 2$. Using Lemma 4 we obtain that

$$\frac{1}{x}\sum_{m\leqslant x}\left|f_x(m) - \frac{1}{x}\sum_{m\leqslant x} f_x(m)\right|^l$$

$$\ll_l \left(\sum_{p\leqslant x}{}^{*}\frac{1}{p}\left(1-\frac{1}{p}\right)^{-1}\right)^{l/2} + \sum_{p\leqslant x}{}^{*}\frac{1}{p}\left(1-\frac{1}{p}\right)^{-1}.$$

We deduce from relations (5), (21), and the inequality

$$(a+b)^l \leqslant 2^l(a^l+b^l), \qquad a,b>0, \quad l\in \mathbf{N},$$

that

$$\frac{1}{x}\sum_{m\leqslant x} f_x^l(m) \ll_l \frac{1}{x}\sum_{m\leqslant x}\left|f_x(m)-\frac{1}{x}\sum_{m\leqslant x} f_x(m)\right|^l + \left|\frac{1}{x}\sum_{m\leqslant x} f_x(m)\right|^l \ll_l 1.$$

Let now $\{x_k\}$ be an arbitrary unbounded increasing sequence of real numbers. The last estimates show that there exists a subsequence $\{x_k^*\}$ for which

$$\lim_{k\to\infty}\frac{1}{x_k^*}\sum_{m\leqslant x_k^*} f_{x_k^*}^l(m) = \beta_l, \qquad l=1,2,\dots . \tag{22}$$

Here β_l, $l=1,2\dots$, are real numbers. By means of Lemma 5 we have that there exists a subsequence $\{x'_k\}$ of the sequence $\{x_k^*\}$ for which

$$\nu_{x'_k}\left(f_{x'_k}<u\right) \underset{k\to\infty}{\Longrightarrow} F(u),$$

where the limiting distribution function $F(u)$ has the moments β_l, $l=1,2,\dots$. According to the hypothesis (1),

$$F(u)=\Pi_\lambda(u).$$

Hence

$$\beta_l=\alpha_l \qquad \text{for} \quad l=1,2,\dots,$$

and

$$\lim_{k\to\infty}\frac{1}{x_k^*}\sum_{m\leqslant x_k^*} f_{x_k^*}^l(m)=\alpha_l, \qquad l=1,2,\dots .$$

Since $\{x_k\}$ is an arbitrary unbounded increasing sequence of real numbers, the assertion (20) holds.
Step Four. In this part we prove that for every natural number $l\geqslant 2$ the relation

$$\lim_{x\to\infty}\sideset{}{^*}\sum_{p_1\leqslant x}\frac{1}{p_1}\sideset{}{^*}\sum_{p_2\leqslant x}\frac{1}{p_2}\cdots\sideset{}{^*}\sum_{p_{l-1}\leqslant x}\frac{1}{p_{l-1}}\sideset{}{^*}\sum_{x/P_l<p_l\leqslant x}\frac{1}{p_l}=0 \tag{23}$$

holds.
The factorial moments of the Poisson law with parameter λ equal to λ^l for $l=1,2,\dots$. Hence, using the assertion (20) we deduce that

$$\lim_{x\to\infty}\frac{1}{x}\sum_{m\leqslant x} f_x(m)(f_x(m)-1)\dots(f_x(m)-l+1)=\lambda^l, \tag{24}$$

where $l=1,2,\dots$.

We shall prove the relation (23) by induction on l.
Suppose that $l = 2$. From the equality (24) we have

$$\frac{1}{x}\sum_{m\leqslant x} f_x(m)\big(f_x(m)-1\big) = \lambda^2 + \varepsilon(x). \tag{25}$$

On the other hand,

$$\frac{1}{x}\sum_{m\leqslant x} f_x(m)\big(f_x(m)-1\big) = \frac{1}{x}\sum_{m\leqslant x}\ \sum_{\substack{p_1|m\\ f_x(p_1)=1}}\ \sum_{\substack{p_2|m\\ f_x(p_2)=1\\ p_2\neq p_1}} 1$$

$$= \sum_{p_1\leqslant x}{}^{*}\ \sum_{\substack{p_2\leqslant x\\ p_2\neq p_1}}{}^{*}\ \frac{1}{x}\sum_{\substack{m\leqslant x\\ p_1p_2|m}} 1 = \sum_{p_1\leqslant x}{}^{*}\ \sum_{p_2\leqslant x/p_1}{}^{*}\ \frac{1}{p_1p_2}$$

$$+ B\sum_{p\leqslant x}{}^{*}\frac{1}{p^2} + \frac{B}{x}\sum_{p_1\leqslant x}{}^{*}\ \sum_{p_2\leqslant x/p_1}{}^{*} 1.$$

Applying the relations (2) and (3) we see that

$$\sum_{p\leqslant x}{}^{*}\frac{1}{p^2} \leqslant \max_{p\leqslant x}{}^{*}\frac{1}{p}\sum_{p\leqslant x}{}^{*}\frac{1}{p} = \varepsilon(x),$$

$$\frac{1}{x}\sum_{p_1\leqslant x}{}^{*}\ \sum_{p_2\leqslant x/p_1}{}^{*} 1 \ll \frac{1}{x}\sum_{p\leqslant x^{\psi(x)}}{}^{*}\frac{x/p}{\log(x/p)} + \frac{1}{x}\sum_{x^{\psi(x)}<p\leqslant x}\frac{x}{p}$$

$$\ll \frac{1}{\sqrt{\log x}}\sum_{p\leqslant x}{}^{*}\frac{1}{p} + \log\frac{1}{\psi(x)} + \frac{1}{\psi(x)\log x} = \varepsilon(x).$$

Therefore, we deduce the equality

$$\frac{1}{x}\sum_{m\leqslant x} f_x(m)\big(f_x(m)-1\big) = \sum_{p_1\leqslant x}{}^{*}\ \sum_{p_2\leqslant x/p_1}{}^{*}\ \frac{1}{p_1p_2} + \varepsilon(x)$$

$$= \lambda^2 - \sum_{p_1\leqslant x}{}^{*}\frac{1}{p_1}\sum_{x/p_1<p_2\leqslant x}{}^{*}\frac{1}{p_2} + \varepsilon(x).$$

The last equality and the relation (25) show that the assertion (23) holds for $l = 2$.
Suppose now that the equality (23) is valid for $l = 1, 2, \ldots, k$. It is clear that

$$\frac{1}{x}\sum_{m\leqslant x} f_x(m)\big(f_x(m)-1\big)\ldots\big(f_x(m)-k\big)$$

$$= \frac{1}{x}\sum_{m\leqslant x}\ \sum_{\substack{p_1|m\\ f_x(p_1)=1}}\ \sum_{\substack{p_2|m\\ f_x(p_2)=1\\ p_2\neq p_1}}\ \cdots \sum_{\substack{p_{k+1}|m\\ f_x(p_{k+1})=1\\ p_{k+1}\neq p_1,p_2,\ldots,p_k}} 1$$

$$= \frac{1}{x}\sum_{p_1\leqslant x}{}^{*}\ \sum_{p_2\leqslant x/p_1}{}^{*}\cdots \sum_{\substack{p_{k+1}\leqslant x/P_{k+1}\\ p_{k+1}\neq p_1,p_2,\ldots,p_k}}{}^{*}\left(\frac{x}{P_{k+2}} + B\right).$$

From (2) and (3) we see that for every pair $i, j,\ 1 \leqslant i < j \leqslant k+1$, the estimate

$$\sum_{p_1 \leqslant x}^{*} \sum_{p_2 \leqslant x}^{*} \cdots \sum_{p_i \leqslant x}^{*} \cdots \sum_{\substack{p_j \leqslant x \\ p_j = p_i}}^{*} \cdots \sum_{p_{k+1} \leqslant x}^{*} \frac{1}{P_{k+2}} = \varepsilon(x)$$

is valid.

Similarly to the case of $l = 2$, we have

$$\frac{1}{x} \sum_{p_1 \leqslant x}^{*} \sum_{p_2 \leqslant x/p_1}^{*} \cdots \sum_{p_{k+1} \leqslant x/P_{k+1}}^{*} 1$$

$$\leqslant \frac{1}{x} \sum_{p_1 \leqslant x}^{*} \sum_{p_2 \leqslant x/p_1}^{*} \cdots \sum_{p_k \leqslant \frac{\Psi(x)}{P_k}}^{*} \sum_{p_{k+1} \leqslant \frac{x}{P_{k+1}}}^{*} 1$$

$$+ \frac{1}{x} \sum_{p_1 \leqslant x}^{*} \sum_{p_2 \leqslant x/p_1}^{*} \cdots \sum_{\frac{\Psi(x)}{P_k} \leqslant p_k \leqslant \frac{x}{P_k}}^{*} \sum_{p_{k+1} \leqslant \frac{x}{P_{k+1}}}^{*} 1$$

$$\ll \frac{1}{x} \sum_{p_1 \leqslant x}^{*} \sum_{p_2 \leqslant x/p_1}^{*} \cdots \sum_{p_k \leqslant \frac{\Psi(x)}{P_k}}^{*} \frac{x/P_{k+1}}{\log(x/P_{k+1})}$$

$$+ \frac{1}{x} \sum_{p_1 \leqslant x}^{*} \sum_{p_2 \leqslant x/p_1}^{*} \cdots \sum_{\frac{\Psi(x)}{P_k} < p_k \leqslant \frac{x}{P_k}}^{*} \frac{x}{P_{k+1}}$$

$$\ll \varepsilon(x) + \sum_{p_1 \leqslant x}^{*} \frac{1}{p_1} \cdots \sum_{p_{k-1} \leqslant \frac{\Psi_3(x)}{P_{k-1}}}^{*} \frac{1}{p_{k-1}} \sum_{\frac{\Psi(x)}{P_k} < p_k \leqslant \frac{x}{P_k}}^{*} \frac{1}{p_k}$$

$$+ \sum_{p_1 \leqslant x}^{*} \frac{1}{p_1} \cdots \sum_{\frac{\Psi_3(x)}{P_{k-1}} < p_{k-1} \leqslant \frac{x}{P_{k-1}}}^{*} \frac{1}{p_{k-1}} \sum_{\frac{\Psi(x)}{P_k} < p_k \leqslant \frac{x}{P_k}}^{*} \frac{1}{p_k}$$

$$= \varepsilon(x) + \sum_{p_1 \leqslant x}^{*} \frac{1}{p_1} \cdots \sum_{p_{k-1} \leqslant \frac{\Psi_3(x)}{P_{k-1}}}^{*} \frac{1}{p_{k-1}}$$

$$\times \left(\Lambda_2(x, p_1, p_2, \ldots, p_{k-1}) + \frac{B}{\Theta_2(x, p_1, p_2, \ldots, p_{k-1})} \right)$$

$$+ \sum_{p_1 \leqslant x}^{*} \frac{1}{p_1} \sum_{p_2 \leqslant x/p_1}^{*} \frac{1}{p_2} \cdots \sum_{\frac{\Psi_3(x)}{P_{k-1}} < p_{k-1} \leqslant \frac{x}{P_{k-1}}}^{*} \frac{1}{p_{k-1}} \sum_{\frac{\Psi(x)}{P_k} < p_k \leqslant \frac{x}{P_k}}^{*} \frac{1}{p_k}$$

$$\ll \varepsilon(x) + \sum_{p_1 \leqslant x}^{*} \frac{1}{p_1} \cdots \sum_{p_{k-2} \leqslant \frac{\Psi_4(x)}{P_{k-2}}}^{*} 1$$

$$\times\left(\Lambda_3(x,p_1,p_2,\ldots,p_{k-2})+\frac{B}{\Theta_3(x,p_1,p_2,\ldots,p_{k-2})}\right)$$
$$+\sum_{p_1\leqslant x}^{*}\frac{1}{p_1}\sum_{p_2\leqslant x/p_1}^{*}\frac{1}{p_2}\cdots\sum_{\frac{\Psi_4(x)}{P_{k-2}}<p_{k-2}\leqslant\frac{x}{P_{k-2}}}^{*}\frac{1}{p_{k-2}}\sum_{\frac{\Psi_3(x)}{P_{k-1}}<p_{k-1}\leqslant\frac{x}{P_{k-1}}}^{*}\frac{1}{p_{k-1}}$$
$$\times\sum_{\frac{\Psi(x)}{P_k}<p_k\leqslant\frac{x}{P_{k+1}}}^{*}\frac{1}{p_k}=\ldots=\varepsilon(x)$$
$$+\sum_{\Psi_{k+1}(x)<p_1\leqslant x}^{*}\frac{1}{p_1}\sum_{\frac{\Psi_k(x)}{p_1}<p_2\leqslant\frac{x}{p_1}}^{*}\frac{1}{p_2}\cdots\sum_{\frac{\Psi(x)}{P_k}<p_k\leqslant\frac{x}{P_k}}^{*}\frac{1}{p_k}=\varepsilon(x).$$

Hence,

$$\frac{1}{x}\sum_{m\leqslant x}f_x(m)(f_x(m)-1)\ldots(f_x(m)-k)$$
$$=\varepsilon(x)+\sum_{p_1\leqslant x}^{*}\frac{1}{p_1}\sum_{p_2\leqslant\frac{x}{p_1}}^{*}\frac{1}{p_2}\cdots\sum_{p_{k+1}\leqslant x/P_{k+1}}^{*}\frac{1}{p_{k+1}}.$$

From the induction hypothesis we obtain that

$$\frac{1}{x}\sum_{m\leqslant x}f_x(m)(f_x(m)-1)\ldots(f_x(m)-k)$$
$$=\sum_{p_1\leqslant x}^{*}\sum_{p_2\leqslant x}^{*}\cdots\sum_{p_{k+1}\leqslant x}^{*}\frac{1}{p_1p_2\cdots p_{k+1}}+\varepsilon(x)$$
$$-\sum_{p_1\leqslant x}^{*}\frac{1}{p_1}\sum_{p_2\leqslant x}^{*}\frac{1}{p_2}\cdots\sum_{p_k\leqslant x}^{*}\frac{1}{p_k}\sum_{\frac{x}{P_{k+1}}<p_{k+1}\leqslant x}^{*}\frac{1}{p_{k+1}}.$$

Using the relation (24) for $l=k+1$ we see that the assertion (23) is valid if $l=k+1$. Hence, the relation (23) is proved.

Step Five. In this section we prove the assertion (4) of our theorem. It is clear that

$$\sum_{x^{1/l}<p_1\leqslant x}^{*}\frac{1}{p_1}\cdots\sum_{x^{1/l}<p_{l-1}\leqslant x}^{*}\frac{1}{p_{l-1}}\sum_{x^{1/l}<p_l\leqslant x}^{*}\frac{1}{p_l}$$
$$\leqslant\sum_{p_1\leqslant x}^{*}\frac{1}{p_1}\cdots\sum_{p_{l-1}\leqslant x}^{*}\frac{1}{p_{l-1}}\sum_{\frac{x}{P_l}<p_l\leqslant x}^{*}\frac{1}{p_l},$$

where $l\geqslant 2$. Therefore, we have for every natural number $l\geqslant 2$

$$\lim_{x\to\infty}\sum_{x^{1/l}<p\leqslant x}^{*}\frac{1}{p}=0.$$

Now the assertion (4) follows from the inequality

$$\lim_{x\to\infty}\frac{1}{\log x}\sum_{p\leqslant x}{}^{*}\frac{\log p}{p}\ll\varepsilon+\lim_{x\to\infty}\sum_{x^{\varepsilon}<p\leqslant x}{}^{*}\frac{1}{p},$$

which is valid for an arbitrary small real positive ε.

THE PROOF OF SUFFICIENCY

From the assertions (2), (3), and the Theorem M we see that

$$\mathbf{P}\left(\sum_{p\leqslant x}\xi_{xp}<u\right)\underset{x\to\infty}{\Longrightarrow}\Pi_{\lambda}(u).$$

Here ξ_{xp} are the random variables defined in the Step One of the proof of necessity of our theorem. The assertion (4) shows that

$$\lim_{x\to\infty}\sum_{x^{\varepsilon}<p\leqslant x}{}^{*}\frac{1}{p}=0$$

for an arbitrary small positive number ε. According to Lemma 6 we have

$$\rho\Big(\nu_x(f_x(m)<u),\mathbf{P}\Big(\sum_{p\leqslant x}\xi_{xp}<u\Big)\Big)$$
$$\ll\sum_{x^{\varepsilon}<p\leqslant x}{}^{*}\frac{1}{p}+\varepsilon+\exp\left\{-\frac{1}{8\varepsilon}\log\frac{1}{\varepsilon}\right\}+x^{-1/15}.$$

Since ε may be chosen arbitrary small, we have completed the proof of our theorem.

REFERENCES

Elliott, P. D. T. A. (1980). *Probabilistic Number Theory*, II. Springer-Verlag, Berlin, Heidelberg, New York.

Fréchet, M. and Shohat, J. (1931). A proof of the generalized central limit theorem. *Trans. Amer. Math. Soc.* **33**, 533–543.

Halász, G. (1975). On the distribution of additive arithmetical functions. *Acta Arithm.* **27**, 143–152.

Levin, B. V. and Timofeev, N. M. (1975). On the distribution of values of additive function. *Acta Arithm.* **26**, 4, 333–364.

Mačys, J. (1979). On generalized Bernoulli trials. *Liet. Matem. Rink.* **19**, 4, 145–151 (in Russian).

von Mises, R. (1921). Über Warscheinlichkeit seltener Ereignisse. *Z. angew. Math. Mech.* **1**, 121–124.

Raikov, D. A. (1937). On a decomposition of the Poisson law. *Dokl. Akad. Nauk SSSR* **14**, 9–12 (in Russian).

Rusza, I. Z. (1984). Generalized moments of additive functions. *J. Number Theory* **18**, 27–33.

New Trends in Probab. and Statist., pp. 311–315
F. Schweiger and E. Manstavičius (Eds.)
1992 VSP/TEV

THE LAW OF LARGE NUMBERS IN THE CASE OF SHIFTED PRIMES

N. M. TIMOFEEV
Lenin avenue 41–63,
600015 Vladimir, Russia

ABSTRACT

The weak law of large numbers for a normalized additive arithmetical function on the sequence of shifted primes is considered. In the case when the normalizing sequence satisfies some extra requirement, the paper presents the necessary and sufficient conditions for the law of large numbers to hold.

Let $g(n)$ be a real-valued additive function, $A(x)$ and $B(x)$ denote real-valued functions such that $A(x) = A([x])$, $B(x) = B([x]) > 0$, $x \geqslant 1$. The central problem in the probabilistic theory of numbers is to determine when the additive function $g(n)$ can be normalized by the functions $A(x)$ and $B(x)$ so that the frequencies

$$[x]^{-1}\operatorname{card}\{n\colon\ n \leqslant x,\ g(n) - A(x) \leqslant uB(x)\} \tag{1}$$

converge weakly as $x \to \infty$. We recall that distribution functions $F_x(u)$ converge weakly ($\Rightarrow$) to a distribution function $F(u)$ if $F_x(u) \to F(u)$ as $x \to \infty$ for every point u at which $F(u)$ is continuous. The problem of characterization of the class of limiting distributions is of great interest too.

The necessary and sufficient conditions for the existence of a limiting distribution, in the cases when $B(x) \not\to \infty$ or $B(x) \to \infty$ and $B(x^y)/B(x) \to 1$ as $x \to \infty$ for each $y > 0$, had been found before 1980. In (Hildebrand, 1989; Timofeev, 1990, 1991) these results were extended for additive functions on the set of shifted primes.

Let $\pi(x)$ be the number of primes not exceeding x, let p denote a prime number and $a \neq 0$ be an integer. Let $u^* = u$ if $|u| \leqslant 1$ and $u^* = 1$ if $|u| > 1$. Denote by $E(u)$ the distribution function of the improper law. In the present paper we shall prove the following theorem.

THEOREM. *Let $B(x)$ be a positive unbounded function satisfying the condition*

$$\max_{x \leqslant y \leqslant x^2} B(y) \leqslant cB(x) \qquad \text{for} \quad x \geqslant 1 \tag{2}$$

with some constant $c > 0$. In order that there be a function $A(x)$ such that

$$\pi(x)^{-1}\operatorname{card}\{p\colon\ p \leqslant x,\ g(p+a) - A(x) \leqslant uB(x)\} \Longrightarrow E(u) \tag{3}$$

it is both necessary and sufficient that there exist a sequence of real numbers $\lambda(x)$ such that

$$\lim_{x\to\infty}\sum_{p\leqslant x}\frac{1}{p}\Big(\frac{g(p)-\lambda(x)\log p}{B(x)}\Big)^{*^2}=0. \tag{4}$$

When the conditions (2) *and* (4) *are satisfied, one may choose*

$$A(x)=\lambda(x)\log x+B(x)\sum_{p\leqslant x}\frac{1}{p}\Big(\frac{g(p)-\lambda(x)\log p}{B(x)}\Big)^{*}.$$

The necessary and sufficient conditions on a real additive function $g(n)$ for the frequencies (1) to approach the improper law as $x\to\infty$, with the side condition $B(x^y)=O(B(x))$ for each fixed positive y, had been obtained by P. D. T. A. Elliott (1980) (see also (Manstavičius, 1976)). In complete generality, i.e. for an arbitrary sequence of additive functions $g_x(n)$, this problem was solved by I. Z. Ruzsa (1979). Nevertheless, for an additive function on the set of shifted primes we are able to prove substantial results concerning the convergence in (3) only when $B(x)$ satisfies the condition (2).

We shall use the following auxiliary results.

LEMMA 1 (Timofeev, 1990, 1991). *For every additive function $g(n)$, we have*

$$\sup_b \pi(x)^{-1}\mathrm{card}\,\{p\colon\ p\leqslant x,\ b<g(p+a)\leqslant b+1\}\leqslant c_1\frac{\log^2(W(x)+2)}{\sqrt{W(x)}},$$

where

$$W(x)=\min_\lambda\Big(\lambda^2+\sum_{p\leqslant x}\frac{1}{p}(g(p)-\lambda\log p)^{*^2}\Big).$$

The constant c_1 depends on a only.

LEMMA 2 (Timofeev, 1983). *Let $f(n)$ be a complex-valued multiplicative function such that $|f(n)|\leqslant 1$ and*

$$\sum_{p\leqslant x}|f(p)-1|\frac{\log p}{p}\leqslant\varepsilon(x)\log x, \tag{5}$$

where $\varepsilon(x)\downarrow 0$ as $x\to\infty$. Then

$$\frac{1}{\pi(x)}\sum_{p\leqslant x}f(p+a)=\prod_{\substack{p\leqslant x\\ p\nmid a}}\Big(1+\sum_{r=1}^{\infty}\frac{f(p^r)-f(p^{r-1})}{p^r-p^{r-1}}\Big)+O\Big(\sqrt{\varepsilon(x)}+\frac{1}{\sqrt{\log x}}\Big).$$

The constant implied in $O(\ldots)$ depends on a only.

Proof of Theorem. Suppose that the relation (3) holds. Then we have

$$\lim_{x\to\infty}\pi(x)^{-1}\mathrm{card}\,\{p\colon\ p\leqslant x,\ |g(p+a)-A(x)|>\varepsilon B(x)\}=0 \tag{6}$$

for each $\varepsilon > 0$. This implies that (6) holds with $\varepsilon = \eta(x) \downarrow 0$ as $x \to \infty$. We may assume that $\eta(x)B(x) \to \infty$ as $x \to \infty$.
Let

$$z = z(x) = \sup\left\{y\colon\ y \geqslant x,\ \sqrt{\eta(y)}B(y) \leqslant B(x)\right\}.$$

In virtue of the condition (2) we have $(\log z)/\log x \to \infty$ as $x \to \infty$. From the definitions of $\eta(x)$ and $z(x)$ we see that

$$\lim_{y\to\infty} \pi(y)^{-1}\operatorname{card}\{p\colon\ p \leqslant y,\ |g(p+a) - A(y)| \leqslant \delta B(x)\} = 1 \tag{7}$$

for each $\delta > 0$ and $x \leqslant y \leqslant z$. Hence, applying Lemma 1 we deduce that there are the constant $c_2 > 0$ and the function $\lambda(x)$ such that

$$\sum_{p\leqslant z} \frac{1}{p}\left(\frac{g(p) - \lambda(x)\log p}{B(x)}\right)^{*^2} \leqslant c_2. \tag{8}$$

Now we shall show that $\lambda(x)/B(x) \to 0$ as $x \to \infty$. While $B(x) \to \infty$ as $x \to \infty$, there is a function $\rho(x) \to \infty$ such that

$$\sum_{p\leqslant\rho(x)} \frac{1}{p}\left(\frac{g(p)}{B(x)}\right)^{*^2} \longrightarrow 0 \qquad \text{as} \quad x \to \infty.$$

Using the inequality

$$(u_1 + u_2)^{*^2} \leqslant 2u_1^{*^2} + 2u_2^{*^2}, \qquad u_1, u_2 \in \mathbf{R},$$

and (8) we obtain

$$\sum_{p\leqslant\rho(x)} \frac{1}{p}\left(\frac{\lambda(x)}{B(x)}\log p\right)^{*^2} \leqslant 3c_2.$$

Thus, $\lambda(x) = o(B(x))$ as $x \to \infty$.
Suppose $z_0 = z_0(x)$ is such that

$$\frac{\log z_0}{\log x} \to \infty \quad \text{and} \quad \frac{\log z}{\log z_0} \to \infty \qquad \text{as} \quad x \to \infty.$$

Let $\Delta = \log((\log z)/\log z_0)$. We divide the interval $(z_0, z]$ into subintervals

$$(z_i, z_{i+1}] \qquad \text{where} \quad z_{i+1} = z_i^{\Delta}, \quad i = 0, 1, \dots, s-1.$$

The number s of the subintervals $(z_i, z_{i+1}] \subset (z_0, z]$ equals to $[(\log z)/(\Delta \log z_0)]$ and tends to infinity as $x \to \infty$. Hence, applying (8) we obtain that there are functions $y_1 = y_1(x)$ and $y_2 = y_2(x)$ such that

$$z_0 \leqslant y_1 < y_2 \leqslant z, \qquad (\log y_2)/\log y_1 \to \infty,$$

and

$$\sum_{y_1<p\leqslant y_2} \frac{1}{p}\left(\frac{g(p) - \lambda(x)\log p}{B(x)}\right)^{*^2} \longrightarrow 0 \tag{9}$$

as $x \to \infty$.
Let us consider the multiplicative function

$$f(n) = \exp\left\{i\xi \frac{g(n) - \lambda(x)\log n}{B(x)}\right\}.$$

Using the inequality

$$\left|e^{i\xi u} - 1\right| = 2|\sin(\xi u/2)| \leqslant 2(|\xi| + 1)|u|^{*}$$

and (9), we have

$$\sum_{p\leqslant y_2} |f(p) - 1|\frac{\log p}{p} \leqslant 4(|\xi|+1)\left(\sum_{p\leqslant y_2}\frac{\log p}{p}\left(\frac{g(p)-\lambda(x)\log p}{B(x)}\right)^{*^2}\right)^{1/2}\sqrt{\log y_2}$$

$$\leqslant 4(|\xi|+1)\left(\frac{2\log y_1}{\log y_2} + \sum_{y_1<p\leqslant y_2}\frac{1}{p}\left(\frac{g(p)-\lambda(x)\log p}{B(x)}\right)^{*^2}\right)^{1/2}\log y_2$$

$$= o(\log y_2(x)), \qquad x \to \infty.$$

Therefore the condition (5) of Lemma 2 is satisfied. Since $B(x) \to \infty$ as $x \to \infty$, this lemma implies

$$\frac{1}{\pi(y_2)}\sum_{p\leqslant y_2}\exp\left\{i\xi\frac{g(p+a)-A(y_2)}{B(x)}\right\}$$

$$= \exp\left\{\sum_{p\leqslant y_2}\frac{1}{p}\left(\exp\left\{i\xi\frac{g(p)-\lambda(x)\log p}{B(x)}\right\}-1\right)\right. \tag{10}$$

$$\left. - i\xi\frac{A(y_2)-\lambda(x)\log x}{B(x)}\right\} + o(1)$$

uniformly in ξ on each interval $[-C, C]$, $C > 0$. In virtue of (7) the left-hand side of (10) is equal to $1 + o(1)$ uniformly in ξ on each interval $[-C, C]$. Thus,

$$\sum_{p\leqslant y_2}\frac{1}{p}\big(1 - \cos\xi\alpha(x,p)\big) = o(1), \qquad x \to \infty,$$

uniformly in ξ, where

$$\alpha(x,p) = \big(g(p) - \lambda(x)\log p\big)/B(x).$$

Since $1 - \alpha^{-1}\sin\alpha \geqslant c_3\alpha^{*^2}$, $c_3 > 0$, we obtain

$$c_3\sum_{p\leqslant y_2}\frac{1}{p}\alpha(x,p)^{*^2} \leqslant \int_0^1\sum_{p\leqslant y_2}\frac{1-\cos\xi\alpha(x,p)}{p}\,d\xi = o(1)$$

where $y_2 \geqslant x$. The equality (4) is proved.
When the condition (4) is satisfied, applying Lemma 2 for the same multiplicative function $f(n)$, we deduce (10) with x in the place of y_2. The choice of the function $A(x)$ given in Theorem, due to the inequality

$$\left|e^{i\xi\alpha} - 1 - i\xi\alpha\right| \leqslant c_4\left(|\xi|^2 + 1\right)\alpha^{*^2}, \qquad \alpha \in \mathbf{R},$$

leads us to the estimate

$$\frac{1}{\pi(x)} \sum_{p \leqslant x} \exp\left\{i\xi \frac{g(p+a) - A(x)}{B(x)}\right\} = 1 + o(1)$$

uniformly in ξ, $|\xi| \leqslant C$. The proof of Theorem is completed.

Acknowledgment. The author is indebted to the referee for many valuable corrections.

REFERENCES

Elliott, P. D. T. A. (1980). *Probabilistic Number Theory, II.* Springer-Verlag, Berlin, Heidelberg, New York.

Hildebrand, A. (1989). Additive and multiplicative functions on shifted primes. *Proc. London Math. Soc.* **53**, 209–232.

Manstavičius, E. (1976). The degenerate distribution for additive arithmetic functions. *Liet. Matem. Rink.* **16**, 194–195 (in Russian).

Ruzsa, I. Z. (1979). The law of large numbers for additive functions. *Stud. Scient. Math. Hung.* **14**, 247–253.

Timofeev, N. M. (1983). Distribution of values of additive functions on the sequence $\{p+1\}$. *Math. Notes* **33**, 478–483.

Timofeev, N. M. (1990). On the distribution of additive arithmetical functions on the set of shifted primes. In: *Probab. Theory and Math. Statist. Proc. Fifth Vilnius Conf., II.* VSP BV/Mokslas, Utrecht, Vilnius, pp. 495–504.

Timofeev, N. M. (1991). The Erdös–Kubilius conjecture on the distribution of the values of additive functions on the sequence of shifted primes. *Acta Arithm.* **58**, 2, 113–131 (in Russian).

Part Six

ZETA-FUNCTIONS AND ALGEBRAIC NUMBER THEORY

New Trends in Probab. and Statist., pp. 319–334
F. Schweiger and E. Manstavičius (Eds.)
1992 VSP/TEV

ON SOME FORMULAE IN ANALYTIC NUMBER THEORY. II

T. FRYSKA
Institute of Mathematics, A. Mickiewicz University,
Matejki 48/49, Poznań, Poland

ABSTRACT

Exact formulae for functions related to the nontrivial zeros of the Hecke-Landau function in an algebraic number field are considered. The paper continues author's investigation in (Fryska, 1989).

1. INTRODUCTION

In this paper an application of the theory developed in (Fryska, 1989) is presented. We deal with so-called complex explicit formulae of the analytic number theory. An arithmetical description of some function defined in terms of the nontrivial zeros of the Hecke function $\zeta_{\mathbf{K}}(s,\chi)$ is obtained.
The classical exact Riemann–Mangoldt formula is one of examples in the subject. The Riemann–Mangoldt formula reads as

$$\psi_0(x,\chi) = x - \sum_{\rho} \frac{x^{\rho}}{\rho} - \log(2\pi) - \frac{1}{2}\log\left(1 - x^{-2}\right), \tag{1.1}$$

where ρ runs over all complex zeros of the Riemann ζ-function, and

$$\sum_{\rho} \frac{x^{\rho}}{\rho} = \lim_{T\to\infty} \sum_{|\mathrm{Im}\,\rho|\leqslant T} \frac{x^{\rho}}{\rho}.$$

The function

$$f(x) = \sum_{\rho} \frac{x^{\rho}}{\rho}$$

also falls into the scope of our interest. On the left-hand side of (1.1) we have the Chebyshev function

$$\psi_0(x) = \frac{1}{2}\left(\psi(x+0) + \psi(x-0)\right),$$

where

$$\psi(x) = \sum_{n\leqslant x} \Lambda(n) = \sum_{p^k\leqslant x} \log p \tag{1.2}$$

and p is a prime number.
Now it follows from the formula (1.1) that the function $f(x)$ is continuous for $x > 1$ except the points $x_k = p^k$ where it has the discontinuities of the first kind with the corresponding jumps equal to $\log p$. It turns out that similar description (expressed in arithmetical terms) of singularities can be given for a fairly large class of functions of the complex variable. Such description is just what we call "complex exact formula". The fact that the variable in these formulae is complex leads us to new applications and some interesting results.
One can find the introduction to this approach as well as its applications in the series of papers of J. Kaczorowski (see (Kaczorowski, 1990) and forthcoming papers). J. Kaczorowski deals with functions which are defined in terms of the zeros of the Dirichlet L-functions. Other exact formulae related to the Möbius function have been investigated by K. Bartz (see *Acta Arithmetica* **56**). In (Fryska, 1989) I have considered a formula of such kind in the case of the Hecke–Landau function $\zeta(s,\chi)$ in an algebraic number field $\mathbf{K}$. Now I shall show how the classical expression (1.1) follows from this formula. Such formulae related to the Dedekind zeta-function are considered in (Fryska, 1991).

2. NOTATIONS

In the paper the following notations are used.

$\mathbf{K}$ is a fixed algebraic number field,

r_1 and r_2 are numbers of real and complex fields conjugate to $\mathbf{K}$,

$n = r_1 + 2r_2$ is the degree of the field $\mathbf{K}$,

Δ is the discriminant of $\mathbf{K}$,

$\mathfrak{a}$ is an integer ideal in $\mathbf{K}$,

$\mathfrak{p}$ is a prime ideal,

$\mathfrak{f}$ is a fixed integer ideal,

$N(\mathfrak{a})$ is the norm of the ideal a,

Λ is the generalized Mangoldt function on $\mathbf{K}$,

$$\Lambda(a) = \begin{cases} \log N(\mathfrak{p}) & \text{for } a = p^k, \quad k = 1, 2 \dots, \\ 0 & \text{otherwise,} \end{cases}$$

χ is a character mod $\mathfrak{f}$ proper in the narrow sense,

$d_0 = d_0(\chi)$ is equal to 1 for the principal character and to 0 for others,

$q = q(\chi)$ is an integer, $0 \leqslant q \leqslant r_1$,

$s = \sigma + it$ is a complex number, as usual in the analytic number theory,

$\zeta(s,\chi)$ is the Hecke–Landau zeta-function which is defined for $\sigma > 1$ by the uniformly convergent series

$$\zeta(s,\chi) = \sum_{\mathfrak{a}} \chi(\mathfrak{a})(N(\mathfrak{a}))^{-s}$$

(see (Landau, 1919)),

$\rho = \beta + i\gamma$ is a nontrivial zero of $\zeta(s, \chi)$, $0 < \beta < 1$, and so are ρ_χ, β_χ, and γ_χ,

γ is the Euler constant,

Γ is the Euler function,

$\lambda(w)$ is some function (not the same in different places) regular on the domain considered.

3. FORMULATION OF THE MAIN RESULTS

By analogy to (1.2) we introduce the notations

$$\psi_0(x, \chi) = \sum_{N(\mathfrak{a}) \leqslant x}{}' \chi(\mathfrak{a})\Lambda(\mathfrak{a})$$

and

$$\tilde{\psi}(x, \chi) = \sum_{N(\mathfrak{a}) \leqslant x}{}' \frac{\chi(\mathfrak{a})\Lambda(\mathfrak{a})}{N(\mathfrak{a})},$$

where the 'prime' at the sum means that in case of x being integer only a half of the last term should be taken.
In the notations of § 2 we have the following two formulae for $x > 1$:

$$\begin{aligned}\psi_0(x, \chi) = \sum_\rho \frac{1 - x^\rho}{\rho} &+ d_0(x + \log x - 1) - (r_1 + r_2 - q)\log x \\ &+ \left(q - \frac{r_1}{2}\right)\log\frac{1 + x^{-1}}{2} - \frac{n}{2}\log\left(1 - x^{-1}\right) \\ &- \frac{n}{2}\gamma - \frac{1}{2}\log\frac{(2\pi)^n}{|\Delta| N(\mathfrak{f})},\end{aligned} \tag{A}$$

$$\begin{aligned}\tilde{\psi}(x, \bar{\chi}) = \sum_\rho \frac{x^{-\rho} - 1}{\rho} &+ d_0\left(\log x + 1 - \frac{1}{x}\right) \\ &- \left(q - \frac{r_1}{2}\right)\log\frac{1 + x^{-1}}{2} - \frac{n}{2}\log\left(1 - x^{-1}\right) \\ &- \frac{n}{2}\gamma - \frac{1}{2}\log\frac{(2\pi)^n}{|\Delta| N(\mathfrak{f})}.\end{aligned} \tag{B}$$

The sums in the above formulae are taken over all zeros $\rho = \beta + i\gamma$, with $0 < \beta < 1$, of the zeta-function $\zeta(s, \chi)$ mod $\mathfrak{f}$. The convergence of these sums is understood in the sense of

$$\lim_{T \to \infty} \sum_{|\rho| \leqslant T}.$$

4. FURTHER NOTATIONS AND AUXILIARY RESULTS

We need for the proofs of (A) and (B) some properties of the function $k(w,\chi)$ proved in (Fryska, 1989).
Let M be the Riemann surface of $\log s$ with the sheets H_k,

$$\mathsf{H}_k = \{re^{i\varphi}\colon\ r>0,\ (2k+1)\pi \geqslant \varphi > (2k-1)\pi\}, \qquad k\in\mathbf{Z},$$

and let

$$\mathsf{H}^+ = \{re^{i\varphi}\colon\ r>0,\ \pi>\varphi>0\}$$

denote the upper complex half-plane. We put

$$k(w;\chi) = \sum_{\operatorname{Im}(\rho)>0} e^{\rho w}, \qquad \operatorname{Im}(w)>0,$$

for $w\in\mathsf{H}^+$. The series is almost uniformly convergent and represents a function holomorphic on H^+. We have proved the following "exact formula" for the function $k(w,\chi)$.

LEMMA 1.

$$\begin{aligned}2\pi i k(w,\chi) &= \int_{l_\chi} \frac{\zeta'}{\zeta}(s,\chi)e^{sw}\,ds + \frac{e^{-w/2}}{w}\log\frac{(2\pi)^n}{|\Delta|N(\mathfrak{f})}\\
&+\frac{ne^w}{1-e^w}\left\{\int_{1/2}^{3/2}\frac{\Gamma'}{\Gamma}(x)e^{-xw}\,dx + E_1(w/2)\right\}\\
&+\frac{i}{2}(r_1-2q)e^{-\frac{w}{2}}\beta\Big(\frac{w}{2\pi i}+\frac12\Big) - \frac{n}{2}e^{-\frac{w}{2}}\Big(\frac{\pi}{iw}+\beta\Big(\frac{w}{2\pi i}\Big)\Big)\\
&+e^{\frac32 w}\sum_{\mathfrak{a}}\frac{\Lambda(\mathfrak{a})\chi(\mathfrak{a})}{(N(\mathfrak{a}))^{3/2}(w-\log N(\mathfrak{a}))}\\
&+e^{-\frac{w}{2}}\sum_{\mathfrak{a}}\frac{\Lambda(\mathfrak{a})\overline{\chi(\mathfrak{a})}}{(N(\mathfrak{a}))^{3/2}(w+\log N(\mathfrak{a}))},\end{aligned} \tag{4.1}$$

where

$$\begin{aligned}E_1(s) &= \int_s^\infty \frac{e^{-u}}{u}\,du = -\gamma - \log s - \sum_{n=1}^{\infty}\frac{(-s)^n}{n!n},\\
\beta(s) &= \frac12\Big(\frac{\Gamma'}{\Gamma}\Big(\frac{s+1}{2}\Big) - \frac{\Gamma'}{\Gamma}\Big(\frac{s}{2}\Big)\Big),\end{aligned} \tag{4.2}$$

and $l_\chi = l_\chi(-1/2, 3/2)$ *denotes a curve in* H^+ *which joins the points* $-1/2$ *and* $3/2$ *and lies below the all zeros* $\rho=\beta+i\gamma$ *with* $\gamma>0$.

From this formula one can derive the following properties of $k(w,\chi)$.

LEMMA 2 (Fryska, 1989, Thm. 4.2). *The function $k(w,\chi)$ has continuation to a meromorphic function on* M, *that is*

$$k(w,\chi)=\frac{1}{2\pi i}\frac{ne^{w}}{e^{w}-1}\log w+N(w,\chi),\qquad(4.3)$$

where $N(w,\chi)$ is a uniquely defined meromorphic function.

LEMMA 3 (Fryska, 1989, Thm. 5.2). *We have*

$$k(w,\chi)+\overline{k(\overline{w},\bar{\chi})}=e^{w}D(-w,\chi)$$

on the Riemann surface M, *where*

$$D(w,\chi)=\frac{e^{w}}{e^{2w}-1}\left(e^{w}(q+r_2)+r_1+r_2-q\right)+d_0(1+e^{w})-\sum_{0<\beta<1}e^{\beta w}.\qquad(4.4)$$

LEMMA 4 (Fryska, 1989, Thm. 6.1). *The only singularities on H_0 of the function $k(w,\chi)$ are the simple poles*

$$w_{k,\mathfrak{p}}=k\log N(\mathfrak{p}),\qquad k\in\mathbf{Z}\setminus\{0\},\quad \mathfrak{p}\nmid\mathfrak{f},$$

with the residues

$$\operatorname*{res}_{w=w_{k,\mathfrak{p}}}k(w,\chi)=\begin{cases}\dfrac{1}{2\pi i}\displaystyle\sum_{N(\mathfrak{a})=N(\mathfrak{p}^{k})}\Lambda(\mathfrak{a})\chi(\mathfrak{a}) & \text{if}\quad k>0,\\[2ex] \dfrac{N(\mathfrak{p}^{k})}{2\pi i}\displaystyle\sum_{N(\mathfrak{a})=N(\mathfrak{p}^{|k|})}\Lambda(\mathfrak{a})\overline{\chi(\mathfrak{a})} & \text{if}\quad k<0.\end{cases}\qquad(4.5)$$

We observe that the formula (4.1) allows us to calculate the residue for $w=0$ of the function $N(w,\chi)$ in (4.3). From (4.1) and (4.2) we obtain

$$\frac{2\pi i}{n}k(w,\chi)-\frac{e^{w}}{e^{w}-1}\log w=\frac{e^{w}}{1-e^{w}}\left\{\int_{1/2}^{3/2}\frac{\Gamma'}{\Gamma}(x)e^{-xw}dx+\log 2-\gamma+w\lambda(w)\right\}$$

$$+\frac{e^{-\frac{w}{2}}}{nw}\log\frac{(2\pi)^{n}}{|\Delta|N(\mathfrak{f})}-\frac{1}{2}e^{-\frac{w}{2}}\left(\frac{\pi}{iw}+\beta\left(\frac{w}{2\pi i}\right)\right)+\lambda(w).$$

Since $\operatorname{res}_{s=0}\beta(s)=1$, the following lemma is true.

LEMMA 5. *The function $k(w,\chi)$ has the representation*

$$k(w,\chi)=\frac{n}{2\pi i}\frac{e^{w}}{e^{w}-1}\log w+\frac{nA(\chi)}{2\pi i w}+\lambda(w,\chi)\qquad(4.6)$$

in the neighbourhood of $\omega=0$, where

$$A(\chi)=\gamma+\frac{1}{n}\log\frac{(2\pi)^{n}}{|\Delta|N(\mathfrak{f})}-\pi i.\qquad(4.7)$$

5. SKETCH OF THE CONSIDERATIONS

The singularities of $k(w,\chi)$ just described in (4.5) lead us to the idea that we should use the exact formulae of the Riemann type. We consider as follows.
Let u, v, and $0 \in [u,v]$ be points of regularity of $k(w,\chi)$. We enclose the interval $[u,v]$ between two symmetric parts m and m' of the curve $m(t)$,

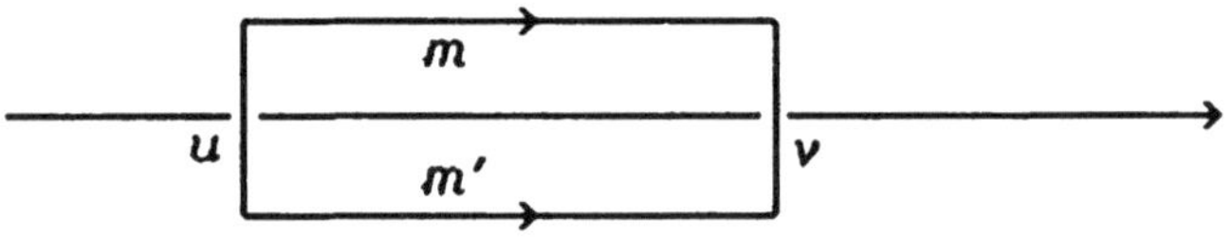

where

$$m(\alpha) = m'(\alpha) = u, \quad m(\beta) = m'(\beta) = v, \quad t \in [\alpha,\beta], \quad \operatorname{Im} m(t) < 2\pi.$$

The Cauchy theorem implies that

$$2\pi i \sum_{w\in[u,v]} \operatorname*{res}_{w=w_{k,p}} k(w,\chi) = \left(\int_{m'} - \int_{m}\right) k(w,\chi)\,dw. \tag{5.1}$$

If $u, v < 0$, then m' lies on the sheet H_1 of the Riemann surface M. Thus, if we take

$$w = w' e^{2\pi i}, \qquad dw = e^{2\pi i}\, dw',$$

for $w \in m'$, then $w' \in \overline{m} \subset \mathsf{H}_0$, and

$$k(w'e^{2\pi i},\chi) = k(w',\chi) + \frac{ne^w}{e^w - 1}$$

in virtue of (4.3). It follows that

$$\int_{m'} k(w,\chi)\,dw = \int_{\overline{m}} k(w'e^{2\pi i},\chi)e^{2\pi i}\,dw' = \int_{\overline{m}} k(w,\chi)\,dw + \int_u^v \frac{ne^x}{e^x - 1}\,dx,$$

because $0 \notin [u,v]$ and $\operatorname{Im} m(t) < 2\pi$.
If $v, u > 0$, then the curves m' and $\overline{m}$ coincide. So, in both cases we obtain

$$\begin{aligned} 2\pi i \sum_{w\in[u,v]} \operatorname*{res}_{w=w_{k,p}} k(w,\chi) &= \left(\int_{\overline{m}} - \int_{m}\right) k(w,\chi)\,dw \\ &\quad + \frac{1-\operatorname{sgn} u}{2} \int_u^v \frac{ne^x}{e^x - 1}\,dx. \end{aligned} \tag{5.2}$$

Furthermore, Lemma 3 yields

$$\begin{aligned}\int_{\overline{m}} k(w,\chi)\,dw &= \int_{\alpha}^{\beta} k(x(t)-iy(t),\chi)\big(x'(t)-iy'(t)\big)\,dt\\ &= \int_{\alpha}^{\beta} [-\overline{k(x(t)+iy(t),\bar\chi)}\\ &+ e^{x(t)+iy(t)}D(-x(t)-iy(t),\chi)]\big(x'(t)-iy'(t)\big)\,dt\\ &= -\overline{\int k(w,\bar\chi)dw} + \overline{\int e^{\bar w}\overline{D(-w,\chi)}\,dw}.\end{aligned} \tag{5.3}$$

From (4.4) one observes that the only singularity of $D(w,\chi)$ is the point $w=0$. Thus,

$$\overline{\int_m e^{\bar w}\overline{D(-w,\chi)}\,dw} = \overline{\int_u^v e^{x}\overline{D(-x,\chi)}\,dx} = \int_u^v e^{x}D(-x,\chi)\,dx.$$

Whence and from (5.1)–(5.3) we obtain

$$\begin{aligned}2\pi i \sum_{w_{k,p}\in[u,v]} \operatorname*{res}_{w=w_{k,p}} k(w,\chi) &= -\left(\int_m k(w,\chi)\,dw + \int_m \overline{k(w,\bar\chi)\,dw}\right)\\ &+ \int_u^v e^{x}D(-x,\chi)\,dx + \frac{1-\operatorname{sgn} u}{2}\int_u^v \frac{ne^{x}}{e^{x}-1}\,dx.\end{aligned} \tag{5.4}$$

The Cauchy theorem implies that

$$\int_{m(u,v)} k(w,\chi)\,dw = \left(-\int_{u+i\infty}^{u} + \int_{v+i\infty}^{v}\right)k(w,\chi)\,dw = \mathcal{K}(v,\chi)-\mathcal{K}(u,\chi), \tag{5.5}$$

where

$$\mathcal{K}(z,\chi) = \int_{x+i\infty}^{x+iy} k(w,\chi)\,dw, \qquad z = x+iy. \tag{5.6}$$

It follows from the definition, that $\mathcal{K}(z,\chi)$ is regular on H^+ and has an analytic continuation at each point of regularity of $k(w,\chi)$ on the real axis. The same is true for the function

$$\mathcal{F}(z,\chi) = \mathcal{K}(z,\chi) + \overline{\mathcal{K}(z,\bar\chi)}. \tag{5.7}$$

Thus, the relation (5.4) takes the form

$$
\begin{aligned}
2\pi i \sum_{w_{k,p}\in[u,v]} \operatorname*{res}_{w=w_{k,p}} k(w,\chi) &= -\Big(\mathcal{K}(v,\chi)-\mathcal{K}(u,\chi)+\overline{\mathcal{K}(v,\bar\chi)}-\overline{\mathcal{K}(u,\bar\chi)}\Big) \\
&+\int_u^v e^x D(-x,\chi)\,dx + \frac{1-\operatorname{sgn} u}{2}\int_u^v \frac{ne^x}{e^x-1}\,dx \\
&= \mathcal{F}(u,\chi)-\mathcal{F}(v,\chi)+\int_u^v e^x D(-x,\chi)\,dx \\
&+\frac{1-\operatorname{sgn} u}{2}\int_u^v \frac{ne^x}{e^x-1}\,dx.
\end{aligned} \tag{5.8}
$$

Now we put

$$
u=\begin{cases} x & \text{if } \ u<0, \\ a & \text{if } \ u>0, \end{cases} \qquad v=\begin{cases} a & \text{if } \ v<0, \\ x & \text{if } \ v>0. \end{cases}
$$

From (5.8) one has

$$
\begin{aligned}
2\pi i \sum_{w_{k,p}\in[a,x]} \operatorname*{res}_{w=w_{k,p}} k(w,\chi) &= \mathcal{F}(a,\chi)-\mathcal{F}(x,\chi) \\
&+\int_a^x e^t D(-t,\chi)\,dt + \frac{1-\operatorname{sgn} x}{2}\int_a^x \frac{ne^t}{e^t-1}\,dt
\end{aligned}
$$

in virtue of the equality $\operatorname{sgn} a = \operatorname{sgn} x$.
We introduce the abbreviations

$$
\psi_\varepsilon(y,\chi) = \sideset{}{'}\sum_{\substack{N(\mathfrak{a})\leqslant y \\ (\mathfrak{a},\mathfrak{f})=1}} \frac{\Lambda(\mathfrak{a})\chi(\mathfrak{a})^{1-2\varepsilon}}{(N(\mathfrak{a}))^{\varepsilon}}, \qquad \varepsilon = \frac{1-\operatorname{sgn} x}{2}, \tag{5.9}
$$

and obtain

$$
\begin{aligned}
\psi_\varepsilon(e^{|x|},\chi)\operatorname{sgn} x &= -\mathcal{F}(x,\chi) \\
&+\lim_{|a|\to 0}\left\{\mathcal{F}(a,\chi)+\int_a^x e^t D(-t,\chi)\,dt + \frac{1-\operatorname{sgn} x}{2}\int_a^x \frac{ne^t}{e^t-1}\,dt\right\}
\end{aligned} \tag{5.10}
$$

by Lemma 4. As usual, the 'prime' at the sum in (5.9) means that

$$
\psi_\varepsilon(y,\chi) = \frac{1}{2}\{\psi_\varepsilon(y-0,\chi)+\psi_\varepsilon(y+0,\chi)\}.
$$

If the above written limits exist, the formula (5.10) is evident at points of regularity of $k(w,\chi)$, because both sides of (5.10) are continuous at these points. It remains to prove that the function

$$\mathcal{F}(x,\chi)=\lim_{y\to 0+}\mathcal{F}(x+iy,\chi)$$

exists at other points of the real axis and satisfies the desired equality

$$\mathcal{F}(x,\chi)=\frac{1}{2}\big(\mathcal{F}(x-0,\chi)+\mathcal{F}(x+0,\chi)\big).$$

6. INVESTIGATION OF THE FUNCTION $\mathcal{K}(z,\chi)$

We present the results of this chapter as the following lemma.

LEMMA 6.
a) *The function $\mathcal{K}(x,\chi)$ is analytically continuable at each point $x\in\mathbf{R}$ of regularity of $k(w,\chi)$ and*

$$\mathcal{K}(z,\chi)=\sum_{\operatorname{Im}\rho>0}\frac{e^{\rho z}}{\rho},\qquad \operatorname{Im}z>0. \tag{6.1}$$

Moreover, the series converges uniformly on every interval $[a,b]$ containing only the points of regularity of $k(w,\chi)$.
b) *The points $x=w_{k,\mathfrak{p}}$ (see Lemma 4) are logarithmic branch points for $\mathcal{K}(z,\chi)$. In their neighbourhoods*

$$\mathcal{K}(z,\chi)=c(\chi)\log(z-w_{k,\mathfrak{p}})+h(z,\chi), \tag{6.2}$$

and

$$\sum_{0<\operatorname{Im}\rho\leqslant T}\frac{e^{\rho x}}{\rho}=-c\Big(\log T+\gamma-\frac{\pi i}{2}\Big)+h(x,\chi)+o(1), \tag{6.3}$$

where $h(z,\chi)$ is regular on a neighbourhood of x, and $c=c(\chi)$ is defined by the formula (4.5), i.e.

$$c(\chi)=\operatorname*{res}_{z=w_{k,\mathfrak{p}}}k(z,\chi)=\begin{cases}\dfrac{1}{2\pi i}\displaystyle\sum_{N(\mathfrak{a})=N(\mathfrak{p}^k)}\Lambda(\mathfrak{a})\chi(\mathfrak{a}) & \text{if}\quad k>0,\\[2ex] \dfrac{(N(\mathfrak{p}))^k}{2\pi i}\displaystyle\sum_{N(\mathfrak{a})=N(\mathfrak{p}^{|k|})}\Lambda(\mathfrak{a})\overline{\chi(\mathfrak{a})} & \text{if}\quad k<0.\end{cases}$$

The first part of a) is an immediate corollary of the definition. For the second part of a), we need the following.

THEOREM I (Kaczorowski, 1990, Thm 4.2). *Let $w_n=a_n+ib_n$, $n=1,2,\ldots$, be a sequence with*

$$|a_n|\leqslant A,\qquad n\geqslant 1,\qquad b_1\leqslant b_2\leqslant\ldots,\qquad \lim_{n\to\infty}b_n=\infty,$$

and let the series

$$f(z) = \sum_{n=1}^{\infty} A_n e^{w_n z}, \qquad A_n \in \mathbf{C},$$

converge for $y = \operatorname{Im} z > 0$. *Suppose that*

$$\left| \sum_{n=N+1}^{\infty} A_n e^{(w_n - w_N)z} \right| = o\big(|y|^{-1}\big), \qquad N \to \infty, \tag{6.4}$$

as $y \to 0+$, *and*

$$\left| \sum_{n=1}^{N} A_n e^{(w_n - w_N)z} \right| = o\big(|y|^{-1}\big), \qquad N \to \infty, \tag{6.5}$$

as $y \to 0-$ *almost uniformly in* $x = \operatorname{Re} z$. *If* f *is holomorphic at the boundary point* $x \in \mathbf{R}$, *then the series*

$$\sum_{n=1}^{\infty} A_n e^{w_n x}$$

converges to $f(x)$ *and the convergence is uniform on each closed interval* $[a, b]$ *which consists of points of regularity of* f *only.*

To prove the remaining part of Lemma 6, a), we arrange the zeros in the ascending order of their absolute values,

$$|\rho_1| \leqslant |\rho_2| \leqslant \dots,$$

and verify the conditions (6.4) and (6.5). For this purpose we split the sum

$$S_N = \sum_{n=N+1}^{\infty} \frac{1}{\rho_n} e^{(\rho_n - \rho_N)z}$$

into three parts:

$$S_N = \Bigg(\sum_{\gamma_N < \gamma_n \leqslant \gamma_N + \sqrt{\gamma_N}} + \sum_{\gamma_N + \sqrt{\gamma_N} < \gamma_n \leqslant 2\gamma_N} + \sum_{2\gamma_N < \gamma_n} \Bigg) \frac{e^{(\rho_n - \rho_N)z}}{\rho_n} = S_1 + S_2 + S_3.$$

We suppose that $y = \operatorname{Im} z > 0$. Let $N(T, \chi)$ be the number of zeros ρ with $T < \operatorname{Im} \rho < T + 1$. Then

$$N(T + 1, \chi) - N(T, \chi) = O(\log T).$$

(see (Landau, 1919, S. 107)). Hence,

$$\begin{aligned} |S_1| &\leqslant \frac{1}{|\rho_N|} \sum_{\gamma_N < \gamma_n \leqslant \gamma_N + \sqrt{\gamma_N}} e^{(\beta_n - \beta_N)x - y(\gamma_n - \gamma_N)} \\ &\ll \frac{1}{\gamma_N} e^{B} \big(N(\gamma_N + \sqrt{\gamma_N}, \chi) - N(\gamma_N, \chi)\big) \ll \frac{\log \gamma_N}{\sqrt{\gamma_N}}. \end{aligned}$$

Analogously, we obtain

$$|S_2| \leqslant \frac{1}{\gamma_N + \sqrt{\gamma_N}} e^{|x| - y\sqrt{\gamma_N}} N(2\gamma_N, \chi) \ll e^{-y\sqrt{\gamma_N}} \log \gamma_N \ll \frac{\log \gamma_N}{y\sqrt{\gamma_N}}$$

and

$$|S_3| \ll \frac{1}{y} \sum_{2\gamma_N < \gamma_n} \frac{1}{\gamma_n(\gamma_n - \gamma_N)} \ll \frac{1}{y} \sum_{\gamma_N < \gamma_n} \frac{1}{\gamma_n^2} \ll \frac{1}{y\sqrt{\gamma_N}}.$$

Hence,

$$S_N = O\Big(\frac{1}{y\sqrt{\gamma_N}}\Big) = o\Big(\frac{1}{y}\Big)$$

as $y \to 0+$ and $N \to \infty$. Thus the condition (6.3) is satisfied. For the second sum we have

$$\begin{aligned} S'_N &= \sum_{n=1}^{N} \frac{1}{\rho_n} e^{(\rho_n - \rho_N)z} \ll \sum_{n=1}^{[\gamma_N]} \sum_{n \leqslant \gamma < n+1} \frac{e^{-y(\gamma_N - \gamma)}}{\gamma} \\ &\ll \sum_{n=1}^{[\gamma_N]} \frac{e^{-y(\gamma_N - n - 1)}}{n} \log(n+1) \ll \sum_{n=1}^{[\gamma_N]} \frac{\log n}{ny(\gamma_N - n + 1)} \\ &= \frac{1}{y(\gamma_N + 1)} \sum_{n=1}^{[\gamma_N]} \Big(\frac{1}{n} + \frac{1}{\gamma_N - n + 1}\Big) \log n \ll \frac{(\log \gamma_N)^2}{y\gamma_N}. \end{aligned}$$

It means that (6.4) is also satisfied, and the proof of the assertion a) is finished. The second part of Lemma rests on the following result.

THEOREM II (Kaczorowski, 1990, Thm 4.3). *In the notations of Theorem* I, *suppose that for some* $x_0 \in \mathbf{R}$, $g \in \mathbf{C}$, *and* $r_0 > 0$,

$$f(z) = g \log(z - x_0) + h(z) \tag{6.6}$$

on the region $|z - x_0| < r_0$, $\operatorname{Im} z > 0$, *where* h *is holomorphic on the circle* $|z - x_0| < r_0$. *Then*

$$\sum_{\operatorname{Im} w_n \leqslant T} A_n e^{w_n x_0} = -g \log T - g\gamma + h(x_0) + \frac{g\pi i}{2} + o(1), \qquad T \to \infty.$$

From the definition of $\mathcal{K}(z, \chi)$ it follows immediately that having gone around the point $x_0 = w_{k,p}$ the value of $\mathcal{K}(z, \chi)$ changes into

$$K(z, \chi) + \operatorname*{res}_{z = w_{k,p}} k(z, \chi).$$

This means that the condition (6.6) is satisfied with

$$g = \operatorname*{res}_{z = w_{k,p}} k(z, \chi).$$

Hence, Theorem II implies the part b) of Lemma 6.

COROLLARY. *The series*

$$\sum_{\rho} \frac{e^{\rho x}}{\rho},$$

where the sum runs over all nontrivial zeros of $\zeta(s,\chi)$, *converges for each* $x \in \mathbf{R}$. *If* $x = w_{k,\mathfrak{p}}$ *or* $x = 0$, *then the convergence is understood in the sense*

$$\lim_{T\to\infty} \sum_{|\rho|\leqslant T} \frac{e^{\rho x}}{\rho}.$$

Proof. We have

$$\overline{\sum_{0<\gamma_{\bar\chi}<T} \frac{e^{\rho_{\bar\chi} x_0}}{\rho_{\bar\chi}}} = \sum_{-T<\gamma<0} \frac{e^{\rho x_0}}{\rho}. \tag{6.7}$$

From Lemma 4 it follows that the points of regularity of $k(w,\chi)$ and of $k(w,\bar\chi)$ are the same. So, the convergence of the series

$$\sum_{\gamma>0} \frac{e^{\rho x}}{\rho}$$

implies that of the series

$$\sum_{\gamma<0} \frac{e^{\rho x}}{\rho}.$$

For $x_0 = w_{k,\mathfrak{p}}$, it follows from (6.2) that

$$\begin{aligned}
\sum_{\beta\neq 0} \frac{e^{\rho x_0}}{\rho} &= \lim_{T\to\infty}\left(\sum_{0<\gamma<T} \frac{e^{\rho x_0}}{\rho} + \sum_{-T<\gamma<0} \frac{e^{\rho x_0}}{\rho}\right) \\
&= \lim_{T\to\infty}\left(\sum_{0<\gamma<T} \frac{e^{\rho x_0}}{\rho} + \overline{\sum_{0<\gamma_{\bar\chi}<T} \frac{e^{\rho_{\bar\chi} x_0}}{\rho_{\bar\chi}}}\right) \\
&= c(\chi)\pi i + \left(h(x_0,\chi) + \overline{h(x_0,\bar\chi)}\right),
\end{aligned} \tag{6.8}$$

since $\overline{c(\bar\chi)} = -c(\chi)$ due to (4.5). The case $x = 0$ is well-known (see (Landau, 1919)).

7. INVESTIGATION OF THE FUNCTION $\mathcal{F}(x,\chi)$

This function is defined by the limit

$$\mathcal{F}(x,\chi) = \lim_{y\to 0+} \mathcal{F}(x+iy,\chi) = \lim_{y\to 0+} \mathcal{K}(x+iy,\chi) + \overline{\mathcal{K}(x+iy,\bar\chi)}. \tag{7.1}$$

LEMMA 7.

a) *The function* $\mathcal{F}(x,\chi)$ *is correctly defined for all* $x \in \mathbf{R}$.

b) *We have*

$$\mathcal{F}(x,\chi)=\sum_{\gamma\neq 0}\frac{e^{\rho x}}{\rho}$$

on each closed interval $[a,b]$ *consisting of points of regularity of* $k(w,\chi)$ *only. The series converges uniformly on each interval* $[a,b]$ *and represents a continuous function.*
c) *Each point* $x=w_{k,p}$ *is a point of discontinuity of the first kind for the function* $\mathcal{F}(x,\chi)$ *with the jump*

$$\mathcal{F}(w_{k,p}+0,\chi)-\mathcal{F}(w_{k,p}-0,\chi)=2i\pi c(\chi)$$

(see (4.5)). Moreover,

$$\mathcal{F}(x,\chi)=\lim_{T\to\infty}\sum_{0<|\gamma|<T}\frac{e^{\rho x}}{\rho}$$

and

$$\mathcal{F}(w_{k,p},\chi)=\frac{1}{2}(\mathcal{F}(w_{k,p}+0,\chi)+\mathcal{F}(w_{k,p}-0,\chi)).$$

d)

$$\mathcal{F}(0,\chi)=\lim_{T\to\infty}\sum_{0<|\gamma|<T}\frac{1}{\rho}$$

and

$$\mathcal{F}(0,\chi)=\frac{1}{2}\lim_{\varepsilon\to 0+}(\mathcal{F}(\varepsilon,\chi)+\mathcal{F}(-\varepsilon,\chi)).$$

Proof. The part b) follows from (6.1), (6.7), and (7.1). We use (6.2) to obtain the equality

$$\mathcal{F}(x+iy,\chi)=2ic(\chi)\mathrm{Im}\log(x+iy-w_{k,p})+h(x+iy,\chi)+\overline{h(x+iy,\bar\chi)}$$

in the neighbourhood of $x=w_{k,p}$.
Then taking the limit as $y\to 0+$ we have

$$\mathcal{F}(x,\chi)=\begin{cases}2c(\chi)i\pi+h(x,\chi)+\overline{h(x,\bar\chi)} & \text{if } x<w_{k,p},\\ c(\chi)i\pi+h(w_{k,p},\chi)+\overline{h(w_{k,p},\bar\chi)} & \text{if } x=w_{k,p},\\ h(x,\chi)+\overline{h(x,\bar\chi)} & \text{if } x>w_{k,p}.\end{cases}$$

Whence the part c) follows if we additionally use the formula (6.8). In the neighbourhood of $w=0$ the Cauchy theorem implies

$$\mathcal{K}(a,\chi)=\mathcal{K}(i|a|,\chi)-\int_a^{i|a|}k(w,\chi)\,dw. \tag{7.2}$$

Lemma 5 and the equality

$$\frac{e^w}{e^w-1}=\frac{1}{w}+\frac{1}{2}+w\lambda(w)$$

give that

$$k(w,\chi)=\frac{n}{2\pi i}\Big(\frac{1}{w}+\frac{1}{2}+\lambda(w)\Big)\log w+\frac{nA(\chi)}{2\pi i w}+\lambda(w), \tag{7.3}$$

which after some calculations implies the equality

$$\begin{aligned}\int_a^{i|a|} k(w,\chi)\,dw &= i|a|\int_\alpha^\pi \Big\{\frac{n}{2\pi i}\Big(\frac{1}{|a|e^{i\varphi}}+\lambda(|a|e^{i\varphi})\Big)\big(\log|a|+i\varphi\big)\\ &+\frac{nA(\chi)}{2\pi i|a|e^{i\varphi}}+\lambda(|a|e^{i\varphi})\Big\}e^{i\varphi}\,d\varphi=\frac{n\operatorname{sgn}a}{4}\big(\log|a|+A(\chi)\big)\\ &+\frac{i\pi^2}{8}(2\operatorname{sgn}a-1)+o(1).\end{aligned} \tag{7.4}$$

From (7.2) and (7.4) one can derive that

$$\mathcal{F}(a,\chi)=\mathcal{F}(i|a|,\chi)-\frac{n\operatorname{sgn}a}{2}\Big(\log|a|+\gamma+\frac{1}{n}\log\frac{(2\pi)^n}{|\Delta|N(\mathfrak{f})}\Big)+o(1),\qquad a\to 0. \tag{7.5}$$

Whence the second part of the statement d) follows, provided that

$$\mathcal{F}(0,\chi)=\lim_{y\to 0+}\mathcal{F}(iy,\chi).$$

But this can be easily proved using the formula (7.3). It remains to prove the formula

$$\mathcal{F}(0,\chi)=\lim_{T\to\infty}\sum_{0<|\gamma|<T}\frac{1}{\rho}.$$

Now we use some reduction. At first,

$$\sum_{\gamma>0}\frac{e^{\rho a i}}{\rho}=\sum_{\gamma>0}\frac{e^{-a\gamma}\cos a\beta}{\rho}+i\sum_{\gamma>0}\frac{e^{-a\gamma}\sin a\beta}{\rho},$$

where the second series converges uniformly in $a\geqslant 0$. The same is true for the series

$$\sum_{\gamma>0}\frac{e^{-a\gamma}\cos a\beta}{\rho}-\frac{e^{-a\gamma}\cos a\beta}{i\gamma}.$$

So, it is enough to investigate the series

$$\sum_{\gamma>0}\frac{e^{-a\gamma}\cos a\beta}{\rho}$$

more exactly. Since

$$\sum_{\gamma>0}\frac{e^{-a\gamma}(\cos a\beta-1)}{\gamma}=2\sum_{\gamma>0}\frac{e^{-a\gamma}(\sin a\beta/2)^2}{\gamma},$$

the relation

$$i\mathcal{F}(ia,\chi)=\sum_{\gamma>0}\frac{e^{-a\gamma}}{\gamma}+\sum_{\gamma<0}\frac{e^{a\gamma}}{\gamma}+\lambda(a,\chi)$$

is valid. In order to determine the principal value we arrange the imaginary parts $\gamma>0$ and $\gamma'<0$ with respect to their absolute values and obtain that

$$S_N=\sum_{n\leqslant N}\frac{e^{-a\gamma_n}}{\gamma_n}+\frac{e^{a\gamma'_n}}{\gamma'_n}$$

with $\gamma_1\leqslant\gamma_2\leqslant\ldots,\ \gamma'_1\geqslant\gamma'_2\geqslant\ldots$ and $\gamma_n+\gamma'_n=O(1)$.
The mean-value theorem implies that

$$\frac{e^{-a\gamma_n}}{\gamma_n}+\frac{e^{a\gamma'_n}}{\gamma'_n}=-\frac{e^{-at_n}(at_n+1)}{t_n^2}(\gamma_n+\gamma'_n)\ll t_n^{-2}$$

for some $t_n\in[\gamma_n,-\gamma'_n]$. Thus,

$$S_N\ll\sum_{n\leqslant N}t_n^{-2}\ll\sum_{n\leqslant N}(\gamma_n+O(1))^{-2}\ll\sum\frac{\log n}{n^2}.$$

Whence the uniformity of the convergence of our series to the principal value follows, and, finally, the proposition of the part d) is valid.

8. CONCLUSION

With the Lemma 7, we have realized the program which has been sketched at the end of § 5. It remains only to calculate the limit in (5.10). Elementary calculations yield that

$$\begin{aligned}\int_a^x e^tD(-t,\chi)\,dt+\frac{1-\operatorname{sgn}x}{2}\int_a^x\frac{ne^t}{e^t-1}\,dt=\int_a^x\Big\{&\frac{e^{-t}(q+r_2)+(r_1+r_2-q)}{e^{-2t}-1}\\&+\frac{n(1-\operatorname{sgn}x)}{2}\frac{e^t}{e^t-1}+d_0(1+e^t)-\sum_{0<\beta<1}e^{(1-\beta)t}\Big\}\,dt\\=\Big(q-\frac{r_1}{2}\Big)\log\frac{1}{2}(1+e^{-|x|})-\frac{n\operatorname{sgn}x}{2}&\log(1-e^{-|x|})\\+\frac{n\operatorname{sgn}x}{2}\log|a|-x\frac{r_1+r_2-q}{2}&(1+\operatorname{sgn}x)+d_0(e^x+x-1)\\-\sum_{0<\beta<1}\frac{e^{\beta x}-1}{\beta}+o(1),\qquad a\to 0.&\end{aligned}$$

Taking into account (7.5), we obtain from (5.10) that

$$\begin{aligned}\psi_\varepsilon\big(e^{|x|},\chi\big)\operatorname{sgn} x &= -\mathcal{F}(x,\chi) + \lim_{|a|\to 0}\big(\mathcal{F}(a,\chi) + I(a)\big)\\ &= -\mathcal{F}(x,\chi) + \mathcal{F}(0,\chi) - \frac{n\operatorname{sgn} x}{2}\Big(\gamma + \frac{1}{n}\log\frac{(2\pi)^n}{|\Delta| N(\mathfrak{f})}\Big)\\ &+ \Big(q - \frac{r_1}{2}\Big)\log\frac{1}{2}\big(1 + e^{-|x|}\big) - \frac{n\operatorname{sgn} x}{2}\log\big(1 - e^{-|x|}\big)\\ &- x\frac{r_1 + r_2 - q}{2}(1 + \operatorname{sgn} x) + d_0\big(e^x + x - 1\big)\\ &- \sum_{0<\beta<1}\frac{e^{\beta x} - 1}{\beta}.\end{aligned}$$

The formula (A) follows from the last equality with $x < 0$, $e^x = y$, and so does the formula (B) with $x < 0$, $e^{|x|} = y$, and $\psi_1 = \tilde{\psi}$.

REFERENCES

Fryska T., (1989). Über einige Formeln in analytischer Zahlentheorie, I. *Funct. et Approx. Comment. Math.* **18**, 121–133. Berichtigung. *Ibidem*, **19**.

Fryska T., (1991). Von der Reihe $\sum x^\rho/\rho$. *Liet. Matem. Rink.* **31**, 187–204.

Kaczorowski J., (1990). The k-functions in multiplicative number theory, I. *Acta Arithm.* **56**, 3, 195–211.

Landau E., (1919). Über Ideale und Primideale in Idealklassen. *Math. Z.* **2**, 52–154.

New Trends in Probab. and Statist., pp. 335–354
F. Schweiger and E. Manstavičius (Eds.)
1992 VSP/TEV

THE ATKINSON FORMULA NEAR THE CRITICAL LINE

A. LAURINČIKAS
Department of Mathematics, Vilnius University,
Naugarduko 24, 2006 Vilnius, Lithuania

ABSTRACT

The asymptotics of the square mean of the Riemann zeta-function near the critical line is considered. The analogue of the Atkinson formula is obtained.

Let $s = \sigma + it$ be a complex variable and let $\zeta(s)$ denote the Riemann zeta-function the classical problems in the zeta-function theory. The investigation of the asymptotic behaviour of its square mean

$$I_T(\sigma, T_0) = \int_{T_0}^{T} |\zeta(\sigma + it)|^2\, dt$$

where T_0 is a fixed number is one of the classical problems in the zeta-function theory. It is well known (Titchmarsh, 1951) that on the half-plane $\sigma > 1/2$

$$\lim_{T\to\infty} \frac{I_T(\sigma, 1)}{T} = \zeta(2\sigma)$$

and as $T \to \infty$

$$I_T\left(\frac{1}{2}, 0\right) = (1 + o(1))T \log T.$$

Let

$$E(T) = I_T\left(\frac{1}{2}, 0\right) - T \log \frac{T}{2\pi} - (2\gamma - 1)T$$

where γ denotes the Euler's constant. The function $E(T)$ has been studied by many authors. The best known estimate for $E(t)$ is due to D. R. Heath-Brown and M. H. Huxley (1990) who proved that

$$E(T) = BT^{\frac{7}{22}+\varepsilon} \qquad \text{for every} \quad \varepsilon > 0.$$

Here and further, B denotes a quantity (not always the same) which is bounded by a constant.

F. V. Atkinson (1949) has obtained the famous explicit formula for $E(t)$. See also (Ivić, 1985, Chap. XV).
In the proofs of the probabilistic limit theorems for the function $\zeta(s)$ near the critical line a crucial role is played by the moments whose asymptotics has an interesting dependence on the closeness to the line $\sigma = 1/2$ (Laurinčikas, 1988, 1989). This stimulated us to consider the square mean of the Riemann zeta-function near the critical line.
Let $l_T > 0$ and let l_T tend to infinity monotonically as $T \to \infty$. We put

$$\sigma_T = \frac{1}{2} + \frac{1}{l_T}.$$

First we shall obtain some results which give the principal term in the asymptotic formula for $I_T(\sigma_T, 0)$.

THEOREM 1. *Let $l_T = o(\log T)$ as $T \to \infty$. Then*

$$I_T(\sigma_T, 0) = \frac{1}{2}(1 + o(1))Tl_T.$$

THEOREM 2. *Let $\log T = o(l_T)$ as $T \to \infty$. Then*

$$I_T(\sigma_T, 0) = (1 + o(1))T \log T.$$

THEOREM 3. *Let $l_T = \varkappa \log T$ where $\varkappa > 0$. Then as $T \to \infty$*

$$I_T(\sigma_T, 0) = \frac{\varkappa}{2}(1 - e^{-2/\varkappa})T \log T + o(T \log T).$$

We shall suppose further that $0 < l_T \leqslant c \log T$, $c > 0$. For brevity we write

$$\delta_T = 2/l_T.$$

Let

$$E_T(\sigma_T) = I_T(\sigma_T, 0) - T\zeta(1 + \delta_T) - (1 - \delta_T)^{-1}(2\pi e)^{\delta_T} T^{1-\delta_T} \zeta(1 - \delta_T).$$

The main aim of our paper is to obtain an explicit formula for $E_T(\sigma_T)$.

THEOREM 4. *Let c_1, c_2 be any fixed constants such that $0 < c_1 < c_2$ and $c_1 T < N < c_2 T$, and let*

$$N_1 = N_1(T) = \frac{T}{2\pi} + \frac{N}{2} - \sqrt{\frac{N^2}{4} + \frac{NT}{2\pi}}.$$

Then

$$E_T(\sigma_T) = \frac{1}{\sqrt{2}}\left(\frac{2\pi}{T}\right)^{\delta_T/2} \sum_{m \leqslant N} (-1)^m \frac{\sigma_{\delta_T}(m)}{m^{(1+\delta_T)/2}} \frac{\cos f(T, m)}{\left(\frac{T}{2\pi m} + \frac{1}{4}\right)^{1/4} \sinh^{-1}\left(\sqrt{\frac{\pi m}{2T}}\right)}$$
$$- 2\left(\frac{2\pi}{T}\right)^{\delta_T/2} \sum_{m \leqslant N_1} \frac{\sigma_{\delta_T}(m)}{m^{(1+\delta_T)/2}} \left(\log \frac{T}{2\pi m}\right)^{-1} \cos f_1(T, m)$$
$$+ B \log^2 T.$$

Here

$$f(T,m) = 2T\sinh^{-1}\left(\sqrt{\frac{\pi m}{2T}}\right) + \sqrt{2\pi m T + \pi^2 m^2} - \frac{\pi}{4},$$

$$f_1(T,m) = T\log\frac{T}{2\pi m} - T + \frac{\pi}{4},$$

and, as usual,

$$\sinh^{-1}(x) = \log\left(x + \sqrt{1+x^2}\right), \qquad \sigma_a(m) = \sum_{d|m} d^a.$$

For the proof of Theorems 1–3 we shall use the following well-known propositions.

LEMMA 1. *Let* $0 \leqslant \sigma \leqslant 1$, $x, y, t > c > 0$, $2\pi xy = t$, *and*

$$\chi(s) = \frac{(2\pi)^s}{2\Gamma(s)\cos(\pi s/2)}.$$

Then uniformly in σ

$$\zeta(s) = \sum_{m\leqslant x} m^{-s} + \chi(s)\sum_{m\leqslant y} m^{s-1} + Bx^{-\sigma} + Bt^{(1/2)-\sigma}y^{\sigma-1}.$$

Lemma 1 is called the approximation equation for the function $\zeta(s)$. Its proof may be found, for example, in (Ivić, 1985, p. 97–104).

LEMMA 2. *On the fixed strip* $\alpha \leqslant \sigma \leqslant \beta$

$$\chi(s) = \left(\frac{2\pi}{t}\right)^{s-(1/2)} e^{i(t+(\pi/4))}\left(1 + Bt^{-1}\right) \qquad \text{for} \quad t > t_0 > 0.$$

For the proof see (Titchmarsh, 1951, Chap. IV).

LEMMA 3. *The estimate*

$$\sum\sum_{1\leqslant m<k<T} m^{-\sigma}k^{-\sigma}\left(\log\frac{k}{m}\right)^{-1} = BT^{2-2\sigma}\log T$$

holds for $1/2 \leqslant \sigma < 1$, *uniformly in* σ, $1/2 \leqslant \sigma \leqslant \sigma_0 < 1$.

The proof is contained in (Titchmarsh, 1951, Chap. VII).

LEMMA 4. *Let* $s \neq 1$ *and* $x \geqslant 1$. *Then*

$$\sum_{m\leqslant x} m^{-s} = \frac{s}{s-1} - \frac{x^{1-s}}{s-1} - s\int_1^x \frac{u-[u]}{u^{s+1}}\,du - \frac{x-[x]}{x^s}.$$

Lemma 4 may be obtained on summing by parts.

Proof of Theorem 1. In Lemma 1 we put

$$\sigma = \sigma_T, \qquad x = \frac{t}{2\pi\sqrt{l_T}}, \qquad y = \sqrt{l_T}.$$

From Lemma 2 the estimate

$$\chi(\sigma_T + it) = B$$

follows. Consequently, applying Lemma 1, we find that

$$\begin{aligned}\zeta(\sigma_T + it) &= \sum_{m\leqslant x} m^{-\sigma_T - it} \\ &\quad + B\sum_{m\leqslant y} m^{\sigma_T - 1} + Bt^{-\sigma_T} l_T^{\sigma_T/2} + Bt^{(1/2)-\sigma_T} l_T^{(\sigma_T-1)/2} \\ &= \sum_{m\leqslant x} m^{-\sigma_T - it} + Bl_T^{\sigma_T/2}.\end{aligned} \tag{1}$$

Let the quantity l_T be defined for $T \geqslant T_0$. Then

$$\int_{T_0}^{T} l_t^{\sigma_T}\, dt = BTl_T^{\sigma_T} = o(Tl_T), \tag{2}$$

and taking into account the equality (1) we see that it suffices to prove the equality of the theorem for the quantity

$$J_T(\sigma_T) \stackrel{\text{def}}{=} \int_0^T \Big| \sum_{m\leqslant x} m^{-\sigma_T - it} \Big|^2\, dt.$$

Let $X = T/(2\pi\sqrt{l_T})$. Then taking

$$\max(m, k) = \frac{T_1}{2\pi\sqrt{l_{T_1}}}$$

we find from Lemma 3

$$\begin{aligned}J_T(\sigma_T) &= \sum_{m\leqslant X}\sum_{k\leqslant X}\int_{T_1}^{T} m^{-\sigma_T - it} k^{-\sigma_T + it}\, dt \\ &= \sum_{m\leqslant X} m^{-2\sigma_T}(T - T_1(m,m)) + B\sum\sum_{1\leqslant m<k\leqslant X} m^{-\sigma_T} k^{-\sigma_T}\left(\log\frac{k}{m}\right)^{-1} \\ &= T\sum_{m\leqslant X} m^{-2\sigma_T} + B\sum_{m\leqslant X} m^{1-2\sigma_T}\sqrt{l_T} + B\Big(\frac{T}{\sqrt{l_T}}\Big)^{2-2\sigma_T}\log T.\end{aligned} \tag{3}$$

From Lemma 4 it follows that

$$\sum_{m\leqslant X} m^{-2\sigma_T} = (2\sigma_T - 1)^{-1} + o(2\sigma_T - 1)^{-1} = \frac{1}{2}l_T + o(l_T).$$

Hence in view of the equality (3) we have

$$J_T(\sigma_T) = \frac{1}{2}(1+o(1))Tl_T \qquad \text{as} \quad T \to \infty.$$

This proves the theorem.

Proof of Theorem 2. Repeating the proof of Theorem 1 with

$$x = \frac{t}{2\pi\sqrt{\log t}}, \qquad y = \sqrt{\log t}$$

we obtain

$$I_T(\sigma_T, 0) = T \sum_{m \leqslant T/\left(2\pi\sqrt{\log T}\right)} m^{-2\sigma_T} + o(T\log T). \tag{4}$$

From (4) and from the estimate $\log T = o(l_T)$ the assertion of the theorem follows.

Proof of Theorem 3. In a way similar to that used above, the equality (4) follows. In virtue of the conditions of the theorem, applying Lemma 4, we find that

$$\sum_{m \leqslant T/\left(2\pi\sqrt{\log T}\right)} m^{-2\sigma_T} = 2\sigma_T(2\sigma_T - 1)^{-1} - \left(\frac{T}{2\pi\sqrt{\log T}}\right)^{1-2\sigma_T}(2\sigma_T - 1)^{-1} + B$$
$$= \frac{\varkappa}{2}\log T - \frac{\varkappa}{2}e^{-2/\varkappa}\log T + o(\log T).$$

From this equality and from (4) we have the equality of the theorem.

Proof of Theorem 4. For the beginning we note that we shall base ourselves on the proof of Atkinson's formula for the case $\sigma = 1/2$. From the definition of the modulus, in virtue of the well-known properties of the Riemann zeta-function, we have

$$I_T(\sigma_T, 0) = \int_0^T \zeta(\sigma_T + it)\zeta(\sigma_T - it)\,dt = \frac{1}{2i}\int_{\sigma_T - iT}^{\sigma_T + iT} \zeta(z)\zeta(2\sigma_T - z)\,dz. \tag{5}$$

For the calculation of the above integral we shall need the following Lemma 5. Let, as usual, $\Gamma(s)$ denote the gamma-function,

$$f(u,v) = \sum_{m=1}^{\infty}\sum_{k=1}^{\infty} m^{-u}(m+k)^{-v} \qquad \text{for} \quad \operatorname{Re} u > 1,\ \operatorname{Re} v > 1,$$

and let $g(u,v)$ be the analytic continuation of the function given by the equality

$$g(u,v) = f(u,v) - \frac{\Gamma(u+v-1)\Gamma(1-u)}{\Gamma(v)}\zeta(u+v-1)$$
$$= 2\sum_{m=1}^{\infty} \sigma_{1-u-v}(m)\int_0^{\infty} y^{-u}(1+y)^{-v}\cos(2\pi m y)\,dy.$$

LEMMA 5. *Let* $0 < \operatorname{Re} u < 1$, $0 < \operatorname{Re} v < 1$, *and* $u + v \neq 1$. *Then*

$$\begin{aligned}\zeta(u)\zeta(v) &= \zeta(u+v) + \zeta(u+v-1)\Gamma(u+v-1)\\ &\times \left(\frac{\Gamma(1-u)}{\Gamma(v)} + \frac{\Gamma(1-v)}{\Gamma(u)}\right)\\ &+ g(u,v) + g(v,u).\end{aligned}$$

For the proof see (Ivić, 1985, p. 443–446).

In view of Lemma 5 the integrand in the formula (5) may be written as follows

$$\begin{aligned}\zeta(z)\zeta(2\sigma_T - z) &= \zeta(2\sigma_T) + \zeta(2\sigma_T - 1)\Gamma(2\sigma_T - 1)\\ &\times \left(\frac{\Gamma(1-z)}{\Gamma(2\sigma_T - z)} + \frac{\Gamma(1-2\sigma_T+z)}{\Gamma(z)}\right)\\ &+ (g(z, 2\sigma_T - z) + g(2\sigma_T - z, z))\\ &\overset{\text{def}}{=} F_1(z) + F_2(z).\end{aligned} \tag{6}$$

LEMMA 6. *We have*

$$\int_{\sigma_T - iT}^{\sigma_T + iT} F_1(z)\,dz = 2iT\zeta(1+\delta_T) + 2i(2\pi e)^{\delta_T} T^{1-\delta_T}\zeta(1-\delta_T)(1-\delta_T)^{-1}$$

$$+ B\delta_T^{-1}\min(\delta_T^{-1}, \log T).$$

Proof. We shall use Stirling's formula

$$\log\Gamma(s) = \left(s - \frac{1}{2}\right)\log s - s + \log\sqrt{2\pi} + B|s|^{-1}, \tag{7}$$

which is valid on the region $|\arg s| \leqslant \pi - \varepsilon$, $\varepsilon > 0$. Here the constant bounding B depends only on ε. From (7) we deduce that

$$\begin{aligned}&\frac{\Gamma(1-z)}{\Gamma(2\sigma_T - z)}\\ &= \exp\left\{\left(\frac{1}{2} - z\right)\log(1-z) - \left(2\sigma_T - \frac{1}{2} - z\right)\log(2\sigma_T - z)\right.\\ &\left. -(1-z) + (2\sigma_T - z) + B|1-z|^{-1} + B|2\sigma_T - z|^{-1}\right\}\\ &= \exp\left\{-\delta_T\log(1-z) + \delta_T + B|1-z|^{-1} + B|2\sigma_T - z|^{-1}\right\}.\end{aligned} \tag{8}$$

Thus, from the equality (8)

$$\begin{aligned}&\int_{\sigma_T - iT}^{\sigma_T + iT} \frac{\Gamma(1-z)}{\Gamma(2\sigma_T - z)}\,dz\\ &= e^{\delta_T}\int_{\sigma_T - iT}^{\sigma_T + iT} \frac{dz}{(1-z)^{\delta_T}} + B\min\left(\delta_T^{-1}, \log T\right)\\ &= \frac{e^{\delta_T}}{(1-\delta_T)}\left[(1-\sigma_T+iT)^{1-\delta_T} - (1-\sigma_T-iT)^{1-\delta_T}\right]\\ &+ B\min\left(\delta_T^{-1}, \log T\right).\end{aligned} \tag{9}$$

It is easy to see that the term in brackets of the formula (9) is equal to

$$T^{1-\delta_T}\left(i^{1-\delta_T}-(-i)^{1-\delta_T}\right)+BT^{-\delta_T}. \tag{10}$$

Since

$$i^{1-\delta_T}=\sin\frac{\pi\delta_T}{2}+i\cos\frac{\pi\delta_T}{2},$$
$$(-i)^{1-\delta_T}=\sin\frac{\pi\delta_T}{2}-i\cos\frac{\pi\delta_T}{2},$$

it follows, in view of (9) and (10), that

$$\int\limits_{\sigma_T-iT}^{\sigma_T+iT}\frac{\Gamma(1-z)}{\Gamma(2\sigma_T-z)}\,dz=2i\,e^{\delta_T}T^{1-\delta_T}(1-\delta_T)^{-1}\cos\frac{\pi\delta_T}{2}\quad\cdot \tag{11}$$
$$+\,B\min\left(\delta_T^{-1},\log T\right).$$

Obviously,

$$\int\limits_{\sigma_T-iT}^{\sigma_T+iT}\frac{\Gamma(1-2\sigma_T+z)}{\Gamma(z)}\,dz=\int\limits_{\sigma_T-iT}^{\sigma_T+iT}\frac{\Gamma(1-z)}{\Gamma(2\sigma_T-z)}\,dz. \tag{12}$$

Using the functional equation for the Riemann zeta-function and the properties of the gamma-function we obtain

$$\zeta(2\sigma_T-1)\Gamma(2\sigma_T-1)=\frac{(2\pi)^{\delta_T}}{2\cos(\pi\delta_T/2)}\zeta(1-\delta_T). \tag{13}$$

Besides, it is well known that

$$\zeta(s)=(s-1)^{-1}+\gamma+B|s-1|. \tag{14}$$

Consequently, the assertion of the lemma follows from (6) and (11)–(14).
Let x be a noninteger positive number and let

$$D_{-\delta_T}(x)=\sum_{m\leqslant x}\sigma_{-\delta_T}(m),$$

$$\Delta_{-\delta_T}=D_{-\delta_T}(x)-x\zeta(1+\delta_T)-x^{1-\delta_T}\zeta(1-\delta_T)(1-\delta_T)^{-1}+\frac{\zeta(\delta_T)}{2}. \tag{15}$$

For the proof of Theorem 4 we shall need the explicit formula for the function $\Delta_{-\delta_T}(x)$ as well as its estimate. Let $J_\nu(z)$, $Y_\nu(z)$, and $K_\nu(z)$ denote the Bessel functions.

LEMMA 7. *We have*

$$\Delta_{-\delta_T}(x)=\frac{x(2\pi)^{1+\delta_T}}{2\sin(\pi\delta_T/2)}\sum_{m=1}^{\infty}\sigma_{\delta_T}(m)\left(2\pi\sqrt{mx}\right)^{-1-\delta_T}\Big(J_{-1+\delta_T}(4\pi\sqrt{mx})$$
$$+\,J_{1-\delta_T}\left(4\pi\sqrt{mx}\right)-\frac{2}{\pi}K_{1-\delta_T}\left(4\pi\sqrt{mx}\right)\sin\pi\delta_T\Big).$$

LEMMA 8. *For any $\varepsilon > 0$*

$$\Delta_{-\delta_T}(x) = Bx^{\frac{1}{3}-\frac{\delta_T}{6}+\varepsilon}.$$

Proofs of Lemmas 7 and 8 are given in (Laurinčikas, 1992).
Let, as in (Ivić, 1985, p. 447),

$$h(s,x) = 2\int_0^\infty y^{-s}(1+y)^{s-2\sigma_T}\cos(2\pi xy)\,dy.$$

It is clear that the function $h(s,x)$ is analytic on the half-plane $\sigma < 1$. On the half-plane $\sigma < 0$ we have

$$g(s, 2\sigma_T - s) = 2\sum_{m=1}^\infty \sigma_{-\delta_T}(m)\int_0^\infty y^{-s}(1+y)^{s-2\sigma_T}\cos(2\pi my)\,dy.$$

Now we must give an analytic continuation of the function $g(s, 2\sigma_T - s)$ to the region containing the line $\sigma = \sigma_T$. Let N be a natural number, restricted in the range $c_1T < N < c_2T$, and let $X = N + 1/2$. Then, taking into account the equality (15) and Lemma 8, we find

$$\begin{aligned}\sum_{m>N}\sigma_{-\delta_T}(m)h(s,m) &= \int_X^\infty h(s,x)\,dD_{-\delta_T}(x)\\ &= \zeta(1+\delta_T)\int_X^\infty h(s,x)\,dx + \zeta(1-\delta_T)\int_X^\infty x^{-\delta_T}h(s,x)\,dx\\ &\quad - \Delta_{-\delta_T}(X)h(s,X) - \int_X^\infty \Delta_{-\delta_T}(x)\frac{\partial h(s,x)}{\partial x}\,dx,\end{aligned}$$

since (Ivić, 1985, p. 445)

$$h(s,x) = Bx^{\sigma-1}|s-1|^{-1} \qquad \text{for} \quad -1 < \sigma < 1.$$

Thus,

$$\begin{aligned} g(s, 2\sigma_T - s) &= \sum_{m\leqslant N}\sigma_{-\delta_T}(m)h(s,m)\\ &\quad + \zeta(1+\delta_T)\int_X^\infty h(s,x)\,dx + \zeta(1-\delta_T)\int_X^\infty x^{-\delta_T}h(s,x)\,dx\\ &\quad - \Delta_{-\delta_T}(X)h(s,X) - \int_X^\infty \Delta_{-\delta_T}(x)\frac{\partial h(s,x)}{\partial x}\,dx\\ &\stackrel{\text{def}}{=} g_1(s) + g_2(s) + g_3(s) - g_4(s) - g_5(s).\end{aligned} \tag{16}$$

Since $h(s,x)$ is an analytic function on the strip $0 < \sigma < 1$, the functions $g_1(s)$ and $g_4(s)$ are also analytic on this region. It remains to consider $g_2(s)$, $g_3(s)$, and $g_5(s)$. In a similar manner as in (Ivić, 1985, p. 448), we obtain the estimate

$$\frac{\partial h(s,x)}{\partial x} = Bx^{\sigma-2} \qquad \text{on the strip} \quad -1 < \sigma < 1.$$

Hence, according to Lemma 8, $g_5(s)$ is an analytic function on the strip

$$0 < \sigma < \frac{2}{3} + \frac{\sigma_T}{6}.$$

Now we shall consider $g_2(s)$. When $\sigma < 0$, we see that

$$g_2(s) = \zeta(1+\delta_T) \int_X^\infty \left(\int_0^{i\infty} y^{-s}(1+y)^{s-2\sigma_T} e^{2\pi i x y}\, dy + \int_0^{-i\infty} y^{-s}(1+y)^{s-2\sigma_T} e^{-2\pi i x y}\, dy \right) dx. \tag{17}$$

Since for $\sigma < 0$

$$\int_X^\infty \left(\int_0^{i\infty} y^{-s}(1+y)^{s-2\sigma_T} e^{2\pi i x y}\, dy \right) dx = -\frac{1}{2\pi i} \int_0^\infty \frac{(1+y)^{s-2\sigma_T}}{y^{1+s}} e^{2\pi i X y}\, dy,$$

$$\int_X^\infty \left(\int_0^{-i\infty} y^{-s}(1+y)^{s-2\sigma_T} e^{-2\pi i x y}\, dy \right) dx = \frac{1}{2\pi i} \int_0^\infty \frac{(1+y)^{s-2\sigma_T}}{y^{1+s}} e^{-2\pi i X y}\, dy,$$

we have in view of the equality (17) that

$$g_2(s) = -\frac{1}{\pi}\zeta(1+\delta_T) \int_0^\infty \frac{(1+y)^{s-2\sigma_T}}{y^{1+s}} \sin(2\pi X y)\, dy. \tag{18}$$

The ' ι:·ι integral is uniformly convergent when $\sigma \leqslant 1-\varepsilon$. Consequently, the equality (18) gives us an analytic continuation of the function $g_2(s)$ to the region which contains the line $\sigma = \sigma_T$.

Assuming $\sigma < 0$ and using integration by parts we find, that

$$\begin{aligned} J^+ &\stackrel{\text{def}}{=} \int_X^\infty x^{-\delta_T} \int_0^{i\infty} \frac{(1+y)^{s-2\sigma_T}}{y^s} e^{2\pi i x y}\, dx\, dy \\ &= -\frac{X^{-\delta_T}}{2\pi i} \int_0^{i\infty} \frac{(1+y)^{s-2\sigma_T}}{y^{1+s}} e^{2\pi i X y}\, dy \\ &\quad + \frac{\delta_T}{2\pi i} \int_X^\infty \int_0^{i\infty} \frac{(x+y)^{s-2\sigma_T}}{y^{1+s}} e^{2\pi i x y}\, dx\, dy. \end{aligned} \tag{19}$$

It is easy to see that

$$\int_X^\infty \int_0^{i\infty} \frac{(x+y)^{s-2\sigma_T}}{y^{1+s}} e^{2\pi i x y}\, dx\, dy = -\frac{X^{-\delta_T}}{s-\delta_T} \int_0^\infty \frac{(1+y)^{s-\delta_T}}{y^{1+s}} e^{2\pi i X y}\, dy.$$

From this and from (19) we obtain that for $\sigma < 0$

$$\begin{aligned} J^+ = &-\frac{X^{-\delta_T}}{2\pi i} \int_0^\infty \frac{(1+y)^{s-2\sigma_T}}{y^{1+s}} e^{2\pi i X y}\, dy \\ &- \frac{\delta_T X^{-\delta_T}}{2\pi i(s-\delta_T)} \int_0^\infty \frac{(1+y)^{s-\delta_T}}{y^{1+s}} e^{-2\pi i X y}\, dy. \end{aligned} \tag{20}$$

Reasoning similarly as above we find that for $\sigma < 0$

$$\begin{aligned} J^- &\stackrel{\text{def}}{=} \int_X^\infty x^{-\delta_T} \Big(\int_0^{-i\infty} y^{-s}(1+y)^{s-2\sigma_T} e^{-2\pi i x y}\, dy \Big)\, dx \\ &= \frac{X^{-\delta_T}}{2\pi i} \int_0^\infty \frac{(1+y)^{s-2\sigma_T}}{y^{1+s}} e^{-2\pi i X y}\, dy \\ &+ \frac{\delta_T X^{-\delta_T}}{2\pi i(s-\delta_T)} \int_0^\infty \frac{(1+y)^{s-\delta_T}}{y^{1+s}} e^{-2\pi i X y}\, dy. \end{aligned} \tag{21}$$

Since on the half-plane $\sigma < 0$

$$g_3(s) = \zeta(1-\delta_T)(J^+ + J^-),$$

it follows from (20) and (21) that

$$\begin{aligned} g_3(s) = &-\zeta(1-\delta_T)\Big(\frac{X^{-\delta_T}}{\pi} \int_0^\infty \frac{(1+y)^{s-2\sigma_T}}{y^{1+s}} \sin(2\pi X y)\, dy \\ &+ \frac{\delta_T X^{-\delta_T}}{\pi(s-\delta_T)} \int_0^\infty \frac{(1+y)^{s-\delta_T}}{y^{1+s}} \sin(2\pi X y)\, dy \Big). \end{aligned} \tag{22}$$

In the last equality the integrals are uniformly convergent when $\sigma \leqslant 1-\varepsilon$, and they define analytic functions on the region $\sigma < 1$. Thus, the equality (22) gives us an analytic continuation of the function $g_3(s)$ which is valid when $\sigma = \sigma_T$. Now, by the properties of $g_i(s)$, $1 \leqslant i \leqslant 5$, in virtue of (16), we obtain that the function $g(s, 2\sigma_T - s)$ is analytic on the region which contains the line $\sigma = \sigma_T$.

Thus, the integration over this line is possible. Moreover, it can be easily seen that

$$\int_{\sigma_T-iT}^{\sigma_T+iT} g(s,2\sigma_T-s)\,ds+\int_{\sigma_T-iT}^{\sigma_T+iT} g(2\sigma_T-s,s)\,ds=2\int_{\sigma_T-iT}^{\sigma_T+iT} g(s,2\sigma_T-s)\,ds. \quad (23)$$

We have from the formula (16) that

$$\int_{\sigma_T-iT}^{\sigma_T+iT} g(s,2\sigma_T-s)\,ds \stackrel{\text{def}}{=} I_1+I_2+I_3-I_4-I_5 \quad (24)$$

where

$$I_j=\int_{\sigma_T-iT}^{\sigma_T+iT} g_j(s)\,ds, \qquad j=1,\dots,5.$$

For the evaluation of the integrals I_j we shall apply, as in (Ivić, 1985), the following lemma.

LEMMA 9. *Let α, β, γ, a, b, k, T be real numbers such that α, β, γ are positive and bounded, $\alpha\neq 1$, $0<a<1/2$, $a<T/8\pi k$, $b\geqslant T$, $k\geqslant 1$, and $T\geqslant 1$. Then*

$$\begin{aligned}\int_a^b & y^{-\alpha}(1+y)^{-\beta}(\lambda(y))^{-\gamma}\exp\Big\{iT\lambda(y)+2\pi iky\Big\}\,dy\\ &=\big(2k\sqrt{\pi}\big)^{-1}\sqrt{T}V^{-\gamma}U^{-1/2}\Big(U-\frac12\Big)^{-\alpha}\Big(U+\frac12\Big)^{-\beta}\\ &\times\exp\Big\{i\Big(TV+2\pi kU-\pi k+\frac{\pi}{4}\Big)\Big\}\\ &+Ba^{1-\alpha}T^{-1}+Bb^{\gamma-\alpha-\beta}k^{-1}+R(T,k)\end{aligned}$$

uniformly for $|\alpha-1|>\varepsilon$, where

$$\lambda(y)=\log\frac{1+y}{y},\qquad U=\Big(\frac{T}{2\pi k}+\frac14\Big)^{1/2},\qquad V=2\sinh^{-1}\Big(\sqrt{\frac{\pi k}{2T}}\Big),$$

$$R(T,k)=B\begin{cases}T^{\frac{\gamma-\alpha-\beta}{2}-\frac14}k^{-\frac{\gamma-\alpha-\beta}{2}-\frac54} & \text{for}\quad 1\leqslant k\leqslant T,\\ T^{-\frac12-\alpha}k^{\alpha-1} & \text{for}\quad k\geqslant T.\end{cases}$$

A similar result holds for the corresponding integral with $-k$ in place of k, except that in that case the principal term on the right-hand side of the equality vanishes.

For the proof see (Ivić, 1985, p. 452–453)

Evaluation of I_1. From the definitions of the function $g_1(s)$ (the equality (16)) and $h(s,x)$ we have that

$$
\begin{aligned}
I_1 &= 2\sum_{m\leqslant N}\sigma_{-\delta_T}(m)\int_0^\infty \cos(2\pi my)\int_{\sigma_T-iT}^{\sigma_T+iT}(1+y)^{-2\sigma_T}\left(\frac{1+y}{y}\right)^s ds\,dy\\
&= 2\sum_{m\leqslant N}\sigma_{-\delta_T}(m)\int_0^\infty \cos(2\pi my)(1+y)^{-2\sigma_T}\left(\frac{1+y}{y}\right)^s(\lambda(y))^{-1}\Big|_{\sigma_T-iT}^{\sigma_T+iT}dy \qquad (25)\\
&= 4i\sum_{m\leqslant N}\sigma_{-\delta_T}(m)\int_0^\infty \cos(2\pi my)\frac{\sin(T\lambda(y))}{y^{\sigma_T}(1+y)^{\sigma_T}\lambda(y)}\,dy.
\end{aligned}
$$

For the integral in (25) we shall apply Lemma 9 with $\alpha,\beta=\sigma_T$, $\gamma=1$, $k=m$, and $k=-m$. Thus since $2\sigma_T>1$, we have for $m\leqslant cT$

$$
\begin{aligned}
&\int_0^\infty \frac{\sin(T\lambda(y))\cos(2\pi my)}{y^{\sigma_T}(1+y)^{\sigma_T}\lambda(y)}\,dy\\
&= \frac{1}{4m}\left(\frac{T}{\pi}\right)^{1/2}\frac{\sin\left(TV+2\pi mU-\pi m+\frac{\pi}{4}\right)}{VU^{1/2}\left(U-\frac12\right)^{\sigma_T}\left(U+\frac12\right)^{\sigma_T}}+BT^{-\frac{\delta_T}{2}-\frac14}m^{\frac{\delta_T}{2}-\frac54}\\
&= \frac{(-1)^m 2^{\sigma_T-1}\sin\left(f(T,m)+\frac{\pi}{2}\right)}{4m^{1-\sigma_T}T^{\sigma_T-(1/2)}\pi^{(1/2)-\sigma_T}\left(\frac{T}{2\pi m}+\frac14\right)^{1/4}\sinh^{-1}\left(\sqrt{\frac{\pi m}{2T}}\right)}+BT^{-\frac{\delta_T}{2}-\frac14}m^{\frac{\delta_T}{2}-\frac54}.
\end{aligned}
$$

From this and from (25), taking into account the inequality $c_1T<N<c_2T$, we obtain

$$
\begin{aligned}
I_1 &= \frac{i}{\sqrt2}\left(\frac{2\pi}{T}\right)^{\delta_T/2}\sum_{m\leqslant N}\frac{(-1)^m\sigma_{\delta_T}(m)}{m^{(1+\delta_T)/2}}\frac{\cos(f(T,m))}{\left(\frac{T}{2\pi m}+\frac14\right)^{1/4}\sinh^{-1}\left(\sqrt{\frac{\pi m}{2T}}\right)} \qquad (26)\\
&\quad + BT^{-\frac{\delta_T}{2}-\frac14}.
\end{aligned}
$$

Estimation of I_4. Reasoning similarly as in the case of the integral I_1, since $c_3T<X<c_4T$, we find

$$
\begin{aligned}
I_4 &= 2\Delta_{-\delta_T}(X)\int_0^\infty \cos(2\pi Xy)\int_{\sigma_T-iT}^{\sigma_T+iT}y^{-s}(1+y)^{s-2\sigma_T}\,dy\,ds\\
&= 4i\Delta_{-\delta_T}(X)\int_0^\infty \frac{\sin(T\lambda(y))\cos(2\pi Xy)}{y^{\sigma_T}(1+y)^{\sigma_T}\lambda(y)}\,dy\\
&= i\Delta_{-\delta_T}(X)\left(\frac{\sin\left(f(T,X)+\frac{\pi}{2}-\pi X\right)}{X^{1-\sigma_T}T^{\sigma_T-(1/2)}\left(\frac{T}{2\pi X}+\frac14\right)^{1/4}\sinh^{-1}\left(\sqrt{\frac{\pi X}{2T}}\right)}\frac{2^{\sigma_T-1}}{\pi^{-\delta_T/2}}\right.\\
&\quad \left.+BT^{-\frac{\delta_T}{2}-\frac14}X^{\frac{\delta_T}{2}-\frac54}\right).
\end{aligned}
$$

Hence in view of Lemma 8 the estimate

$$I_4 = BT^{-\frac{1}{6}-\frac{\delta_T}{6}+\varepsilon} \tag{27}$$

follows.

Estimation of I_2. The equality (18) gives us

$$\begin{aligned} I_2 &= -\frac{1}{\pi}\zeta(1+\delta_T)\int\limits_0^\infty \int\limits_{\sigma_T-iT}^{\sigma_T+iT} \frac{(1+y)^{s-2\sigma_T}}{y^{1+s}}\sin(2\pi Xy)\,ds\,dy \\ &= -2i\frac{1}{\pi}\zeta(1+\delta_T)\int\limits_0^\infty \frac{\sin(T\lambda(y))\sin(2\pi Xy)}{y^{1+\sigma_T}(1+y)^{\sigma_T}\lambda(y)}\,dy. \end{aligned} \tag{28}$$

We write integral in (28) as a sum of two integrals

$$\int\limits_0^\infty \ldots dy = \int\limits_0^{(2X)^{-1}} \ldots dy + \int\limits_{(2X)^{-1}}^\infty \ldots dy. \tag{29}$$

Applying twice the second mean value theorem we obtain that

$$\begin{aligned} \int\limits_0^{(2X)^{-1}} \ldots dy &= 2\pi X\int\limits_0^{\xi_1} \frac{\sin(T\lambda(y))}{y(1+y)}\,\frac{y^{1-\sigma_T}(1+y)^{1-\sigma_T}}{\lambda(y)}\,dy \\ &= 2\pi X\xi_1^{1-\sigma_T}(1+\xi_1)^{1-\sigma_T}(\lambda(\xi_1))^{-1}\int\limits_{\xi_2}^{\xi_1}\frac{\sin(T\lambda(y))}{y(1+y)}\,dy \\ &= 2\pi X\xi_1^{1-\sigma_T}(1+\xi_1)^{1-\sigma_T}(\lambda(\xi_1))^{-1}\frac{1}{T}\cos(T\lambda(y))\Big|_{\xi_2}^{\xi_1} \\ &= BT^{\sigma_T-1}. \end{aligned} \tag{30}$$

Here $0 \leqslant \xi_2 \leqslant \xi_1 \leqslant (2X)^{-1}$. Besides, by Lemma 9

$$\int\limits_{(2X)^{-1}}^\infty \ldots dy = BT^{\sigma_T-1}. \tag{31}$$

Thus, from (28)–(31) we deduce the estimate

$$I_2 = BT^{\sigma_T-1}|\zeta(1+\delta_T)|. \tag{32}$$

Evaluation of I_3. By applying the formula (22) we find that

$$\begin{aligned} I_3 = &-\frac{1}{\pi}\zeta(1-\delta_T)X^{-\delta_T}\int\limits_0^\infty \int\limits_{\sigma_T-iT}^{\sigma_T+iT} \frac{(1+y)^{s-2\sigma_T}}{y^{1+s}}\sin(2\pi Xy)\,ds\,dy \\ &+\frac{1}{\pi}\zeta(1-\delta_T)\delta_T X^{-\delta_T}\int\limits_0^\infty \int\limits_{\sigma_T-iT}^{\sigma_T+iT} \frac{(1+y)^{s-\delta_T}}{(s-\delta_T)y^{1+s}}\sin(2\pi Xy)\,ds\,dy. \end{aligned} \tag{33}$$

The first integral in (33) is similar to I_2, therefore it is estimated as

$$BT^{\sigma_T-1-\delta_T}|\zeta(1-\delta_T)|. \tag{34}$$

The second integral in (33) will be denoted by I_3', and we shall write it as a sum of two terms

$$I_3' = \int_0^1 \dots ds\, dy + \int_1^\infty \dots ds\, dy \stackrel{\text{def}}{=} I_{31}' + I_{32}'. \tag{35}$$

The integrand in I_{31}' has a simple pole at $s = \delta_T$, hence by the residue theorem

$$\int_{\sigma_T-iT}^{\sigma_T+iT} \left(\frac{1+y}{y}\right)^{s-\delta_T} \frac{ds}{s-\delta_T} = 2\pi i - \left(\int_{\sigma_T+iT}^{-\infty+iT} + \int_{-\infty-iT}^{\sigma_T-iT}\right)\left(\frac{1+y}{y}\right)^{s-\delta_T} \frac{ds}{s-\delta_T}. \tag{36}$$

Since for $0 < y \leqslant 1$

$$\int_{\sigma_T\pm iT}^{-\infty\pm iT} \left(\frac{1+y}{y}\right)^{s-\delta_T} \frac{ds}{s-\delta_T} = \frac{B}{T}\int_{-\infty}^{\sigma_T-\delta_T} \left(\frac{1+y}{y}\right)^t dt = BT^{-1}y^{-\sigma_T+\delta_T},$$

it follows from (35) and (36) that

$$I_{31}' = 2\pi i \int_0^1 \frac{\sin(2\pi Xy)\, dy}{y^{1+\delta_T}} + \frac{B}{T}\int_0^T \frac{|\sin(2\pi Xy)|}{y^{1+\sigma_T}}\, dy. \tag{37}$$

Let us write

$$v(y) = \int_0^y \frac{\sin(2\pi Xu)}{u}\, du.$$

Integrating by parts we find that

$$\begin{aligned}\int_0^1 \frac{\sin(2\pi Xy)}{y^{1+\delta_T}}\, dy &= \int_0^1 y^{-\delta_T}\, dv(y) = v(1) + \delta_T \int_0^1 y^{-1-\delta_T} v(y)\, dy \\ &= \frac{\pi}{2} + \frac{B}{X} + B\delta_T \int_0^{X^{-1}} y^{-1-\delta_T}|v(y)|\, dy \\ &\quad + B\delta_T \int_{X^{-1}}^1 y^{-1-\delta_T}|v(y)|\, dy \\ &= \frac{\pi}{2} + \frac{B}{X} + B\delta_T T^{\delta_T}\min\left(\delta_T^{-1}, \log T\right).\end{aligned} \tag{38}$$

Moreover, in a similar manner we obtain that

$$\int_0^1 y^{-1-\sigma_T} |\sin(2\pi X y)|\, dy = BT^{\sigma_T}. \tag{39}$$

Now from (37)–(39) we conclude that

$$I'_{31} = \pi^2 i + B\delta_T T^{\delta_T} \min(\delta_1^{-1}, \log T) + BT^{\sigma_T - 1}. \tag{40}$$

It remains to consider I'_{32}. Using the integration by parts we have

$$\begin{aligned} I'_{32} &= -\int_1^\infty y^{-1-\delta_T} \int_{\sigma_T - iT}^{\sigma_T + iT} \left(\frac{1+y}{y}\right)^{s-\delta_T} \frac{ds}{s-\delta_T} \frac{d\cos(2\pi X y)}{2\pi X} \\ &= \left(-\frac{\cos(2\pi X y)}{2\pi X y^{1+\delta_T}} \int_{\sigma_T - iT}^{\sigma_T + iT} \left(\frac{1+y}{y}\right)^{s-\delta_T} \frac{ds}{s-\delta_T} \right)\Bigg|_1^\infty \\ &\quad - (1+\delta_T) \int_1^\infty \frac{\cos(2\pi X y)}{2\pi X y^{2+\delta_T}}\, dy \int_{\sigma_T - iT}^{\sigma_T + iT} \left(\frac{1+y}{y}\right)^{s-\delta_T} \frac{ds}{s-\delta_T} \\ &\quad - \int_1^\infty \frac{\cos(2\pi X y)}{2\pi X y^{1+\delta_T}}\, dy \int_{\sigma_T - iT}^{\sigma_T + iT} \left(\frac{1+y}{y}\right)^{s-1-\delta_T} \frac{ds}{y^2}. \end{aligned} \tag{41}$$

Since for $y \geqslant 1$

$$\int_{\sigma_T - iT}^{\sigma_T + iT} \left(\frac{1+y}{y}\right)^{s-\delta_T} \frac{ds}{s-\delta_T} = B \int_{\sigma_T - iT}^{\sigma_T + iT} \frac{|ds|}{|s-\delta_T|} = B \log T,$$

$$\int_{\sigma_T - iT}^{\sigma_T + iT} \left(\frac{1+y}{y}\right)^{s-1-\delta_T} ds = By,$$

we have from (41) the estimate

$$I'_{32} = BT^{-1} \log T. \tag{42}$$

Consequently, from (33), (34), (40), and (42) it is seen that

$$\begin{aligned} I_3 &= -\pi i \zeta(1-\delta_T)\delta_T X^{-\delta_T} \\ &\quad + B\delta_T^2 |\zeta(1-\delta_T)| \min(\delta_T^{-1}, \log T) + BT^{\sigma_T - 1 - \delta_T} |\zeta(1-\delta_T)|. \end{aligned} \tag{43}$$

Evaluation of I_5. Taking into account the equalities (16) and (24) we can write that

$$I_5 = \int_X^\infty \Delta_{-\delta_T}(x) \int_{\sigma_T - iT}^{\sigma_T + iT} \frac{\partial h(s,x)}{\partial x}\, dx\, ds. \tag{44}$$

From the definition of $h(s,x)$ it follows that (with $\lambda_1(x,y) = \log((x+y)/y)$)

$$\begin{aligned}
\int_{\sigma_T - iT}^{\sigma_T + iT} \frac{\partial h(s,x)}{\partial x}\, ds &= 4i \frac{\partial}{\partial x} \int_0^\infty \frac{\sin(T\lambda(y))\cos(2\pi xy)}{y^{\sigma_T}(1+y)^{\sigma_T}\lambda(y)}\, dy \\
&= 4i \frac{\partial}{\partial x} \int_0^\infty \frac{\sin(T\lambda_1(x,y))\cos(2\pi y) x^{\delta_T}}{y^{\sigma_T}(x+y)^{\sigma_T}\lambda_1(x,y)}\, dy \\
&= 4i \int_0^\infty \frac{x^{\delta_T}\cos(2\pi y)}{y^{\sigma_T}(x+y)^{\sigma_T+1}\lambda_1(x,y)} \big(T\cos(T\lambda_1(x,y)) \\
&\quad - \big(\sigma_T + (\lambda_1(x,y))^{-1}\big)\sin(T\lambda_1(x,y))\big)\, dy \\
&\quad + 4i \int_0^\infty \frac{\delta_T x^{\delta_T - 1}\sin(T\lambda_1(x,y))\cos(2\pi y)}{y^{\sigma_T}(x+y)^{\sigma_T}\lambda_1(x,y)}\, dy.
\end{aligned}$$

From this and from (44), after the change of the variable $y = xy'$, we obtain that

$$\begin{aligned}
I_5 &= 4i \int_X^\infty \frac{\Delta_{-\delta_T}(x)}{x} \int_0^\infty \frac{\cos(2\pi xy)}{y^{\sigma_T}(1+y)^{1+\sigma_T}\lambda(y)} \\
&\quad \times \big(T\cos(T\lambda(y)) - \big(\sigma_T + (\lambda(y))^{-1}\big)\sin(T\lambda(y))\, dx\, dy \\
&\quad + 4i\delta_T \int_X^\infty \frac{\Delta_{-\delta_T}(x)}{x} \int_0^\infty \frac{\cos(2\pi xy)\sin(T\lambda(y))}{y^{\sigma_T}(1+y)^{\sigma_T}\lambda(y)}\, dx\, dy \\
&\stackrel{\text{def}}{=} 4i(I_5' + \delta_T I_5'').
\end{aligned} \tag{45}$$

Using Lemma 9 with $k = x > cT$ we find

$$\begin{aligned}
&\int_0^\infty \frac{\cos(2\pi xy)\sin(T\lambda(y))}{y^{\sigma_T}(1+y)^{\sigma_T}\lambda(y)}\, dy \\
&= \frac{1}{4x}\left(\frac{T}{\pi}\right)^{1/2} \frac{\sin\left(TV + 2\pi xU - \pi x + \frac{\pi}{4}\right)}{VU^{1/2}\left(U - \frac{1}{2}\right)^{\sigma_T}\left(U + \frac{1}{2}\right)^{\sigma_T}} + BT^{-\sigma_T - (1/2)} x^{\sigma_T - 1},
\end{aligned} \tag{46}$$

$$\begin{aligned}
&\int_0^\infty \frac{\cos(T\lambda(y))\cos(2\pi xy)}{y^{\sigma_T}(1+y)^{1+\sigma_T}\lambda(y)}\, dy \\
&= \frac{1}{4x}\left(\frac{T}{\pi}\right)^{1/2} \frac{\cos\left(TV + 2\pi xU - \pi x + \frac{\pi}{4}\right)}{VU^{1/2}\left(U - \frac{1}{2}\right)^{\sigma_T}\left(U + \frac{1}{2}\right)^{\sigma_T + 1}} + BT^{-\sigma_T - (1/2)} x^{\sigma_T - 1}.
\end{aligned} \tag{47}$$

Similarly, in virtue of the same lemma and the properties of U and V the estimate

$$\int_0^\infty \frac{\sin(T\lambda(y))\cos(2\pi xy)}{y^{\sigma_T}(1+y)^{1+\sigma_T}(\lambda(y))^r}\,dy = BT^{-\delta_T/2}x^{\sigma_T-1} \tag{48}$$

is valid for $r = 1, 2$.

Thus, from (47), (48) and from the definition of U and V we have

$$\begin{aligned} I_5' &= \frac{1}{4}\left(\frac{\pi}{T}\right)^{\delta_T/2}\frac{1}{2^{1-\sigma_T}} \\ &\times \int_X^\infty \left(\frac{\Delta_{-\delta_T}(x)T\cos\left(f(T,x)-\pi x+\frac{\pi}{2}\right)}{x^{2-\sigma_T}\left(\left(\frac{T}{2\pi x}+\frac{1}{4}\right)^{1/2}+\frac{1}{2}\right)\left(\frac{T}{2\pi x}+\frac{1}{4}\right)^{1/4}\sinh^{-1}\left(\sqrt{\frac{\pi x}{2T}}\right)}\right. \\ &\left. + B\Delta_{-\delta_T}(x)x^{\sigma_T-2}\right)dx. \end{aligned} \tag{49}$$

In view of known formulae (see (Watson, 1945, p. 361)), the equality

$$\left(2\sin\frac{\pi\delta_T}{2}\right)^{-1}(J_{-1+\delta_T}(z)+J_{1-\delta_T}(z)) = \sin\frac{\pi\delta_T}{2}J_{1-\delta_T}(z) - \cos\frac{\pi\delta_T}{2}Y_{1-\delta_T}(z)$$

is valid. We shall also use the following asymptotic formulae for the Bessel functions (Watson, 1945, p. 222–226):

$$\begin{aligned} J_\nu(z) &\sim \left(\frac{2}{\pi z}\right)^{1/2}\left(\cos\left(z-\frac{\pi\nu}{2}-\frac{\pi}{4}\right)\sum_{m=0}^\infty \frac{(-1)^m(\nu,2m)}{(2z)^{2m}}\right. \\ &\quad \left. -\sin\left(z-\frac{\pi\nu}{2}-\frac{\pi}{4}\right)\sum_{m=0}^\infty\frac{(-1)^m(\nu,2m+1)}{(2z)^{2m+1}}\right), \\ Y_\nu(z) &\sim \left(\frac{2}{\pi z}\right)^{1/2}\left(\sin\left(z-\frac{\pi\nu}{2}-\frac{\pi}{4}\right)\sum_{m=0}^\infty \frac{(-1)^m(\nu,2m)}{(2z)^{2m}}\right. \\ &\quad \left. +\cos\left(z-\frac{\pi\nu}{2}-\frac{\pi}{4}\right)\sum_{m=0}^\infty\frac{(-1)^m(\nu,2m+1)}{(2z)^{2m+1}}\right), \\ K_\nu(z) &\sim \left(\frac{\pi}{2z}\right)^{1/2}e^{-z}\sum_{m=0}^\infty\frac{(\nu,m)}{(2z)^m}, \end{aligned}$$

where

$$(\nu,m) = \frac{\Gamma(\nu+m+1/2)}{m!\Gamma(\nu-m+1/2)},$$

$|z| \to \infty$, and $|\arg z| < \pi$. We observe that these asymptotic relations are uniform in ν, $1/2 \leqslant \nu \leqslant 1$, as it follows from their proofs. Applying Lemma 7 and using the above listed formulae for the Bessel functions, we find

$$\begin{aligned}\Delta_{-\delta_T}(x) &= x(2\pi)^{1+\delta_T}\sum_{m=1}^{\infty}\sigma_{\delta_T}(m)\Big\{\big(2\pi\sqrt{mx}\big)^{-1-\delta_T}\Big(\frac{1}{2\pi^2\sqrt{mx}}\Big)^{1/2}\\ &\times\Big[\Big(\sin\frac{\pi\delta_T}{2}\cos\varkappa(m,x,\delta_T)+\cos\frac{\pi\delta_T}{2}\sin\varkappa(m,x,\delta_T)\Big)\\ &-\Big(\sin\frac{\pi\delta_T}{2}\sin\varkappa(m,x,\delta_T)+\cos\frac{\pi\delta_T}{2}\cos\varkappa(m,x,\delta_T)\Big)\\ &\times\frac{(3-2\delta_T)(1-2\delta_T)}{32\pi\sqrt{mx}}\Big]+\frac{B}{mx}\Big\}\\ &= \frac{x^{(1-2\delta_T)/4}}{\pi\sqrt{2}}\sum_{m=1}^{\infty}\frac{\sigma_{\delta_T}(m)}{m^{(3+2\delta_T)/4}}\Big[\cos\Big(4\pi\sqrt{mx}-\frac{\pi}{4}\Big)\\ &= -\frac{(3-2\delta_T)(1-2\delta_T)}{32\pi\sqrt{mx}}\sin\Big(4\pi\sqrt{mx}-\frac{\pi}{4}\Big)\Big]+Bx^{-(3+2\delta_T)/4},\end{aligned}\tag{50}$$

where

$$\varkappa(m,x,\delta_T)=4\pi\sqrt{mx}-\frac{(1-\delta_T)\pi}{2}-\frac{\pi}{4}.$$

Now Lemma 8, (49), and (50) give us

$$\begin{aligned}I_5' &= \frac{2^{\sigma_T}}{4}T^{1-(\delta_T/2)}\pi^{\delta_T/2}\\ &\times\int_{\sqrt{X}}^{\infty}\frac{\Delta_{-\delta_T}(x^2)\cos\big(f(T,x^2)-\pi x^2+\frac{\pi}{2}\big)\,dx}{x^{3-2\sigma_T}\big(\big(\frac{T}{2\pi x^2}+\frac14\big)^{1/2}+\frac12\big)\big(\frac{T}{2\pi x^2}+\frac14\big)^{1/4}\sinh^{-1}\big(x\sqrt{\frac{\pi}{2T}}\big)}\\ &+BT^{-\delta_T/2}\int_X^{\infty}x^{\sigma_T-\frac53-\frac{\delta_T}{6}+\varepsilon}\,dx\\ &= \Big(\frac{T}{\pi}\Big)^{1-(\delta_T/2)}2^{-(4-\delta_T)/2}\sum_{m=1}^{\infty}\frac{\sigma_{\delta_T}(m)}{m^{(3+2\delta_T)/4}}\\ &\times\int_{\sqrt{X}}^{\infty}\frac{\cos\big(f(T,x^2)-\pi x^2+\frac{\pi}{2}\big)}{x^{3/2}\big(\big(\frac{T}{2\pi x^2}+\frac14\big)^{1/2}+\frac12\big)\big(\frac{T}{2\pi x^2}+\frac14\big)^{1/4}\sinh^{-1}\big(x\sqrt{\frac{\pi}{2T}}\big)}\\ &\times\Big(\cos\Big(4\pi x\sqrt{m}-\frac{\pi}{4}\Big)-\frac{(3-2\delta_T)(1-2\delta_T)}{32\pi x\sqrt{m}}\\ &\times\sin\Big(4\pi x\sqrt{m}-\frac{\pi}{4}\Big)\Big)\,dx+BT^{-(1+\delta_T)/6+\varepsilon}\\ &\overset{\text{def}}{=}\Big(\frac{T}{\pi}\Big)^{1-(\delta_T/2)}2^{-(5-2\sigma_T)/2}\sum_{m=1}^{\infty}\frac{\sigma_{\delta_T}(m)}{m^{(3+2\delta_T)/4}}J_5\\ &+BT^{-(1+\delta_T)/6+\varepsilon}.\end{aligned}\tag{51}$$

The integral J_5 differs from one in the proof of the Atkinson formula on the critical line (Ivić, 1985, p. 458) only by the factor at $\sin(4\pi x\sqrt{m}-\pi/4)$. Therefore, putting

$$Z=\Big(\frac{T}{2\pi}+\frac{X}{2}\Big)-\Big(\frac{X^2}{4}+\frac{XT}{2\pi}\Big)^{1/2}$$

we deduce from (51)

$$\begin{aligned}
I_5' = \frac{1}{2}\Big(\frac{2\pi}{T}\Big)^{\delta_T/2} &\sum_{m<Z} \frac{\sigma_{\delta_T}(m)}{m^{(1+\delta_T)/2}}\Big(\log\frac{T}{2\pi m}\Big)^{-1}\cos\Big(T\log\frac{T}{2\pi m} - T + \frac{\pi}{4}\Big)\\
&+ BT^{-\delta_T/2}\sum_{m<Z}\frac{\sigma_{\delta_T}(m)}{m^{(1+\delta_T)/2}}(T-2\pi m)^{-1}\\
&+ BT^{-(1+\delta_T)/2}\sum_{m<Z}\frac{\sigma_{\delta_T}(m)}{m^{(1+\delta_T)/2}}(T-2\pi m)^{-1/2}\\
&+ BT^{(1-2\delta_T)/4}\sum_{m=1}^{\infty}\frac{\sigma_{\delta_T}(m)}{m^{(3+2\delta_T)/2}}\\
&\times\min\left(1,\left|2\sqrt{m}+\sqrt{X}-\sqrt{X+\frac{2T}{\pi}}\right|^{-1}\right) + BT^{-(1+\delta_T)/6+\varepsilon}\\
&\overset{\mathrm{def}}{=} I_{51}' + I_{52}' + I_{53}' + I_{54}' + BT^{-(1+\delta_T)/6+\varepsilon}
\end{aligned} \tag{52}$$

Since $Z = BT$ and $T/2\pi - Z \geqslant cT$, the evident estimate $\sigma_{\delta_T}(m) = Bm^{\delta_T+\varepsilon}$ yields

$$I_{52}' = BT^{-(2+\delta_T)/2}\sum_{m<Z} m^{-(1-\delta_T)/2+\varepsilon} = BT^{-1/2+\varepsilon}, \tag{53}$$

$$I_{53}' = BT^{-1/2+\varepsilon}. \tag{54}$$

Using the estimates for the sums S_i, $1 \leqslant i \leqslant 4$, from (Ivić, 1985, p. 459–460), we find that

$$\begin{aligned}
&\sum_{m\leqslant 2Z}\frac{\sigma_{\delta_T}(m)}{m^{(3+2\delta_T)/4}}\min\left(1,\left|2\sqrt{m}+\sqrt{X}-\sqrt{X+\frac{2T}{\pi}}\right|^{-1}\right)\\
&\quad = BT^{\delta_T/2}\sum_{m\leqslant 2Z}\frac{d(m)}{m^{3/4}}\min\left(1,\left|2\sqrt{m}+\sqrt{X}-\sqrt{X+\frac{2T}{\pi}}\right|^{-1}\right)\\
&\quad = BT^{(2\delta_T-1)/4}\log^2 T.
\end{aligned} \tag{55}$$

We see that the sum over $m > 2Z$ is bounded by

$$BT^{-(1+2\delta_T)/4}\log^2 T. \tag{56}$$

Now from (52)–(56) we have

$$I_5' = \frac{1}{2}\Big(\frac{2\pi}{T}\Big)^{\delta_T/2}\sum_{m<Z}\frac{\sigma_{\delta_T}(m)}{m^{(1+\delta_T)/2}}\Big(\log\frac{T}{2\pi m}\Big)^{-1}\cos f_1(T,m) + B\log^2 T. \tag{57}$$

Taking into account (48) and applying Lemma 8 we obtain

$$I_5'' = BT^{-\delta_T/2}\int_X^{\infty} x^{\sigma_T - \frac{5}{3} - \frac{\delta_T}{6} + \varepsilon}\,dx = BT^{-(1+\delta_T)/6+\varepsilon}. \tag{58}$$

Consequently, from (45), (57) and (58)

$$I_5 = 2i\left(\frac{2\pi}{T}\right)^{\delta_T/2} \sum_{m \leqslant N_1} \frac{\sigma_{\delta_T}(m)}{m^{(1+\delta_T)/2}} \left(\log \frac{T}{2\pi m}\right)^{-1} \cos f_1(T,m) + B\log^2 T.$$

From the last equality and from the equalities (24), (25), (27), (44), and the estimates (28), (33) we conclude that

$$\begin{aligned}\int_{\sigma_T - iT}^{\sigma_T + iT} F_2(s)\,ds = {} & \frac{2i}{\sqrt{2}}\left(\frac{2\pi}{T}\right)^{\delta_T/2} \sum_{m \leqslant N} \frac{(-1)^m \sigma_{\delta_T}(m)}{m^{(1+\delta_T)/2}} \\ & \times \frac{\cos f(T,m)}{\left(\frac{T}{2\pi m} + \frac{1}{4}\right)^{1/4} \sinh^{-1}\left(\sqrt{\frac{\pi m}{2T}}\right)} \\ & - 4i\left(\frac{2\pi}{T}\right)^{\delta_T/2} \sum_{m \leqslant N_1} \frac{\sigma_{\delta_T}(m)}{m^{(1+\delta_T)/2}} \left(\log \frac{T}{2\pi m}\right)^{-1} \cos f_1(T,m) \\ & + B\log^2 T + B|\zeta(1-\delta_T)|\delta_T T^{-\delta_T} \\ & + B\delta_T^2 |\zeta(1-\delta_T)| \min\left(\delta_T^{-1}, \log T\right) + BT^{\sigma_T - 1}|\zeta(1+\delta_T)|.\end{aligned}$$

Since

$$\zeta(1 \pm \delta_T) = B\delta_T^{-1},$$

the assertion of Theorem 4 follows from (5), (6), and Lemma 6.
Note that the case $l_T > c\log T$ remains open. Theorem 4 can be used as a starting point for further investigations of the function $E_T(\sigma_T)$, for instance, to obtain upper and lower bounds or to study the square mean estimates of $E_T(\sigma_T)$.

Acknowledgement. The author would like to thank the referee for several helpful remarks.

REFERENCES

Atkinson, F. V. (1949). The mean value of the Riemann zeta-function. *Acta Math.* **81**, 353–376.

Heath-Brown, D. R. and Huxley, M. H. (1990). Exponential sums with a difference. *Proc. London Math. Soc.* **61**, 227–250.

Ivić, A. (1985). *The Riemann Zeta-Function.* John Wiley and Sons, New York.

Laurinčikas, A. P. (1988). A limit theorem for the Riemann zeta-function close to the critical line. *Mat. Sb.* (*N.S.*) **135(177)**, 3–11 (in Russian).

Laurinčikas, A. P. (1989). A limit theorem for the Riemann zeta-function close to the critical line. II. *Mat. Sb.* (*N.S.*) **180**, 733–749 (in Russian).

Laurinčikas, A. (1992). Once more on the function $\sigma_a(m)$. *Liet. Matem. Rink.* (to appear).

Tichmarsh, E. C. (1951). *The Theory of the Riemann Zeta-Function.* Oxford University Press, Oxford.

Watson, G. N. (1945). *A Treatise on the Theory of Bessel Functions.* Cambrige University Press, Cambrige.

New Trends in Probab. and Statist., pp. 355–365
F. Schweiger and E. Manstavičius (Eds.)
1992 VSP/TEV

LARGE VALUE ESTIMATES FOR HECKE POLYNOMIALS

M. MAKNYS
Department of Mathematics, Vilnius University,
Naugarduko 24, 2006, Vilnius, Lithuania

ABSTRACT

We give some new 'large' value estimates for the Hecke polynomials

$$\sum_{N(\alpha)\leqslant N}^{*} c(\alpha)\Xi(\alpha)(N(\alpha))^{-s}, \qquad N>1, \quad s=\sigma+it,$$

in an imaginary quadratic number field. The estimates are uniform in all parameters of the Hecke character Ξ and in t.

1. INTRODUCTION AND RESULTS

It is well known (see (Heath-Brown, 1979; Huxley, 1976; Jutila, 1976; Motgomery, 1971)) that mean and 'large' value estimates for the Dirichlet polynomials

$$\sum_{n\leqslant N} a(n)\chi(n)n^{-s} \tag{1.1}$$

are useful in obtaining zero-density results for the Riemann zeta and Dirichlet L-functions.

In this paper we shall consider an analogue of such polynomials in the imaginary quadratic number field $\mathbf{Q}(\sqrt{d})$ (d is a negative and square-free integer) which is extended to an ideal number system $\mathbf{K}$ (see (Hecke, 1918, 1920; Kubilius, 1952)). We shall prove some 'large' value estimates for

$$\sum_{N(\alpha)\leqslant N}^{*} c(\alpha)\Xi(\alpha)(N(\alpha))^{-s}. \tag{1.2}$$

The last sum is called a Hecke polynomial. In (1.2) α is an ideal number from $\mathbf{K}$, $N(\alpha)$ is the norm of α, $s=\sigma+it$, $N>1$, and $\Xi(\alpha)$ is the Hecke character of the second kind (or the so-called Grössencharacter) mod μ with exponent m, $m\in\mathbf{Z}$, where μ is an integer ideal from $\mathbf{Q}(\sqrt{d})$, $\mu\neq(0)$. For the field $\mathbf{Q}(\sqrt{d})$, the character $\Xi(\alpha)$ can be written as

$$\chi(\alpha)\exp(igm\operatorname{arc}\alpha)$$

(see (Hecke, 1918, 1920)). Here $\chi(\alpha)$ is a usual abelian character of the multiplicative group of reduced residue classes mod μ, g is the number of units mod μ, $c(\alpha)$ is a complex number depending on α only. The asterisk, in (1.2) and in other formulas, denotes that α runs over the set of nonassociated nonzero numbers from $\mathbf{K}$. For $\sigma > 1$, Hecke functions $Z(s, \Xi)$ of the field $\mathbf{Q}(\sqrt{d})$ are defined by the Dirichlet series

$$\sum_{\alpha}{}^{*} \Xi(\alpha)(N(\alpha))^{-s} = Z(s, \Xi).$$

The function $Z(s, \Xi)$ is regular on the whole plane $\mathbf{C}$ except for $s = 1$, where it has a simple pole if Ξ is the principal character.
The Hecke functions $Z(s, \Xi)$ satisfy the functional equation of the Riemann–Dirichlet type

$$Z(s, \Xi) = \varkappa(\Xi)\Psi(s, \Xi)Z(1 - s, \overline{\Xi}), \tag{1.3}$$

where $\varkappa(\Xi)$ depends on Ξ only, $|\varkappa(\Xi)| = 1$, and

$$\Psi(s, \Xi) = \Big(\frac{1}{2\pi}\sqrt{|D|N(\mu)}\Big)^{1-2s}\Gamma\Big(\frac{1}{2}g|m| + 1 - s\Big)\Big(\Gamma\Big(\frac{1}{2}g|m| + s\Big)\Big)^{-1}, \tag{1.4}$$

where D is the discriminant of the field $\mathbf{Q}(\sqrt{d})$.
We see that the Hecke polynomials (1.2) depend on μ, m, g, s, $c(\alpha)$, and on the field $\mathbf{Q}(\sqrt{d})$. We want to have some estimates of sums (1.2) uniform in all parameters listed above.
For a fixed μ, the quantity

$$V = V_{mt} = \frac{1}{4}g^2m^2 + t^2,$$

proposed in (Kubilius, 1955), may be a 'measure' of uniformity of estimates of (1.2). Some estimates of (1.2) uniform for V_{mt} are given in (Maknys, 1975, 1985).
In this paper we shall estimate (1.2) uniformly in μ, m, g, and t, and we shall prove a new estimate for a fixed μ.
We propose the following 'measures' of uniformity:

$$L_1 = L_1(\mu, m, t) = \Big(\frac{1}{4}m^2g^2 + t^2\Big)N(\mu),$$

$$L_2 = L_2(\mu, M, T) = \Big(\frac{1}{4}g^2M^2 + T^2\Big)N(\mu)$$

for $|m| \leqslant M$, $M > 1$, $|t| \leqslant T$, $T > 2$, and

$$L_3 = L_3(Q, M, T) = Q\Big(\frac{1}{4}g^2M^2 + T^2\Big)$$

for the same M and T, and $1 < N(\mu) \leqslant Q$.
The letter ε denotes an arbitrarily small positive constant not the same at every on occurrence; $c_2, c_3, \ldots$ are constants that may depend on $\mathbf{Q}(\sqrt{d})$, ε, and natural numbers l and k. The same is true for the constants implied in the symbols O and $\ll$.

The letter g' denotes the number of units of the field $\mathbf{Q}(\sqrt{d})$, h is the number of ideal classes of it,

$$C_{\mathbf{K}} = \frac{h}{g'|\sqrt{d}|},$$

$$E(\Xi) = \begin{cases} 1 & \text{if} \quad \Xi = \Xi_0, \\ 0 & \text{if} \quad \Xi \neq \Xi_0, \end{cases}$$

where Ξ_0 is the principal character,

$$E_1(\operatorname{arc}\alpha, P) \leqslant \begin{cases} M & \text{always}, \\ 3 & \text{if} \quad \operatorname{arc}\alpha = \operatorname{arc}\beta + \dfrac{2\pi n}{g[M]}, \end{cases}$$

where $n \in \mathbf{Z}$, $\beta \in \mathbf{K}$, $0 < N(\beta) \leqslant P$, and $[M]$ denotes the largest integer less than or equal to M.
Using the methods of (Heath-Brown, 1979; Jutila, 1976) we shall prove the following

THEOREM. *Suppose that we have sequences*

$$\Xi_r(\alpha) \quad \textit{and} \quad s_{mr} = \sigma_{mr} + it_{rm},$$

$$r = 1, 2, \ldots, R, \qquad \sigma_{mr} \geqslant 0, \qquad \sigma = \min \sigma_{mr}.$$

Assume that $\Xi_r \neq \Xi_q$ *for* $r \neq q$ *and*

$$|t_{rm} - t_{qm}| \leqq \log^2 L_2 \qquad \textit{for} \quad |t_{rm}| \leqslant T.$$

Furthermore, assume that there are constants c *and* c_1, $0 < c \leqslant 1/2$, $c_1 \geqslant 1$, *such that*

$$\left|Z\left(\frac{1}{2} + it, \Xi\right)\right| \ll L_2^c (\log L_2)^{c_1} \tag{1.5}$$

for $L_2 \geqslant E(\Xi)$.
Then we have

$$\begin{aligned} &\left(\sum_{r \leqslant R} \left| \sum_{N < N(\alpha) \leqslant 2N}^{*} c(\alpha)\Xi_r(\alpha)(N(\alpha)^{-s_{mr}} \right|\right)^2 \\ &\ll \left(R^2\sqrt{N} + R^2\sqrt{N}L_2^{(4c-1)/4k} + C_{\mathbf{K}}^{1/k}\sqrt{L_2}R^{2-1/2k}\right. \\ &+ C_{\mathbf{K}}RN + \sqrt{N}R^{2-1/2kl}(MTN(\mu))^{1/2kl} \\ &\left.+ \sqrt{N}R^{2-1/2kl}E_1\left(\operatorname{arc}\alpha, L_2^{l/2}\right)\right)L_2^{\varepsilon} \sum_{N < N(\alpha) \leqslant 2N}^{*} |c(\alpha)|^2 (N(\alpha))^{-2\sigma} \end{aligned} \tag{1.6}$$

for $1 < N \leqslant \sqrt{L_2}$ *and for fixed integers* $k \geqslant 1$, $l \geqslant 1$.

COROLLARY 1. *We can put* $c = 1/4$ *and* $c_1 = 1$ *in Theorem.*

COROLLARY 2. *If the conditions of Theorem are satisfied, then we have*

$$\left(\sum_{r\leqslant R}\left|\sum_{N<N(\alpha)\leqslant 2N}^{*} c(\alpha)\Xi_r(\alpha)(N(\alpha)^{-s_{mr}}\right|\right)^2$$
$$\ll \Big(R^2\sqrt{N}+R^2\sqrt{N}V_{MT}^{-1/12k}+RN \tag{1.7}$$
$$+R^{2-1/k}\sqrt{V_{MT}}+R^{2-1/2kl}(MT)^{1/2kl}\sqrt{N}$$
$$+\sqrt{N}R^{2-1/2kl}V_{MT}^{1/4k}\Big)V_{MT}^{\varepsilon}\sum_{N<N(\alpha)\leqslant 2N}^{*}|c(\alpha)|^2(N(\alpha))^{-2\sigma}$$

for fixed μ *and* $1<N\leqslant\sqrt{MT}$.

Remarks. Our Theorem is an analogue of Lemma 6 in (Heath-Brown, 1979) and extends corresponding results of (Maknys, 1975, 1985).
Theorem and Corollaries of this paper may be used in obtaining estimates for the number of nontrivial zeros of Hecke Z-functions in the region $3/4<\sigma<1$.
And at last we remark that Theorem and Corollary 1 are true if we change the quantity L_2 by L_3.

2. LEMMAS

We shall use some lemmas for proving Theorem and Corollaries.

LEMMA 1. *The relation*

$$|\Psi(s,\Xi)|\ll L_1^{1/2-\sigma} \tag{2.1}$$

holds in the region $-1\leqslant\sigma\leqslant 2,\ L_1\geqslant 2$.

Proof. The assertion follows from Stirling's formula for $\log\Gamma(s)$ if we rewrite $\Psi(s,\Xi)$ in the form

$$\Psi(s,\Xi)=\Big(\frac{1}{2\pi}\sqrt{|D|N(\mu)}\Big)^{1-2s}\exp\Big\{\log\Gamma\Big(\frac{1}{2}g|m|+1-s\Big)-\log\Gamma\Big(\frac{1}{2}g|m|+s\Big)\Big\}.$$

For details see (Kubilius, 1955).

LEMMA 2. *Let* $h=\log^2 L_2,\ 0\leqslant\sigma\leqslant 1,\ T>2,\ M>1,\ |t|\leqslant h,\ 1<N\leqslant\sqrt{L_2}$,

$$H(s,\Xi)=\sum_{\alpha}^{*}\Big\{\exp\Big(-\Big(\frac{N(\alpha)}{2N}\Big)^h\Big)-\exp\Big(-\Big(\frac{N(\alpha)}{N}\Big)^h\Big)\Big\}\Xi(\alpha)(N(\alpha))^{-s}. \tag{2.2}$$

Then

$$|H(s,\Xi)|\ll NE(\Xi)C_{\mathbf{K}}+1+\sqrt{N}\int_{-h^2}^{h^2}\Big|\sum_{N(\alpha)\leqslant P}^{*}\Xi(\alpha)(N(\alpha))^{-1/2+it+i\tau}\Big|\,d\tau \tag{2.3}$$

for $P=N^{-1}L_2\log^{c_2}L_2$.

Lemma 2 is proved in (Maknys, 1985).

LEMMA 3. *Let $M > 1$ and $T > 1$. For each Ξ we associate natural number R_Ξ and suppose that*

$$|t_{am} - t_{bm}| \geqslant \log^2 L_2 \qquad for \quad 1 \leqslant a \neq b \leqslant R_\Xi,$$

$$|t_{rm}| \leqslant T, \qquad |m| \leqslant M.$$

Then we have

$$\sum_{\substack{\Xi \\ |m|\leqslant M}} \sum_{r=1}^{R_\Xi} |Z(\sigma + it_{rm}, \Xi_r)|^{2l} \ll (L_2^{l/2} E_1(\operatorname{arc}\alpha, L_2^{l/2}) + MTN(\mu)) \log L_2^{c_3}. \quad (2.4)$$

Lemma 3 is proved in (Maknys, 1986) using the method of (Ramachandra, 1974).

LEMMA 4. *Let $-1/2 \leqslant \sigma_1 \leqslant 0$ and $1 \leqslant \sigma_2 \leqslant 2$. Then, in the region*

$$\sigma_1 - \frac{2}{\log L_2} \leqslant \sigma \leqslant \sigma_2 + \frac{2}{\log L_2}, \qquad |t| \geqslant E(\Xi), \quad (2.5)$$

we have the estimate

$$|Z(s,\Xi) \ll \left(\sqrt{L_2}\right)^{\frac{1-2\sigma_1}{\sigma_2-\sigma_1}(\sigma_2-\sigma)} \log L_2. \quad (2.6)$$

Proof. At first we remark that the assertion of Lemma 4 was proved in (Kubilius, 1955) for fixed μ and d, and for the principal character. Here we have a more general case which is proved by the same method as in (Kubilius, 1955).
Let

$$\Phi_j(s,\Xi) = Z(s,\Xi)(f_j(s))^{-1}, \qquad j = 1,2,$$

where

$$f_j(s) = \exp\left\{\left(\frac{1}{2}\log L_2 + \log(s+3) + \frac{\pi}{2}\sigma_j i\right)\frac{1-2\sigma_1}{\sigma_2-\sigma_1}(\sigma_2 - s)\right\}\log L_2 \quad (2.7)$$

with $\sigma_1 = -1$ and $\sigma_2 = 1$, respectively, and only principal values of logarithms are considered.
If $\Xi \neq \Xi_0$, then the functions $\Phi_j(s,\Xi)$ are regular in the region (2.5) and

$$|f_j(s)| \sim \left(\sqrt{L_2}\right)^{\frac{1-2\sigma_1}{\sigma_2-\sigma_1}(\sigma_2-\sigma)} \exp\left\{\frac{\pi}{2}t(\operatorname{sgn} t + \sigma_j)\frac{1-2\sigma_1}{\sigma_2-\sigma_1}\right\}\log L_2.$$

We estimate $\Phi_j(s,\Xi)$ on the boundary of the region (2.5). If

$$\sigma \geqslant 1 + \frac{2}{\log L_2},$$

then

$$|Z(s,\Xi)| \ll \left|Z\left(1 + \frac{2}{\log L_2}\right)\right| \ll L_2.$$

Using the functional equation (1.3), the last estimate, and Lemma 1, we have

$$\left|Z\left(\sigma_1 - \frac{2}{\log L_2} + it, \Xi\right)\right|$$
$$= \left|\Psi\left(\sigma_1 - \frac{2}{\log L_2} + it, \Xi\right)\right|\left|Z\left(1 - \sigma_1 + \frac{2}{\log L_2} - it, \Xi\right)\right|$$
$$\ll \left(\sqrt{L_2}\right)^{1-2\sigma_1+\frac{4}{\log L_2}}\left|Z\left(1 + \frac{2}{\log L_2}\right)\right| \ll L_2^{1/2-\sigma_1}\log L_2.$$

From the above estimates and from (2.7) it follows that $\Phi_j(s, \Xi) \ll 1$ on the boundary, and the lemma follows by the Phragmén–Lindelöf principle from the expression of $Z(s, \Xi)$ as θ-series.

COROLLARY. *The estimate*

$$|Z(s, \Xi)| \ll \left(\sqrt{L_2}\right)^{1-\sigma} L_2$$

is true for

$$-\frac{2}{\log L - 2} \leqslant \sigma \leqslant 1 + \frac{2}{\log L_2}, \qquad L_2 \geqslant E(\Xi).$$

Proof. It follows from (2.6) if we put $\sigma_1 = 0$ and $\sigma_2 = 1$.

LEMMA 5. *For fixed μ and d, we have*

$$\left|Z\left(\frac{1}{2} + it, \Xi\right)\right| \ll V_{mt}^{1/6} \log^{c_{10}} V_{mt}, \qquad c_{10} > 1,$$

in the region $V_{mt} \geqslant E(\Xi)$.

For the proof see (Kaufman, 1978).

LEMMA 6. *Let $|t| \leqslant T$, $T > 2$, and $a = (\log L_2)^{-1}$. Then we have*

$$\sum_{N(\alpha)\leqslant N}^{*} \Xi(\alpha)(N(\alpha))^{-1/2+it} = E(\Xi)C_{\mathbf{K}}\frac{N^{1/2+it}}{1/2+it}$$
$$+ \frac{1}{2\pi i}\int_{a-i\sqrt{L_2}}^{a+i\sqrt{L_2}} Z\left(\frac{1}{2} - it + w, \Xi\right)\frac{N^w}{w}\, dw + O(\log L_2)$$

for a natural number N, $1 < N \leqslant [L_2]$.

Proof. Let $b = a + 1/2 > 1/2$. We can apply the contour integration formula. Thus

$$\sum_{N(\alpha)\leqslant N}^{*} \Xi(\alpha)(N(\alpha))^{-1/2+it} = \frac{1}{2\pi i}\int_{b-i\sqrt{L_2}}^{b+i\sqrt{L_2}} Z\left(\frac{1}{2} - it + w, \Xi\right)\frac{N^w}{w}\, dw$$
$$+ O\left(\frac{N^b}{\sqrt{L_2}(b - 1/2)}\right) + O\left(\sqrt{N}L_2^{-1/2}\log 2N\right), \tag{2.8}$$

where N is a natural number.
From the Cauchy theorem it follows that

$$\frac{1}{2\pi i}\int_{b-i\sqrt{L_2}}^{b+i\sqrt{L_2}} Z\Big(\frac{1}{2}-it+w,\Xi\Big)\frac{N^w}{w}\,dw = E(\Xi)C_{\mathbf{K}}\frac{N^{1/2+it}}{1/2+it}$$
$$-\frac{1}{2\pi i}\left(\int_{a+i\sqrt{L_2}}^{a-i\sqrt{L_2}}+\int_{b+i\sqrt{L_2}}^{a+i\sqrt{L_2}}+\int_{a-i\sqrt{L_2}}^{b-i\sqrt{L_2}}\right)Z\Big(\frac{1}{2}-it+w,\Xi\Big)\frac{N^w}{w}\,dw \tag{2.9}$$
$$= E(\Xi)C_{\mathbf{K}}\frac{N^{1/2+it}}{1/2+it}+\int_{a-i\sqrt{L_2}}^{a+i\sqrt{L_2}} Z\Big(\frac{1}{2}-it+w,\Xi\Big)\frac{N^w}{w}\,dw+I_1+I_2.$$

But

$$|I_1| \ll |I_2| \ll \left|\int_a^b Z\Big(\frac{1}{2}-it+u+i\sqrt{L_2},\Xi\Big)\frac{N^u}{u+i\sqrt{L_2}}\,du\right|$$
$$\ll \frac{N^b}{a+\sqrt{L_2}}(b-a)\max_{\substack{a\leqslant u\leqslant b\\ |t|\leqslant T}}\left|Z\Big(\frac{1}{2}+u-i(t\pm\sqrt{L_2}),\Xi\Big)\right|. \tag{2.10}$$

We obtain the lemma from (2.8), (2.9), and (2.10) if the maximum there is estimated by Corollary of Lemma 4.

COROLLARY 1. *For fixed μ and d, we have*

$$\sum_{N(\alpha)\leqslant N}^{*}\Xi(\alpha)(N(\alpha))^{-1/2+it} = E(\Xi)C_{\mathbf{K}}\frac{N^{1/2+it}}{1/2+it}$$
$$+\frac{1}{2\pi i}\int_{a-i\sqrt{L_2}}^{a+i\sqrt{L_2}} Z\Big(\frac{1}{2}-it+w,\Xi\Big)\frac{N^w}{w}\,dw$$
$$+O\big(V_{MT}^{-1/12}\log^{c_4}V_{MT}\big)+O(\log V_{MT}).$$

COROLLARY 2. *If conditions of Lemma 6 are satisfied and if the estimate* (1.5) *is true, then*

$$\sum_{N(\alpha)\leqslant N}^{*}\Xi(\alpha)(N(\alpha))^{-1/2+it} = E(\Xi)C_{\mathbf{K}}\frac{N^{1/2+it}}{1/2+it}$$
$$+\frac{1}{2\pi i}\int_{a-i\sqrt{L_2}}^{a+i\sqrt{L_2}} Z\Big(\frac{1}{2}-it+w,\Xi\Big)\frac{N^w}{w}\,dw$$
$$+O\big(L_2^{c-1/4}\log^{c_5}L_2\big)+O(\log L_2).$$

Proofs. We may prove Corollaries by the same way as Lemma 6 if we estimate

$$\max_{\substack{a\leqslant u\leqslant b\\ |t|\leqslant T}} \left|Z\left(\frac{1}{2}+u-i(t\pm\sqrt{L_2}),\Xi\right)\right|$$

applying modifications of Lemma 4 respectively to Lemma 5 for first case and with (1.5) in second case.

LEMMA 7. *Let $a(\alpha)$ be a complex number depending only on $\alpha\in\mathbf{K}$, t_r be real numbers, $r=1,2,\dots,R$, and*

$$\max_{1<N(\alpha)\leqslant N} |a(\alpha)|\leqslant A, \qquad N>1.$$

Then

$$\sum_{r,q=1}^{N}\left|\sum\nolimits^{*}_{1<N(\alpha)\leqslant N} a(\alpha)\Xi_r(\alpha)\overline{\Xi}_q(\alpha)(N(\alpha))^{-1/2+it_r-it_q}\right|^2$$
$$\leqslant A^2\sum_{r,q=1}^{N}\left|\sum\nolimits^{*}_{1<N(\alpha)\leqslant N} \Xi_r(\alpha)\overline{\Xi}_q(\alpha)(N(\alpha))^{-1/2+it_r-it_q}\right|^2.$$

It is an analogue of Lemma 2 in (Jutila, 1977). For the proof see (Maknys, 1985).

LEMMA 8. *Let $T\geqslant 2$, $r,q=1,2,\dots,R$, $|t_r|\leqslant T$, and all characters Ξ be of the same modulus μ. Let*

$$\Xi_r\neq\Xi_q \quad \textit{and} \quad |t_r-t_q|\geqslant \log^2 L_2 \qquad \textit{for} \quad r\neq q.$$

Then

$$\sum_{r,q\leqslant R}\left|\sum\nolimits^{*}_{N(\alpha)\leqslant N} \Xi_r(\alpha)\overline{\Xi}_q(\alpha)(N(\alpha))^{-1/2+it_r-it_q}\right|^2$$
$$\ll \big(R^2+RNC_{\mathbf{K}}^2+R^{2-1/l}((MTN(\mu))^{1/l}$$
$$+E_1^{1/l}(\operatorname{arc}\alpha, L_2^{l/2}))\sqrt{L_2}\big)\log^{c_6} L_2,$$

for a natural number $N\leqslant[L_2]$ and a fixed natural number $l\geqslant 1$.

Proof. From Lemma 6 and the Schwarz inequality we have

$$\sum_{r,q\leqslant R}\left|\sum\nolimits^{*}_{N(\alpha)\leqslant N} \Xi_r(\alpha)\overline{\Xi}_q(\alpha)(N(\alpha))^{-1/2+it_r-it_q}\right|^2$$
$$\ll RNC_{\mathbf{K}}^2+R^2\log^2 L_2$$
$$+\int_{a-i\sqrt{L_2}}^{a+i\sqrt{L_2}}\sum_{r,q\leqslant R}\left|Z\left(\frac{1}{2}-it_r+it_q+w,\Xi\right)\right|^2|dw| \tag{2.11}$$
$$\times\int_{a-i\sqrt{L_2}}^{a+i\sqrt{L_2}}\left|N^w w^{-1}\right|^2|dw|.$$

If we apply the Hölder inequality for the integrand sum, then

$$
\begin{aligned}
&\sum_{r,q\leqslant R}\Big|Z\Big(\frac{1}{2}-it_r+it_q+w,\Xi\Big)\Big|^2\\
&\leqslant\Big(\sum_{\substack{\Xi\\|m|\leqslant M}}\sum_{r,q=1}^{R}\Big|Z\Big(\frac{1}{2}-it_r+it_q+w,\Xi\Big)\Big|^{2l}\Big)^{1/l}\Big(\sum_{r,q\leqslant R}1\Big)^{1-1/l}\\
&\ll R^{2-2/l}\big(R(MTN(\mu)+L_2^{l/2}E_1(\operatorname{arc}\alpha,L_2^{l/2})\big)^{1/l}\log^{c_7}L_2\\
&\ll R^{2-1/l}\Big((MTN(\mu))^{1/l}+E_1^{1/l}(\operatorname{arc}\alpha,L_2^{l/2})\sqrt{L_2}\Big)\log^{c_8}L_2
\end{aligned}
\tag{2.12}
$$

according to Lemma 3. Thus, Lemma 8 follows from (2.11) and (2.12).

COROLLARY 1. *For fixed μ and d, we have*

$$
\begin{aligned}
&\sum_{r,q\leqslant R}\Big|\sum_{N(\alpha)\leqslant N}{}^{*}\,\Xi_r(\alpha)\overline{\Xi}_q(\alpha)(N(\alpha))^{-1/2+it_r-it_q}\Big|^2\\
&\ll\big(R^2+V_{MT}^{-1/6}+RN+R^{2-1/l}((MT)^{1/l}+\sqrt{MT})\big)\log^{c_9}MT.
\end{aligned}
$$

COROLLARY 2. *If conditions of Lemma 8 are satisfied and if the estimate (1.5) is true, then*

$$
\begin{aligned}
&\sum_{r,q\leqslant R}\Big|\sum_{N(\alpha)\leqslant N}{}^{*}\,\Xi_r(\alpha)\overline{\Xi}_q(\alpha)(N(\alpha))^{-1/2+it_r-it_q}\Big|^2\\
&\ll\big(R^2L_2^{2c-1/2}+RNC_{\mathbf{K}}+R^{2-1/l}((MTN(\mu))^{1/l}+E_1^{1/l}(\operatorname{arc}\alpha,L_2^{l/2})\big)\log^{c_{11}}L_2.
\end{aligned}
$$

Proofs. We obtain both Corollaries by the same way as Lemma 8. We must only apply Corollaries 1 and 2 of Lemma 6 instead of Lemma 6.

3. PROOF OF THEOREM

First we prove Corollary 1. We know that

$$
\begin{aligned}
&\Big(\sum_{r\leqslant R}\Big|\sum_{N\leqslant N(\alpha)\leqslant 2N}{}^{*}\,c(\alpha)\Xi_r(\alpha)(N(\alpha))^{-\sigma_r-it_r}\Big|\Big)^2\\
&\ll\sum_{r,q\leqslant R}\big|H(\sigma_r+\sigma_q+it_r-it_q,\Xi_r\overline{\Xi}_q)\big|\sum_{N\leqslant N(\alpha)\leqslant 2N}{}^{*}\,|c(\alpha)|^2(N(\alpha))^{-2\sigma}.
\end{aligned}
\tag{3.1}
$$

This is the Halász–Montgomery lemma for an imaginary quadratic number field; see (Maknys, 1975).

It follows from Lemma 2 that

$$\sum_{r,q\leqslant R}|H(\sigma_r+\sigma_q+it_r-it_q,\Xi_r\overline{\Xi}_q)| \ll \sum_{r,q\leqslant R}(E(\Xi)C_{\mathbf{K}}N+1$$
$$+\sqrt{N}\int\limits_{-h^2}^{h^2}\sum_{r,q\leqslant R}\Big|\sideset{}{^*}\sum_{N(\alpha)\leqslant P}\Xi_r(\alpha)\overline{\Xi}_q(\alpha)(N(\alpha))^{-1/2+it_r-it_q}(N(\alpha))^{-i\tau}\Big|\,d\tau \qquad (3.2)$$
$$\ll RNC_{\mathbf{K}}+R^2+\sqrt{N}\,I\Big(\sum\Big).$$

At first we shall estimate $\sum_{r,q}$ using Lemma 7 with $a(\alpha)=(N(\alpha))^{-i\tau}$; after that we shall apply the Hölder inequality and Lemma 7 once more with

$$a(\alpha)=a_k(\alpha)\leqslant\tau_k(\alpha)\ll L_2^{\varepsilon}$$

where $\tau_k(\alpha)$ is the number of nonassociated solutions of the Diophantine equation $\alpha=\alpha_1\cdot\ldots\cdot\alpha_k$. Thus,

$$\sum\ll\Big(\sum_{r,q\leqslant R}1\Big)^{1-1/2k}\Big(\sum_{r,q\leqslant R}\Big|\Big(\sideset{}{^*}\sum_{N(\alpha)\leqslant P}\Xi_r(\alpha)\overline{\Xi}_q(N(\alpha))^{-1/2+it_r-it_q}\Big)^k\Big|^2\Big)^{1/2k}$$
$$\ll R^{2-1/k}\Big(\sum_{r,q\leqslant R}\Big|\sideset{}{^*}\sum_{N(\alpha)\leqslant P^k}a_k(\alpha)\Xi_r(\alpha)\overline{\Xi}_q(\alpha)(N(\alpha))^{-1/2+it_r-it_q}\Big|^2\Big)^{1/2k}$$
$$\ll L_2^{\varepsilon/k}R^{2-1/k}\Big(\sum_{r,q\leqslant R}\Big|\sideset{}{^*}\sum_{N(\alpha)\leqslant P^k}\Xi_r(\alpha)\overline{\Xi}(\alpha)(N(\alpha))^{-1/2+it_r-it_q}\Big|^2\Big)^{1/2k}$$

We can apply Lemma 8 to the last sum;

$$\sqrt{N}I\Big(\sum\Big)\ll L_2^{\varepsilon}R^{2-1/k}\sqrt{N}\big(RP^kC_{\mathbf{K}}^2+R^{2-1/2kl}((MTN(\mu))^{1/l}$$
$$+E_1^{1/l}(\operatorname{arc}\alpha,L_2^{l/2}))\sqrt{L_2}+R^2\sqrt{N}\big)^{1/2k}; \qquad (3.3)$$

and Corollary 1 of our Theorem follows immediately from (3.1), (3.2), and (3.3). Theorem and Corollary 2 may be proved by the same way using Corollary 1 and Corollary 2 of Lemma 8, respectively.

REFERENCES

Bulota, K. (1964). On the Hecke Z-functions and the distribution of prime numbers of an imaginary quadratic number field. *Liet. Matem. Rink.* **4**, 3, 309–326 (in Russian).

Hecke, E. (1918, 1920), Eine neue Art von Zetafunktion und ihre Beziehungen zur Verteilung der Primzahlen. I, II. *Math. Z.* **1**, 357–376; *Math. Z.* **6**, 11–50.

Heath-Brown, D. R. (1979). A large value estimate for Dirichlet polynomials. *J. London Math. Soc.* **20**, 8–18.

Huxley, M. N. (1976). An imperfect hybrid zero-density theorem. *J. London Math. Soc.* **13**, 53–56.

Jutila, M. (1976). On large value of Dirichlet polynomials. In: *Coll. Math. Soc. J. Bolyai* **13**. North-Holland, Amsterdam, Oxford, pp. 129–140.

Jutila, M. (1977). Zero-density estimates for L-functions. *Acta Arithm.* **32**, 55–62.

Kaufman, R. M. (1978). A. F. Lavrik's truncated equations. *Zap. Nauchn. Sem. LOMI* **76**, 124–158 (in Russian).

Koval'čik, F. B. (1974). Density theorems and distributions of primes in sectors and progressions. *Dokl. Akad. Nauk SSSR* **219**, 31–34 (in Russian).

Kubilius, J. (1952). On some problems of geometry of primes. *Mat. Sb.* **31**, 507–542 (in Russian).

Kubilius, J. (1955). On some problems of multivariative analytic number theory. *Vilniaus Valst. Univ. Mokslo Darbai, Ser. Mat., Fiz. ir Chem.* **4**, 5–42 (in Lithuanian).

Maknys, M. (1975). Zeros of Hecke Z-functions and the distribution of primes of an imaginary quadratic field. *Lithuanian Math. J.* **15**, 140–149.

Maknys, M. (1976). Density theorems for Hecke Z-functions and the distribution of primes of an imaginary quadratic field. *Lithuanian Math. J.* **16**, 105–110.

Maknys, M. (1985). On the large value estimate of Hecke polynomials. *Ann. Univ. Sci. Budapest, Sect. Math.* **27**, 21–28.

Maknys, M. (1986). Even moments of Hecke Z-fuctions. In: *Analytic Number Theory.* Petrozavodsk State University, Petrozavodsk, pp. 38–46 (in Russian).

Montgomery, H. L. (1968). Mean and large values of Dirichlet polynomials. *Invent. Math.* **8**, 334–345.

Montgomery, H. L. (1971). *Topics in Multiplicative Number Theory.* Springer-Verlag, Berlin, Heidelberg, New York.

Prachar, K. (1957). *Primzahlverteilung.* Springer-Verlag, Berlin, Göttingen, Heidelberg.

Ramachandra, K. (1974). A simple proof of the mean fourth power estimate for $\zeta(1/2+it)$ and $L(1/2+it,\chi)$. *Ann. Sc. Norm. Sup. Pisa* **1**, 81–97.

Part Seven

COMPUTATIONAL NUMBER THEORY AND APPLICATIONS TO NUMERIC ANALYSIS

New Trends in Probab. and Statist., pp. 369–378
F. Schweiger and E. Manstavičius (Eds.)
1992 VSP/TEV

ON NORMAL BASES OF ALGEBRAIC NUMBER FIELDS

S. A. STEPANOV and I. E. SHPARLINSKI
Steklov Institute of Mathematics,
Vavilov 42, 117966 Moscow, Russia

Mosfilmovskaja 2–41
119285 Moscow, Russia

ABSTRACT

In the paper the further development of the method to treat normal bases of a normal extension **K** of a field **L**, due to Schlickewei, Stepanov, and Shparlinski, is given. For instance, it is shown that for an element $\vartheta \in \mathbf{K}$ (under certain natural conditions) such bases are generated by $\vartheta + \alpha$ with a suitable element $\alpha \in \mathbf{K}$ of "small size" (in some sense) and also by ϑ^i, $i = k+1, \dots, k+N$, for some "small" N and any $k \in \mathbf{N}$.

Let the field **K** be a normal extension of a field **L** of finite degree $n \geqslant 2$ with the Galois group

$$G(\mathbf{K}/\mathbf{L}) = \{\sigma_i \colon i = 1, \dots, n\}$$

where $\sigma_1 = \mathrm{id}$ is the identical map.
If $\mathbf{K} = \mathbf{L}(\alpha)$ then **K** has the natural basis

$$1, \alpha, \dots, \alpha^{n-1}$$

over **L** which is called a power basis.
On the other hand, for some applications it is convenient to have a basis of the form

$$\sigma_1(\alpha), \dots, \sigma_n(\alpha),$$

called a normal basis, here α is an element of **K**.
The existence of a normal basis for a Galois algebraic number field (i.e. when $\mathbf{L} = \mathbf{Q}$) and a finite field (i.e. when $\mathbf{L} = \mathbf{F}_q$) was proved for the first time by K. Hensel in 1888. Since then this result was repeated in a number of papers. For the history and details see an excellent survey (Lidl and Niederreiter, 1983).
However, until recent time, there were no fast algorithms to construct such bases. Roughly speaking, the general problem is to find a fast (in some sense) algorithm to obtain a normal basis of **K** over **L** from a given basis

$$\Omega = \{\omega_1, \dots, \omega_n\} \tag{1}$$

of $\mathbf{K}$ over $\mathbf{L}$.
The first explicit construction of the normal bases over $\mathbf{F}_q$ for some very special fields $\mathbf{F}_{q^n}$ was presented in (Sidelnikov, 1987).
Then for q fixed, an algorithm, allowing to construct a normal basis $\mathbf{F}_{q^n}$ over $\mathbf{F}_q$ in time which is polynomial in n, was found by (Semaev, 1988). Independently, a quite different way for finding such bases was proposed in (Stepanov and Shparlinski, 1987a, 1987b, 1991) for all n and q satisfying the condition $(n, q) = 1$. The computing time of this algorithm is polynomial in $n\log q$ (it uses the linear algebra only).
The paper (Lenstra, 1991) contains as an auxiliary result a polynomial algorithm for constructing a normal basis without the restriction $(n, q) = 1$. The main result of that work is the polynomial algorithm for finding isomorphisms between two models of a finite field. Note, that under the Extended Riemann Hypothesis it was made before in (Evdokimov, 1989).
In the case $q > n$, one algorithm in (Stepanov and Shparlinski, 1987a, 1987b, 1991) was presented independently of (von zur Gathen and Giesbrecht, 1990). A probabilistic polynomial algorithm for finding a normal basis is given in that paper also.
In the paper (Stepanov and Shparlinski, 1989) an algorithm for the transition from an arbitrary primitive root $\vartheta \in \mathbf{F}_q$ to some primitive root $\alpha \in \mathbf{F}_{q^n}$ generating the normal basis of $\mathbf{F}_{q^n}$ over $\mathbf{F}_q$ has been given (in this case the corresponding bases are called primitive normal bases). For fixed n its computing time is polynomial in $\log q$.
The crucial point of the algorithm is the following bound for the number of zeros of a linear recurrent sequence over an arbitrary field $\mathbf{K}$.
Define the sequence δ_n by the recurrence

$$\delta_2 = 1, \qquad \delta_l = \delta_{[(l+2)/2]}/[(l+2)/2], \qquad l \geqslant 3.$$

Let $\mathbf{K}$ be an arbitrary field, let

$$\alpha_1, \ldots, \alpha_n, \; \mu_1, \ldots, \mu_n, \qquad n \geqslant 2,$$

be some non-zero elements of $\mathbf{K}$. We consider the sequence

$$v(t) = \alpha_1 \mu_1^t + \ldots + \alpha_n \mu_n^t, \qquad t = 1, 2, \ldots.$$

It was proved in (Stepanov and Shparlinski, 1989) that the number of solutions of the equation

$$v(t) = 0, \qquad 1 \leqslant t \leqslant N,$$

denoted by $R(N)$, does not exceed

$$R(N) \leqslant 2nN\left(N^{-1} + \tau^{-1}\right)^{\delta_n},$$

where τ is the smallest positive integer t satisfying

$$\mu_i^t = \mu_j^t \qquad \text{for some } i \text{ and } j, \quad 1 \leqslant i, j \leqslant n.$$

(If there is no such t then the corresponding term in this bound must be omitted). Then, in (Schlickewei and Stepanov, 1991) the method of (Stepanov and Shparlinski, 1989) was applied to treat the normal bases of the normal algebraic extensions of $\mathbf{Q}$ whose Galois group is cyclic. The crucial point for the results of (Schlickewei and Stepanov, 1991) is a new bound, obtained in (Schlickewei, 1991) for the number of zeros of a linear recurrent sequence in algebraic number fields, used instead of the corresponding bound from the paper (Stepanov and Shparlinski, 1989) for sequences in finite fields. A sharper bound (for the number fields) is obtained now by van der Poorten and Schlickewei. It allows us to improve the results of (Schlickewei and Stepanov, 1991). In our paper we use the bound of (van der Poorten and Schlickewei, 1991) mentioned above. In a special case of recurrent sequences over an algebraic number field $\mathbf{F}$ of degree d over $\mathbf{Q}$ it can be formulated as follows.
Let a nonzero sequence $\omega(t)$ in $\mathbf{F}$ of order N satisfy the linear recurrent equation

$$\omega(t+N) = A_{N-1}\omega(t+N-1) + \ldots + A_o u_v(t), \qquad t = 1, 2, \ldots,$$

with the characteristic polynomial

$$F(x) = x^N + A_{N-1}x^{N-1} + \ldots + A_o \in \mathbf{F}[x]$$

which has $m, m \leqslant N$, different roots $\mu_1, \ldots, \mu_m$. Moreover, let $\mu_1, \ldots, \mu_m \in \mathbf{F}$. If there are no roots of unity among the ratios μ_i/μ_j, $1 \leqslant j < i \leqslant m$, then the sequence $\omega(t)$ is called nondegenerate, and the number of solutions of the equation

$$\omega(t) = 0, \qquad t = 1, 2, \ldots,$$

does not exceed

$$(N-1)(4(d+\omega))^{2(d+1)}, \tag{2}$$

where ω is the number of prime ideal factors in $\mathbf{F}$ occurring in the decomposition of the ideals $(\mu_1), \ldots, (\mu_m)$.
The aim of this paper is a further development of the method of the papers mentioned above that allows us to obtain some new results in this direction. For instance, we show that for an element $\vartheta \in \mathbf{K}$ (under certain natural conditions) such bases are generated by $\vartheta + \alpha$ with a suitable element $\alpha \in \mathbf{K}$ of "small size" (in some sense) and also by

$$\vartheta^i, \qquad i = k+1, \ldots, k+N,$$

for some "small" N and any $k \in \mathbf{N}$.
Let us introduce the following notations.
For an algebraic number α denote by $\varphi_\alpha(x) \in \mathbf{Z}[x]$ its minimal polynomial, i.e. the polynomial of minimal degree satisfying $\varphi_\alpha(\alpha) = 0$ and having integer and coprime coefficients, the leading coefficient being positive. Denote by $H(\alpha)$ the maximal absolute value of all coefficients of $\varphi_\alpha(x)$ called its height.
Then, it is convenient to define the size of an algebraic number α of degree n as

$$S(\alpha) = \max(2, n \log H(\alpha)).$$

It is clear that $S(\alpha)$ is a measure of complexity of the "input size" of α. Indeed, its minimal polynomial has $n+1$ coefficients of binary size at most $\log H(\alpha)+1$. For the basis (1) of algebraic number field $\mathbf{K}$ over $\mathbf{Q}$ we set

$$S(\Omega) = \max\left(S(\omega_1), \ldots, S(\omega_n)\right).$$

First of all, we remark that the algorithm of Lemma 2 from (Stepanov and Shparlinski, 1987b) (see also Lemma 2.1 from (Stepanov and Shparlinski, 1987b) as well) can be extended word-by-word to the algebraic number fields with a cyclic Galois group, i.e. when

$$\sigma_i = \sigma^{i-1}, \qquad i = 1, \ldots, n,$$

where $\sigma \in G(\mathbf{K}/\mathbf{Q})$ is some generating element.

THEOREM 1. *Let* $\mathbf{K}$ *be a finite normal extension of degree* n *over* $\mathbf{Q}$ *with a cyclic Galois group, given by a basis* (1) *over* $\mathbf{Q}$. *Then a normal basis of* $\mathbf{K}$ *over* $\mathbf{Q}$ *can be found in time*

$$T = S(\Omega)^{O(1)}.$$

For a normal extension $\mathbf{K}$ of any fixed degree n over a finite or algebraic number field $\mathbf{L}$ a simpler algorithm may be designed. Moreover, we can omit the restriction on the Galois group.

In fact, let us have any basis (1) of $\mathbf{K}$ over $\mathbf{L}$. If

$$\mu = x_1\omega_1 + \ldots + x_n\omega_n, \qquad x_1, \ldots, x_n \in \mathbf{L},$$

then it is clear that

$$\sigma_i(\mu) = \sum_{j=1}^{n} L_{ij}(x_1, \ldots, x_n)\omega_j, \qquad 1 \leqslant i \leqslant n,$$

where

$$L_{ij}(x_1, \ldots, x_n), \qquad 1 \leqslant i, j \leqslant n,$$

are the homogeneous linear forms in $x_1, \ldots, x_n$. Thus, the element μ generates a normal basis if and only if

$$\Delta(x_1, \ldots, x_n) \neq 0,$$

where

$$\Delta(x_1, \ldots, x_n) = \det\left(L_{ij}(x_1, \ldots, x_n)\right)_{1 \leqslant i,j \leqslant n}.$$

Since there exists a normal basis of $\mathbf{K}$ over $\mathbf{L}$, the determinant $\Delta(x_1, \ldots, x_n)$ is a nonzero homogeneous polynomial of degree n.

Now we assume that either $|\mathbf{L}| > n$ or $|\mathbf{L}|$ is infinite. Let

$$X_1, \ldots, X_n \subseteq \mathbf{L}$$

be any n sets of size $|X_1| = \ldots = |X_n| = n+1$.

It is not difficult to show by induction, that for certain

$$x_1 \in X_1, \ldots, x_n \in X_n$$

we have

$$\Delta(x_1, \ldots, x_n) \neq 0.$$

Therefore, in the case $|\mathbf{L}| > n$ a normal basis can be found in $n^{n+O(1)}$ arithmetical operations in $\mathbf{L}$.
Note, that in the case of $\mathbf{L} = \mathbf{Q}$ the bound for the height of the corresponding element α can be derived from this algorithm. Moreover, we can get a result on the distribution of these bases.

THEOREM 2. *Let* $\mathbf{K}$ *be a finite normal extension of degree* n *over* $\mathbf{Q}$ *given by a basis* (1) *over* $\mathbf{Q}$, *and let* ϑ *be an arbitrary element of* $\mathbf{K}$. *Then there exists an algorithm that in time*

$$T = n^n(S(\Omega) + S(\vartheta))^{O(1)}$$

finds $\alpha \in \mathbf{K}$ *having the size*

$$S(\alpha) = O(n^2 S(\Omega)),$$

such that $\vartheta + \alpha$ *generates the normal basis of* $\mathbf{K}$ *over* $\mathbf{Q}$.

Proof. Let

$$\vartheta = t_1\omega_1 + \ldots + t_n\omega_n, \qquad t_1, \ldots, t_n \in \mathbf{Q}.$$

Then we set

$$X_i = \{t_i + j : j = 0, \ldots, n\}, \qquad i = 1, \ldots, n,$$

and, using the algorithm described above, find μ which generates the normal basis. Set then $\alpha = \mu - \vartheta$.
It is evident that to find the corresponding α the algorithm uses $n^{n+O(1)}$ arithmetical operations in $\mathbf{Q}$ over numbers of logarithmic height at most

$$n^{O(1)}(S(\Omega) + S(\vartheta)).$$

Thus we get the desired bound for the computing time.
For $\lambda \in \mathbf{K}$ denote by $q(\lambda)$ its denominator, i.e. the leading coefficient of its minimal polynomial. Then we set

$$R(\lambda) = \max(|\sigma_1(\lambda)|, \ldots, |\sigma_n(\lambda)|).$$

It is evident, that $R(\lambda) \leqslant H(\lambda)/q(\lambda) + 1$. On the other hand, it is evident that

$$H(\lambda) \leqslant H(\lambda q(\lambda)) \leqslant 2^n R(\lambda q(\lambda))^n \leqslant 2^n R(\lambda)^n q(\lambda)^n.$$

Then define

$$H(\Omega) = \max\big(H(\omega_1), \ldots, H(\omega_n)\big),$$
$$R(\Omega) = \max\big(R(\omega_1), \ldots, R(\omega_n)\big).$$

To obtain the bound for $S(\alpha)$ we note that

$$R(\alpha) \leqslant n^2 R(\Omega) \leqslant n^2 (H(\Omega)+1)^n,$$

and

$$q(\alpha) \leqslant q(\omega_1)\dots q(\omega_n) \leqslant H(\Omega)^n.$$

Thus,

$$H(\alpha) \leqslant n^{2n}(H(\Omega)+1)^{n^2} H(\Omega)^{n^2},$$

and the work is done.

THEOREM 3. *Let* $\mathbf{K}$ *be a finite normal extension of degree* n *over* $\mathbf{Q}$, *and let* ϑ *generate the normal basis of* $\mathbf{K}$ *over* $\mathbf{Q}$. *Then there exists a finite set* $\mathfrak{R}$ *of size*

$$|\mathfrak{R}| \leqslant 2^{6n}$$

such that for each $r \in \mathbf{Q}\backslash\mathfrak{R}$ *there exists a set* $\mathfrak{I}_r$ *of size*

$$|\mathfrak{I}_r| \leqslant \left(6n(\omega_r+1)\right)^{2(n+1)},$$

where ω_r *is the number of prime ideal factors of* $\vartheta + r$ *in* $\mathbf{K}$, *such that all powers*

$$(\vartheta + r)^i, \qquad i \in \mathbf{N}\backslash\mathfrak{I}_r,$$

generate normal bases of $\mathbf{K}$ *over* $\mathbf{Q}$.

Proof. Let v^{T} denote the transposition of a vector v. Using the arguments of (Schlickewei and Stepanov, 1991), for all $i = 1, 2, \dots,$ we get the equality

$$\left(\sigma_1((\vartheta+r)^i), \dots, \sigma_n((\vartheta+r)^i)\right)^{\mathrm{T}} = U_r(i)(\sigma_1(\vartheta), \dots, \sigma_n(\vartheta))^{\mathrm{T}},$$

where the elements $u_{r,k,j}(i)$ of the matrix

$$U_r(i) = \left(u_{r,k,j}(i)\right)_{1 \leqslant k,j \leqslant n}$$

are rational numbers and satisfy the same linear recurrent equation

$$a_{r,n} u_{r,k,j}(i+n) = a_{r,n-1} u_{r,k,j}(i+n-1) + \dots + a_{r,0} u_{r,k,j}(i), \qquad i = 1, 2, \dots,$$

with the characteristic polynomial

$$\varphi_{r,\vartheta}(x) = a_{r,n} x^n + a_{r,n-1} x^{n-1} + \dots + a_{r,0},$$

being the minimal polynomial of $\vartheta + r$.
This implies that any sequence $u_{r,k,j}(i)$ can be represented as a linear combination of power functions

$$\left(\sigma_1(\vartheta + r)\right)^i, \dots, \left(\sigma_n(\vartheta + r)\right)^i.$$

Set

$$N_n = \binom{2n-1}{n}.$$

Thus, from the explicit formula for the determinant of matrix we get that $\det U_r(i)$ can be represented as a linear combination of $N(r)$, $N(r) \leqslant N_n$, power functions

$$(\lambda_v(r))^i, \qquad v = 1, \ldots, N(r),$$

where $\lambda_v(r)$ runs over the set of all the different products of the form

$$\prod_{m=1}^{n} (\sigma_m(\vartheta + r))^{t_m}, \qquad t_1 + \ldots + t_n = n,$$

corresponding to the non-negative integer solutions $(t_1, \ldots, t_n)$ of the equation

$$t_1 + \ldots + t_n = n.$$

Since ϑ generates the normal basis, for all $r \in \mathbf{Q}$, with possible exception of one, say r_0, the element $\vartheta + r$ generates the normal basis too. Indeed, let $\vartheta + r$ do not generate a normal basis, then for some $c_1, \ldots, c_n \in \mathbf{Q}$, not all equal to zero,

$$c_1\sigma_1(\vartheta + r) + \ldots + c_n\sigma_n(\vartheta + r) = 0.$$

Then

$$c_1\sigma_1(\vartheta) + \ldots + c_n\sigma_n(\vartheta) = (c_1 r + \ldots + c_n r) \in \mathbf{Q}.$$

On the other hand,

$$\sigma_1(\vartheta) + \ldots + \sigma_n(\vartheta) = \mathrm{Sp}\,(\vartheta) \in \mathbf{Q}.$$

Taking into account that $\sigma_1(\vartheta), \ldots, \sigma_n(\vartheta)$ form the basis of $\mathbf{K}$ over $\mathbf{Q}$, i.e. they are linearly independent over $\mathbf{Q}$, we get that $\vartheta + r$ does not generate a normal basis only for $r = \mathrm{Sp}\,(\vartheta)/n$.
Let $\rho_1, \ldots, \rho_s$ be all the roots of unity in $\mathbf{K}$. It is known that $s \leqslant 2^{2n}$.
Set $\lambda_0(r) = 1$ and consider at most $sN_n(N_n + 1)/2$ equations (with respect to r) of the form

$$\lambda_k(r) = \rho_j \lambda_l(r), \qquad 0 \leqslant k < l \leqslant N(r), \quad 1 \leqslant j \leqslant s. \tag{3}$$

Since $\lambda_v(r)$, $v = 1, \ldots, N(r)$, are different, for $\rho_1 = 1$ this equation has no solutions. For other $\rho_j \neq 1$ we get for r an equation of degree n (with nonzero leading coefficient).
Let the set $\mathfrak{R}$ consists of the element r_0, $r_0 = \mathrm{Sp}\,(\vartheta)/n$, and all solutions of the system (3). It is clear that

$$|\mathfrak{R}| \leqslant nsN_n(N_n + 1)/2 \leqslant 2^{2n} n N_n^2 \leqslant 2^{6n}.$$

For $r \in \mathbf{Q}/\mathfrak{R}$ we have that $\vartheta + r$ generates the normal basis, i.e. the sequence $\det U_r(i)$ is a nonzero sequence. Moreover, for $r \in \mathbf{Q}/\mathfrak{R}$ we have that there are no roots of unity among the ratios $\lambda_v(r)/\lambda_u(r)$, $u \neq v$, $u, v = 1, \ldots, N(r)$. Then this sequence is nondegenerate in the sense of papers (van der Poorten and Schlickewei, 1991; Schlickewei, 1991), and we can apply the bound (2) with

$$\mathbf{F} = \mathbf{K}, \qquad d = n, \qquad \omega = n\omega_r, \quad \text{and} \quad N \leqslant N_n.$$

It gives the bound

$$|\mathfrak{J}| \leqslant (N_n - 1)\big(4(n + n\omega_r)\big)^{2(n+1)} \leqslant 2^n\big(4n(\omega_r + 1)\big)^{2(n+1)} \leqslant \big(6n(\omega_r + 1)\big)^{2(n+1)}$$

for the size of the exceptional set of those i for which $\det U_r(i) = 0$.
Let $\tau(n)$ denote the number of positive divisors of n and let

$$T_n = \binom{n + \tau(n) - 1}{n}.$$

THEOREM 4. *Let* $\mathbf{K}$ *be a finite normal extension of degree* n *over* $\mathbf{Q}$ *with a cyclic Galois group. If at least one power of* $\alpha \in \mathbf{K}$ *generates the normal basis, then among any* T_n *consecutive powers*

$$\vartheta^{k+i}, \qquad i = 1, \ldots, T_n,$$

at least one power generates the normal basis of $\mathbf{K}$ *over* $\mathbf{Q}$.

Proof. Define the sequence of matrices with rational elements

$$U(i) = \big(u_{k,j}(i)\big)_{1 \leqslant k,j \leqslant n}$$

by the equality

$$\big(\sigma^0(\vartheta^i), \ldots, \sigma^{n-1}(\vartheta^i)\big)^{\mathrm{T}} = U(i)\big(\sigma^0(\vartheta), \ldots, \sigma^{n-1}(\vartheta)\big)^{\mathrm{T}}.$$

For the case of a cyclic Galois group, it was shown in (Schlickewei and Stepanov, 1991) that $U(i)$ is a circulant matrix, i.e.

$$u_{k,j}(i) = u_{k-j}(i), \qquad 1 \leqslant k \leqslant j \leqslant n, \quad i = 1, 2, \ldots,$$

where indices are taken modulo n, and $u_v(i)$, $0 \leqslant v \leqslant n-1$, satisfy the same linear recurring equation

$$a_n u_v(i+n) = a_{n-1} u_v(i+n-1) + \ldots + a_0 u_v(i), \qquad i = 1, 2, \ldots,$$

with the characteristic polynomial

$$\varphi_\vartheta(x) = a_n x^n + a_{n-1} x^{n-1} + \ldots + a_0$$

being the minimal polynomial of ϑ.

It was also shown in (Schlickewei and Stepanov, 1991) that

$$\det U(i) = v_1(i)\dots v_n(i), \qquad i = 1, 2, \dots,$$

where

$$v_j(k) = u_0(k) + \rho_j u_1(k) + \dots + \rho_j^{n-1} u_{n-1}(k), \qquad 1 \leqslant j \leqslant n,$$

and $\rho_1, \dots, \rho_n$ are all roots of the n-th degree of the unity.
Since

$$\rho^n - 1 = \prod_{d|n} Q_d(\rho)$$

where $Q_d(\rho)$ is the d-th cyclotomic polynomial, all sequences $v_1(i), \dots, v_n(i)$ may be divided into $\tau(n)$ subgroups of sequences pairwise conjugate over $\mathbf{Q}$ (in the corresponding algebraic extensions of $\mathbf{Q}$). If we choose the representatives

$$w_t(i), \qquad t = 1, \dots, \tau(n), \tag{$*$}$$

from each the group and define the sequences

$$W(i) = \prod_{t=1}^{\tau(n)} w_t(i),$$

then $\det U(i) = 0$ if and only if $W(i) = 0$. Since all sequences $(*)$ have the same minimal polynomial, $W(i)$ is a linear recurring sequence of order at most

$$T_n = \binom{n + \tau(n) - 1}{n}.$$

Indeed, $W(i)$ can be represented as a linear combination of T, $T \leqslant T_n$, power functions

$$(\mu_v(r))^i, \qquad v = 1, \dots, T,$$

where $\mu_v(r)$ runs over the set of all the different products of the form

$$\prod_{m=1}^{n} (\sigma^m(\vartheta))^{t_m}, \qquad t_1 + \dots + t_n = \tau(n),$$

corresponding to the non-negative integer solutions $(t_1, \dots, t_n)$ of the equation

$$t_1 + \dots + t_n = \tau(n).$$

Since among the powers ϑ^i at least one generates the normal basis, the sequence $W(i)$ is a nonzero one. Thus, $W(i)$ takes at least one nonzero value among any T_n consecutive values, and we have the result.
It follows from the known upper bounds for $\tau(n)$ that for any $\varepsilon > 0$ and sufficiently large $n > n_0(\varepsilon)$ the inequality

$$T_n \leqslant \exp(n^{\varepsilon})$$

holds.

Acknowledgment. The authors would like to express gratitude to Alfred J. van der Poorten and Hans P. Schlickewei for valuable discussions and for the information about their result before it was published.

REFERENCES

Evdokimov, S. A. (1989). Factoring a solvable polynomial over a finite field and the Generalized Riemann Hypothesis. *Zap. Nauchn. Semin. LOMI* **176**, 104–117 (in Russian).

Von zur Gathen, J. and Giesbrecht, M. W. (1990). Constructing normal bases in finite fields. *J. Symb. Comp.* **10**, 547–570.

Lenstra, H. W. (1991). Finding isomorphisms between finite fields. *Math. Comp.* **56**, 329–347.

Lidl, R. and Niederreiter, H. (1983). *Finite Fields*. Addison-Wesley, London, Amsterdam.

Schlickewei, H. P. (1991). Multiplicies of algebraic linear recurrences. *Acta. Math.* (to appear).

Schlickewei, H. P. and Stepanov S. A. (1991). An algorithm to construct normal bases of cyclic number fields. *J. Number Theory* (to appear).

Semaev, I. I. (1988). Construction of irreducible polynomials over finite fields with linearly independent roots. *Mat. Sb. (N. S.)* **135 (177)**, 4, 520–532 (in Russian).

Sidelnikov, V. M. (1987). On the normal bases of a finite field. *Mat. Sb.* **133**, 4, 497–507 (in Russian).

Stepanov, S. A. and Shparlinski, I. E. (1987a). On structure complexity of normal bases of a finite field. *Lecture Notes Comp. Sci.* **278**, 414–416.

Stepanov, S. A. and Shparlinski, I. E. (1987b). On construction of a normal basis of a finite field. *Acta Arithm.* **49**, 2, 189–192.

Stepanov, S. A. and Shparlinski, I. E. (1989). On construction of a primitive normal basis of a finite field. *Mat. Sb.* **180**, 8, 1067–1072 (in Russian).

Stepanov, S. A. and Shparlinski, I. E. (1991). On construction of primitive elements and primitive normal basis in a finite field. In: *Proc. Colloq. Comp. Number Theory, Hungary, 1989*. Walter De Gruyter, New York, pp. 1–24.

Van der Poorten, A. J. and Schlickewei, H. P. (1991). Zeros of recurring sequences. *Bull. Austral. Math. Soc.* **44**, 215–223.

New Trends in Probab. and Statist., pp. 379–386
F. Schweiger and E. Manstavičius (Eds.)
1992 VSP/TEV

NUMBER THEORY AND NUMBER CRUNCHING

P. ZINTERHOF
Universität Salzburg, Institut für Mathematik,
Hellbrunnerstr. 34, A 5020 Salzburg, Austria

ABSTRACT

The problem of numerical inversion of multi-dimensional functional transforms occurs in different fields of applied mathematics. Such fields are, e.g. stochastic processes, PDE, system theory. We give number theoretic methods in the sense of Korobow and Hlawka for the numerical inversion of multivariate Laplace transform which are much more accurate than classical Cartesian-product rules. Similar theorems for other transforms are possible.

Since 1916, when H. Weyl introduced the concept of equidistribution of sequences, this field of research has developed very extensively. The classical result can be summarized in famous Weyl's

THEOREM. *The sequence* (x_n) *is equidistributed* mod 1 *iff*
a)

$$\frac{1}{N}\sum_{k=1}^{N} f(x_k) \longrightarrow \int_0^1 f(x)\,dx \tag{1}$$

for all Riemann-integrable $f(x)$, *or*
b)

$$\frac{1}{N}\sum_{k=1}^{N} e^{2\pi i m x_k} \longrightarrow 0 \tag{2}$$

for all integers $m \neq 0$, *or*
c)

$$D_N = \sup_{0\leqslant\alpha<\beta\leqslant 1}\left|\frac{A_N(\alpha,\beta)}{N} - (\beta-\alpha)\right| \longrightarrow 0 \tag{3}$$

where A_N *is the number of* x_n*'s,* $1 \leqslant n \leqslant N$*, falling into* $[\alpha,\beta)$ mod 1.

The quantity D_N is called the discrepancy of the sequence (x_n).
This theorem has stimulated vast research in abstract spaces up to generalization of equidistribution in metric spaces (Radon measures and equidistribution), and the most important research in multivariate numerics. It also has led to the most important new "mathematical technologies".

This theorem clarifies the strong interconnection between equidistribution of sequences and the theory of trigonometric sums in number theory and the theory of Diophantine approximation. Very interesting results concerning topological groups also were stimulated by the interpretation of (2) as a mean of characters from a dual groups.
This theorem interconnects with probability and probabilistic number theory by the well-known fact that D_N is the Kolmogorow distance between the empirical distribution of $x_1, \ldots, x_N$ and the equidistribution function.
Weyl's theorem may be generalized in a natural way on the s-dimensional unit cube G_s and it leads to the most effective numerical processes in higher dimensions. We mention only some typical results.

THEOREM. *Let* $f(x_1, \ldots, x_s) \in E_\alpha^s(C)$, $\alpha > 1$, *i. e.*

$$f(x_1, \ldots, x_s) = \sum_{m_1, \ldots, m_s = -\infty}^{\infty} C(m_1, \ldots, m_s) e^{2\pi i (m_1 x_1 + \ldots + m_s x_s)}, \tag{4}$$

$$\overline{m} = \max(1, |m|),$$

$$\left|C(m_1, \ldots, m_s)\right| \leqslant \frac{C}{(\overline{m}_1 \ldots \overline{m}_s)^\alpha} \tag{5}$$

and let $a_1, \ldots, a_s \in \mathbf{Z}$ *be optimal coefficients* mod N. *Then*

$$R_N = \frac{1}{N} \sum_{k=1}^{N} f\left((a_1, \ldots, a_s)\frac{k}{N}\right) - \int_{G_s} f(x_1, \ldots, x_s)\, dx_1 \ldots dx_s = O\left(\frac{\log^{\alpha s} N}{N^\alpha}\right). \tag{6}$$

This (essentially the best possible) result belongs to (Korobow, 1963). Optimal coefficients do exist for all dimensions s and all N.

THEOREM. *Let* $f(x_1, \ldots, x_s)$ *be of finite variation* V *in the sense of Hardy and Krause and* D_N *be the discrepancy of the* s*-dimensional sequence* (x_n) mod 1. *Then*

$$\left|\frac{1}{N} \sum_{k=1}^{N} f(x_k) - \int_{G_s} f(x)\, dx\right| \leqslant V(f)\, D_N. \tag{7}$$

This result is due to Hlawka (see (Kuipers and Niederreiter, 1974)).

Remark. The comparison of both results shows that the periodization of a differentiable function f can vastly improve the order of convergence of the integration process. We shall be concerned with this question later.

THEOREM. *Let* G *be a compact abelian metrizable and torsion-free group. Let*

$$f\colon G \longrightarrow \mathbf{C} \tag{8}$$

having an absolutely convergent Fourier series:

$$\hat{f} \in l_1(\hat{G}). \tag{9}$$

Then there always exists a function $x = x(f) \in G$ such that

$$\frac{1}{N^2} \sum_{k=-(N-1)}^{N-1} (N - |k|)\, f(kx) - \int_G f\, d\mu = O\Big(\frac{1}{N}\Big) \tag{10}$$

with μ being the Haar measure on G.

The O-constant is in fact equal to $\|\hat{f}\|_1$. This result is due to the author (see (Zinterhof and Stegbuchner, 1987)) and generalizes and improves $(\log N)/N$ results of Gelfand and Korobow. The result is the best possible even for analytic functions in the one-dimensional case. One gets readily Monte-Carlo methods on such groups using the Borel–Cantelli lemma. More interesting estimate are possible in the case $\hat{f} \in l_p$, $p < 1$.
In this paper we shall concentrate on some new applications of number theoretic methods to functional transforms, i.e. we shall prove some new results on the numerical inversion of the multivariate Laplace transform.

DEFINITION. The function $x = \Psi(\tau)$: $[0,1] \longrightarrow [0,1]$ is called a periodization function of order α (α being a positive integer) if

$$\begin{aligned} &\text{(i) } \Psi'(\tau) > 0 \quad \text{for} \quad 0 < \tau < 1,\\ &\text{(ii) } \Psi(0) = 0, \quad \Psi(1) = 1,\\ &\text{(iii) } \Psi^{(n)}(0) = \Psi^{(n)}(1) = 0 \quad \text{for} \quad n = 1, \ldots, \alpha. \end{aligned} \tag{11}$$

LEMMA 1 (Korobow, 1963). *Suppose that*

$$\frac{\partial^{s\alpha} f(x_1, \ldots, x_s)}{\partial x_1^\alpha \partial x_2^\alpha \ldots \partial x_s^\alpha} = D_\alpha(x_1, \ldots, x_s) \tag{12}$$

is continuous and bounded by C,

$$\max_{G_s} |D_\alpha(x_1, \ldots, x_s)| \leqslant C. \tag{13}$$

Then

$$f(\Psi(\tau_1), \Psi(\tau_2), \ldots, \Psi(\tau_s))\, \Psi'(\tau_1) \ldots \Psi'(\tau_s) \in E_\alpha^s(C)$$

and

$$\int_{G_s} f(x_1, \ldots, x_s)\, dx_1 \ldots dx_s = \int_{G_s} f(\Psi(\tau_1), \ldots, \Psi(\tau_s))\, d\Psi_1 \ldots d\Psi_s, \tag{14}$$

$$d\Psi_l = \Psi'(\tau_l)\, d\tau_l.$$

Korobow (1963) suggested some methods of periodization of functions. We add two other methods slightly more sensitive in the case of $\alpha \notin N$ and for higher dimensions s.

DEFINITION. Using the incomplete Beta-integral the B-periodization is defined by

$$\Psi(\tau) = B(\alpha,\beta,\tau) = \frac{\Gamma(\alpha+\beta+2)}{\Gamma(\alpha+1)\Gamma(\beta+1)} \int_0^\tau t^\alpha (1-t)^\beta \, dt \tag{15}$$

where $\alpha > -1$ and $\beta > -1$.

DEFINITION. The trigonometric periodization of order (α,β) is defined by

$$\Psi(\tau) = \frac{\Gamma\left(\frac{(\alpha+\beta+2)}{2}\right)\pi}{\Gamma\left(\frac{(\alpha+1)}{2}\right)\Gamma\left(\frac{(\beta+1)}{2}\right)} \int_0^\tau \sin\left(\frac{\pi}{2}t\right)^\alpha \cos\left(\frac{\pi}{2}\right)^\beta dt \tag{16}$$

where $\alpha > -1$, and $\beta > -1$.

Remark. Of course both periodizations are closely related. The latter is introduced because the trigonometric functions involved are preferable in many cases. In practice, α and β are taken to be positive in most cases. By means of periodization functions of this kind it is possible to remove certain singularities of integrands too.

Now we suppose that the Laplace transform

$$\tilde{f}(z_1,\dots,z_s) = \int_0^\infty \cdots \int_0^\infty e^{-(t_1 z_1+\dots+t_s z_s)} f(t_1,\dots,t_s)\, dt_1 \dots dt_s \tag{17}$$

of the function $f(t_1,\dots,t_s)$ exists for

$$\mathrm{Re}(z_1) = \mathrm{Re}(x_1+iy_1) > 0, \ \dots, \ \mathrm{Re}(z_s) = \mathrm{Re}(x_s+iy_s) > 0.$$

Let f have continuous partial derivatives at least of order one. The well-known inversion theorem gives:

$$f(t_1,\dots,t_s) = \frac{1}{(2\pi i)^s} \int_{z_1=x_1-i\infty}^{z_1=x_1+i\infty} \cdots \int_{z_s=x_s-i\infty}^{z_s=x_s+i\infty} e^{t_1 z_1+\dots+t_s z_s} \tilde{f}(z_1,\dots,z_s)\, dz_1 \dots dz_s. \tag{18}$$

Let

$$z = \frac{1+w}{1-w} \tag{19}$$

be the conformal mapping which transforms the unit disk $|w| \leqslant 1$ onto the right-hand half-plane $\mathrm{Re}(z) \geqslant 0$, with $z(0) = 1$, and $z(1) = \infty$.

The straight line $\mathrm{Re}(z) = x$ corresponds to the circle

$$w = \frac{x}{1+x} + \frac{e^{i\phi}}{1+x}, \qquad 0 \leqslant \phi \leqslant 2\pi, \tag{20}$$

and we get straightforward that

$$
\begin{aligned}
z &= x - \frac{i(1+x)\sin\phi}{1-\cos\phi} = x + iy, \\
y &= -(1+x)\cot\frac{\phi}{2}, \\
\phi &= -2\operatorname{arccot}\frac{y}{(1+x)}, \\
dz &= \frac{i(1+x)}{2\sin^2\phi/2}\, d\phi.
\end{aligned}
\tag{21}
$$

Now let $\psi(\tau)$ be a periodization of order α and let $0 < \varepsilon < 1$. Let

$$
\begin{aligned}
\phi &= 2\pi\Big(\frac{\varepsilon}{2} + (1-\varepsilon)\,\psi(\tau)\Big), \\
d\phi &= 2\pi\,(1-\varepsilon)\,\psi'(\tau)\,d\tau.
\end{aligned}
\tag{22}
$$

Using these transforms we rewrite the inversion formula (18):

$$
\begin{aligned}
f_\varepsilon(t_1,\dots,t_s) = (1-\varepsilon)^s \int_0^1 \cdots \int_0^1 \prod_{l=1}^{s} e^{t_l(x_l - i(1+x_l))\cot\phi_l/2}\, \frac{1+x_l}{2\sin^2\phi_l/2}\, \psi'(\tau_l) \\
\times \tilde{f}\left(x_1 - i(1+x_1)\cot\frac{\phi_1}{2},\dots,x_s - i(1+x_s)\cot\frac{\phi_s}{2}\right) d\tau_1 \dots d\tau_s.
\end{aligned}
\tag{23}
$$

The equality

$$
\lim_{\varepsilon\to 0} f_\varepsilon(t_1,\dots,t_s) = f_0(t_1,\dots,t_s) = f(t_1,\dots,t_s) \tag{24}
$$

holds obviously. For $\varepsilon = 0$ the singularities at $\tau = 0$, and $\tau = 1$ can be removed by a periodization of order $\alpha \geqslant 2$. Now we derive a numerical process evaluating f for $\varepsilon > 0$.

Remark. It is well known that $\tilde{f}(z_1,\dots,z_s)$ is analytic in $z_1,\dots,z_s$ on the domain $\mathrm{Re}(z_1) > 0, \dots, \mathrm{Re}(z_s) > 0$.

LEMMA 2. *The integrand in* (23) *belongs to* $E^s_\alpha(C)$ *if the used periodization* $\psi(\tau)$ *is at least of order* $2\alpha + 2$.

Proof. Denote the integrand in (23) by $H(\tau_1,\dots,\tau_s)$. The function $\tilde{f}$ and all its partial derivatives are bounded by a suitable constant depending on $x_1,\dots,x_s$ and on the order of derivatives α. We perform the partial integration α-times with respect to each variable $\tau_1,\dots,\tau_s$ of the m-th Fourier coefficient of $H(\tau_1,\dots,\tau_s)$:

$$
C(m_1,\dots,m_s) = \int_0^1 \cdots \int_0^1 H(\tau_1,\dots,\tau_s)\, e^{2\pi i(m_1\tau_1+\dots+m_s\tau_s)}\, d\tau_1 \dots \tau_s, \tag{25}
$$

$$\left|C(m_1,\ldots,m_s)\right| \leqslant CK(\alpha,\tilde{f},x_1,\ldots,x_s)\frac{1}{(\overline{m}_1\ldots\overline{m}_s)^{\alpha}}, \tag{26}$$

with C independing on ε due to the order $2\alpha+2$ of the periodization function $\psi(\tau)$, because the order of periodization removes the possible singularities of the integrand and its derivatives in (23), caused by sin and cot. Being interested in the case $\varepsilon > 0$ one gets a similar Lemma replacing $2\alpha+2$ by α because the integrand in (23) in this case actually has no singularities. The step (25), (26) can be found in (Korobow, 1963, p. 55).

DEFINITION. We say that $f(t_1,\ldots,t_s) \in D_\beta^s(C_1)$ if $\beta > 1$ and

$$\left|\tilde{f}(z_1,\ldots,z_s)\right| \leqslant \frac{C_1}{(|z_1|\ldots|z_s|)^{\beta}}. \tag{27}$$

This holds, e.g. if the derivatives of f of the order β exist and are integrable.

LEMMA 3. *If* $f(t_1,\ldots,t_s) \in D_\beta^s(C_1)$, *then*

$$\left|f_\varepsilon(t_1,\ldots,t_s) - f(t_1,\ldots,t_s)\right| = O\left(\varepsilon^{\beta-1}\right). \tag{28}$$

Proof. We have

$$\begin{aligned}&\left|f_\varepsilon(t_1,\ldots,t_s) - f(t_1,\ldots,t_s)\right|\\ &= \left|\frac{1}{(2\pi i)^s}\int\cdots\int e^{z_1t_1+\ldots+z_st_s}\tilde{f}(z_1,\ldots,z_s)\,dz_1\ldots dz_s\right|.\end{aligned} \tag{29}$$

Here the domain of integration is the compliment of

$$\{y_1,\ldots,y_s\colon\ |y_l| \leqslant \cot\pi\varepsilon,\quad l=1,\ldots,s\},$$

thus $|y_l| > \cot\pi\varepsilon$, for at least one $l = 1,\ldots,s$.
So we get immediately the assertion of the lemma.
Finally we apply an integration rule to the function $H(\tau_1,\ldots,\tau_n)$. Let

$$\left(\tau_1^{(k)},\ldots,\tau_s^{(k)}\right),\qquad k=1,\ldots,N, \tag{30}$$

be a set of nodes in G_s.
There is an extensive literature on the appropriate choice of rules and nodes. Here we shall use only two methods: the "classical" Cartesian product rule and the method of optimal coefficients. We consider the rule

$$\frac{1}{N}\sum_{k=1}^{N} H\left(\tau_1^{(k)},\ldots,\tau_s^{(k)}\right) - \int_0^1\cdots\int_0^1 H\,d\tau_1\ldots d\tau_s = R_N, \tag{31}$$

$$I_N(H) - I(H) = R_N.$$

THEOREM.
a) *Using the method of good lattice points (optimal coefficients) and choosing* $\varepsilon \sim N^{-\alpha/(\beta-1)}$, *the estimate*

$$f(t_1,\ldots,t_s) - I_N(H) = O\left(N^{-\alpha}(\log N)^{\alpha p}\right) \tag{32}$$

holds with $p \leqslant s$.
b) *Using the Cartesian product rule and choosing* $\varepsilon \sim N^{-\alpha/s(\beta-1)}$, *the estimate*

$$f(t_1,\ldots,t_s) - I_N(H) = O\left(N^{-\alpha/s}\right) \tag{33}$$

holds.

Proof.
a) Let $a_1, \ldots, a_s$ be optimal coefficients mod N and $\{x\}$ be the fractional part of x. Consider the integration nodes

$$\left(\tau_1^{(k)}, \ldots, \tau_s^{(k)}\right) = \left(\left\{\frac{a_1 k}{N}\right\}, \ldots, \left\{\frac{a_s k}{N}\right\}\right), \qquad k = 1, \ldots, N. \tag{34}$$

Because

$$H\left(\tau_1, \ldots, \tau_s\right) \in E_\alpha^s(CK) \tag{35}$$

due to Lemma 2, we have (see (Korobow, 1963))

$$R_N = O\left(\frac{(\log N)^{\alpha p}}{N^\alpha}\right). \tag{36}$$

with $p \leqslant s$. Using Lemma 3 we obtain that the equality

$$\begin{aligned} f\left(t_1, \ldots, t_s\right) - I_N\left(H\left(\tau_1, \ldots, \tau_s\right)\right) &= f - f_\varepsilon + I(H) - I_N(H) \\ &= O\left(\varepsilon^{\beta-1} + \frac{(\log N)^{\alpha p}}{N^\alpha}\right) \end{aligned} \tag{37}$$

holds. Choosing ε of the order $N^{-\alpha/(\beta-1)}$ the assertion a) follows.
b) The classical Cartesian rule for periodic functional consists of $N = n^s$ nodes

$$\left(\frac{k_1}{n}, \ldots, \frac{k_s}{n}\right), \qquad k_l = 1, \ldots, n, \quad l = 1, \ldots, s. \tag{38}$$

For $H(\tau_1, \ldots, \tau_s) \in E_\alpha^a(C)$ we have the best possible estimation (see (Korobow, 1963))

$$\frac{1}{n^s} \sum_{k_1, \ldots, k_s = 1}^{n} H\left(\frac{k_1}{n}, \ldots, \frac{k_s}{n}\right) - \int_{G_s} H \, d\tau_1 \ldots d\tau_s = I_N(H) - I(H) = O\left(\frac{1}{N^{\alpha/s}}\right). \tag{39}$$

Using the same arguments as in a) and choosing ε of the order $N^{-\alpha/s(\beta-1)}$ we get the assertion b).

Remark. One can not expect an estimate essentially better as in a) because there exists no integration rule better then $O(1/N^\alpha)$; for good lattice points the lower bound of $E_\alpha^s(C)$ is even $O((\log N)^{s-1}/N^\alpha)$ (see (Korobow, 1963)).

Remark. The method proposed in this paper was stimulated by investigations on Markow renewal processes. The Markow renewal process is described by a matrix-valued integral equation on several variables (generalized renewal equation). We have used multivariate Laplace transforms for analytic solutions. The numerical experiments on parallel computers are promising. So the use of advanced number-theoretic methods and modern computer technologies apparently extends the area of practical computability in a significant way.

Acknowledgement. The author is grateful to the referee for some helpful comments.

REFERENCES

Doetsch, C. (1958). *Einführung in Theorie und Anwendung der Laplace-Transformation*. Birkhäuser, Basel.

Hua, L. K. and Wang, Y. (1981). *Applications of Number Theory to Numerical Analysis*. Springer-Verlag/Sciences Press, Berlin, Beijing.

Korobow, N. M. (1963). *Number Theoretic Methods in Approximate Analysis*. Fizmatgiz, Moscow.

Kuipers, L. and Niederreiter, H. (1974). *Uniform Distribution of Sequences*. John Wiley and Sons, New York.

Zaremba, S. K. (Ed.) (1972). *Application of Number Theory to Numerical Analysis*. Academic Press, New York, London.

Zinterhof, P. (1969). Einige zahlentheoretische Methoden zur numerischen Quadrature und Interpolation. *Sitzungsber. Österr. Akad. Wiss. Math.-naturw. Kl. Abt.* II **177**, 51–77.

Zinterhof, P. (1976). Über einige Abschätzungen bei der Approximation von Funktionen mit Gleichverteilungsmethoden. *Sitzungsber. Österr. Akad. Wiss. Math.-naturw. Kl. Abt.* II **185**, 121–132.

Zinterhof, P. and Stegbuchner, H. (1978). Trigonometrische Approximation mit Gleichverteilungsmethoden. *Studia Scient. Math. Hung.* **13**, 273–289.

Zinterhof, P. (1987a). Über die schnelle Lösung von hochdimensionalen Fredholm Gleichungen vom Faltungstyp mit zahlentheoretichen Methoden. *Sitzungsber. Österr. Akad. Wiss. Math.-naturw. Kl. Abt.* II **196**, 159–169.

Zinterhof, P. (1987b). Gratis lattice points for multidimensional integration. *Computing* **38**, 347–353.

LIST OF CONTRIBUTORS